FOURIER ANALYSIS

Princeton Lectures in Analysis

I Fourier Analysis: An Introduction

II Complex Analysis

III Real Analysis:
Measure Theory, Integration, and
Hilbert Spaces

PRINCETON LECTURES IN ANALYSIS

I

FOURIER ANALYSIS

AN INTRODUCTION

Elias M. Stein

&

Rami Shakarchi

PRINCETON UNIVERSITY PRESS
PRINCETON AND OXFORD

Published by Princeton University Press, 41 William Street,
Princeton, New Jersey 08540
In the United Kingdom: Princeton University Press,
6 Oxford Street, Woodstock, Oxfordshire OX20 1TW

Library of Congress Control Number 2003103688
ISBN 978-0-691-11384-5

British Library Cataloging-in-Publication Data is available

The publisher would like to acknowledge the authors of this volume for
providing the camera-ready copy from which this book was printed

Printed on acid-free paper. ∞

press.princeton.edu

Printed in the United States of America

17 19 20 18 16

To my grandchildren
Carolyn, Alison, Jason

E.M.S.

To my parents
Mohamed & Mireille
and my brother
Karim

R.S.

Contents

Foreword

Beginning in the spring of 2000, a series of four one-semester courses were taught at Princeton University whose purpose was to present, in an integrated manner, the core areas of analysis. The objective was to make plain the organic unity that exists between the various parts of the subject, and to illustrate the wide applicability of ideas of analysis to other fields of mathematics and science. The present series of books is an elaboration of the lectures that were given.

While there are a number of excellent texts dealing with individual parts of what we cover, our exposition aims at a different goal: presenting the various sub-areas of analysis not as separate disciplines, but rather as highly interconnected. It is our view that seeing these relations and their resulting synergies will motivate the reader to attain a better understanding of the subject as a whole. With this outcome in mind, we have concentrated on the main ideas and theorems that have shaped the field (sometimes sacrificing a more systematic approach), and we have been sensitive to the historical order in which the logic of the subject developed.

We have organized our exposition into four volumes, each reflecting the material covered in a semester. Their contents may be broadly summarized as follows:

I. Fourier series and integrals.

II. Complex analysis.

III. Measure theory, Lebesgue integration, and Hilbert spaces.

IV. A selection of further topics, including functional analysis, distributions, and elements of probability theory.

However, this listing does not by itself give a complete picture of the many interconnections that are presented, nor of the applications to other branches that are highlighted. To give a few examples: the elements of (finite) Fourier series studied in Book I, which lead to Dirichlet characters, and from there to the infinitude of primes in an arithmetic progression; the X-ray and Radon transforms, which arise in a number of problems in Book I, and reappear in Book III to play an important role in understanding Besicovitch-like sets in two and three dimensions; Fatou's theorem, which guarantees the existence of boundary values of bounded holomorphic functions in the disc, and whose proof relies on ideas developed in each of the first three books; and the theta function, which first occurs in Book I in the solution of the heat equation, and is then used in Book II to find the number of ways an integer can be represented as the sum of two or four squares, and in the analytic continuation of the zeta function.

A few further words about the books and the courses on which they were based. These courses where given at a rather intensive pace, with 48 lecture-hours a semester. The weekly problem sets played an indispensable part, and as a result exercises and problems have a similarly important role in our books. Each chapter has a series of "Exercises" that are tied directly to the text, and while some are easy, others may require more effort. However, the substantial number of hints that are given should enable the reader to attack most exercises. There are also more involved and challenging "Problems"; the ones that are most difficult, or go beyond the scope of the text, are marked with an asterisk.

Despite the substantial connections that exist between the different volumes, enough overlapping material has been provided so that each of the first three books requires only minimal prerequisites: acquaintance with elementary topics in analysis such as limits, series, differentiable functions, and Riemann integration, together with some exposure to linear algebra. This makes these books accessible to students interested in such diverse disciplines as mathematics, physics, engineering, and finance, at both the undergraduate and graduate level.

It is with great pleasure that we express our appreciation to all who have aided in this enterprise. We are particularly grateful to the students who participated in the four courses. Their continuing interest, enthusiasm, and dedication provided the encouragement that made this project possible. We also wish to thank Adrian Banner and Jose Luis Rodrigo for their special help in running the courses, and their efforts to see that the students got the most from each class. In addition, Adrian Banner also made valuable suggestions that are incorporated in the text.

We wish also to record a note of special thanks for the following individuals: Charles Fefferman, who taught the first week (successfully launching the whole project!); Paul Hagelstein, who in addition to reading part of the manuscript taught several weeks of one of the courses, and has since taken over the teaching of the second round of the series; and Daniel Levine, who gave valuable help in proof-reading. Last but not least, our thanks go to Gerree Pecht, for her consummate skill in typesetting and for the time and energy she spent in the preparation of all aspects of the lectures, such as transparencies, notes, and the manuscript.

We are also happy to acknowledge our indebtedness for the support we received from the 250th Anniversary Fund of Princeton University, and the National Science Foundation's VIGRE program.

Elias M. Stein
Rami Shakarchi

Princeton, New Jersey
August 2002

Preface to Book I

Any effort to present an overall view of analysis must at its start deal with the following questions: Where does one begin? What are the initial subjects to be treated, and in what order are the relevant concepts and basic techniques to be developed?

Our answers to these questions are guided by our view of the centrality of Fourier analysis, both in the role it has played in the development of the subject, and in the fact that its ideas permeate much of the present-day analysis. For these reasons we have devoted this first volume to an exposition of some basic facts about Fourier series, taken together with a study of elements of Fourier transforms and finite Fourier analysis. Starting this way allows one to see rather easily certain applications to other sciences, together with the link to such topics as partial differential equations and number theory. In later volumes several of these connections will be taken up from a more systematic point of view, and the ties that exist with complex analysis, real analysis, Hilbert space theory, and other areas will be explored further.

In the same spirit, we have been mindful not to overburden the beginning student with some of the difficulties that are inherent in the subject: a proper appreciation of the subtleties and technical complications that arise can come only after one has mastered some of the initial ideas involved. This point of view has led us to the following choice of material in the present volume:

- Fourier series. At this early stage it is not appropriate to introduce measure theory and Lebesgue integration. For this reason our treatment of Fourier series in the first four chapters is carried out in the context of Riemann integrable functions. Even with this restriction, a substantial part of the theory can be developed, detailing convergence and summability; also, a variety of connections with other problems in mathematics can be illustrated.

- Fourier transform. For the same reasons, instead of undertaking the theory in a general setting, we confine ourselves in Chapters 5 and 6 largely to the framework of test functions. Despite these limitations, we can learn a number of basic and interesting facts about Fourier analysis in $\mathbb{R}^d$ and its relation to other areas, including the wave equation and the Radon transform.

- Finite Fourier analysis. This is an introductory subject *par excellence*, because limits and integrals are not explicitly present. Nevertheless, the subject has several striking applications, including the proof of the infinitude of primes in arithmetic progression.

Taking into account the introductory nature of this first volume, we have kept the prerequisites to a minimum. Although we suppose some acquaintance with the notion of the Riemann integral, we provide an appendix that contains most of the results about integration needed in the text.

We hope that this approach will facilitate the goal that we have set for ourselves: to inspire the interested reader to learn more about this fascinating subject, and to discover how Fourier analysis affects decisively other parts of mathematics and science.

1 The Genesis of Fourier Analysis

> Regarding the researches of d'Alembert and Euler could one not add that if they knew this expansion, they made but a very imperfect use of it. They were both persuaded that an arbitrary and discontinuous function could never be resolved in series of this kind, and it does not even seem that anyone had developed a constant in cosines of multiple arcs, the first problem which I had to solve in the theory of heat.
>
> *J. Fourier,* 1808-9

In the beginning, it was the problem of the vibrating string, and the later investigation of heat flow, that led to the development of Fourier analysis. The laws governing these distinct physical phenomena were expressed by two different partial differential equations, the wave and heat equations, and these were solved in terms of Fourier series.

Here we want to start by describing in some detail the development of these ideas. We will do this initially in the context of the problem of the vibrating string, and we will proceed in three steps. First, we describe several physical (empirical) concepts which motivate corresponding mathematical ideas of importance for our study. These are: the role of the functions $\cos t$, $\sin t$, and e^{it} suggested by simple harmonic motion; the use of separation of variables, derived from the phenomenon of standing waves; and the related concept of linearity, connected to the superposition of tones. Next, we derive the partial differential equation which governs the motion of the vibrating string. Finally, we will use what we learned about the physical nature of the problem (expressed mathematically) to solve the equation. In the last section, we use the same approach to study the problem of heat diffusion.

Given the introductory nature of this chapter and the subject matter covered, our presentation cannot be based on purely mathematical reasoning. Rather, it proceeds by plausibility arguments and aims to provide the motivation for the further rigorous analysis in the succeeding chapters. The impatient reader who wishes to begin immediately with the theorems of the subject may prefer to pass directly to the next chapter.

1 The vibrating string

The problem consists of the study of the motion of a string fixed at its end points and allowed to vibrate freely. We have in mind physical systems such as the strings of a musical instrument. As we mentioned above, we begin with a brief description of several observable physical phenomena on which our study is based. These are:

- simple harmonic motion,
- standing and traveling waves,
- harmonics and superposition of tones.

Understanding the empirical facts behind these phenomena will motivate our mathematical approach to vibrating strings.

Simple harmonic motion

Simple harmonic motion describes the behavior of the most basic oscillatory system (called the simple harmonic oscillator), and is therefore a natural place to start the study of vibrations. Consider a mass $\{m\}$ attached to a horizontal spring, which itself is attached to a fixed wall, and assume that the system lies on a frictionless surface.

Choose an axis whose origin coincides with the center of the mass when it is at rest (that is, the spring is neither stretched nor compressed), as shown in Figure 1. When the mass is displaced from its initial equilibrium

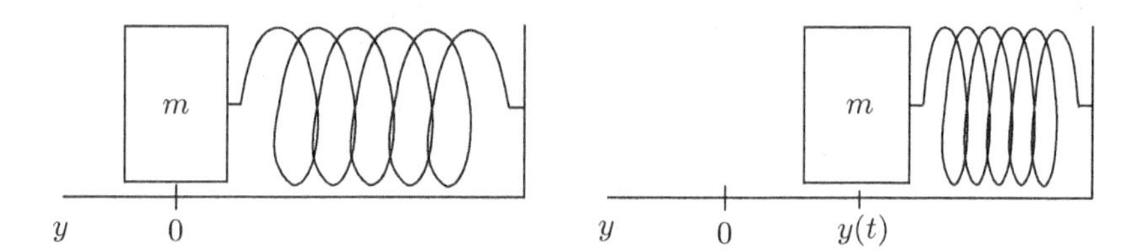

Figure 1. Simple harmonic oscillator

position and then released, it will undergo **simple harmonic motion**. This motion can be described mathematically once we have found the differential equation that governs the movement of the mass.

Let $y(t)$ denote the displacement of the mass at time t. We assume that the spring is ideal, in the sense that it satisfies Hooke's law: the restoring force F exerted by the spring on the mass is given by $F = -ky(t)$. Here

$k > 0$ is a given physical quantity called the spring constant. Applying Newton's law (force = mass × acceleration), we obtain

$$-ky(t) = my''(t),$$

where we use the notation y'' to denote the second derivative of y with respect to t. With $c = \sqrt{k/m}$, this second order ordinary differential equation becomes

$$y''(t) + c^2 y(t) = 0. \tag{1}$$

The general solution of equation (1) is given by

$$y(t) = a \cos ct + b \sin ct\,,$$

where a and b are constants. Clearly, all functions of this form solve equation (1), and Exercise 6 outlines a proof that these are the only (twice differentiable) solutions of that differential equation.

In the above expression for $y(t)$, the quantity c is given, but a and b can be any real numbers. In order to determine the particular solution of the equation, we must impose two initial conditions in view of the two unknown constants a and b. For example, if we are given $y(0)$ and $y'(0)$, the initial position and velocity of the mass, then the solution of the physical problem is unique and given by

$$y(t) = y(0) \cos ct + \frac{y'(0)}{c} \sin ct\,.$$

One can easily verify that there exist constants $A > 0$ and $\varphi \in \mathbb{R}$ such that

$$a \cos ct + b \sin ct = A \cos(ct - \varphi).$$

Because of the physical interpretation given above, one calls $A = \sqrt{a^2 + b^2}$ the "amplitude" of the motion, c its "natural frequency," φ its "phase" (uniquely determined up to an integer multiple of 2π), and $2\pi/c$ the "period" of the motion.

The typical graph of the function $A\cos(ct - \varphi)$, illustrated in Figure 2, exhibits a wavelike pattern that is obtained from translating and stretching (or shrinking) the usual graph of $\cos t$.

We make two observations regarding our examination of simple harmonic motion. The first is that the mathematical description of the most elementary oscillatory system, namely simple harmonic motion, involves

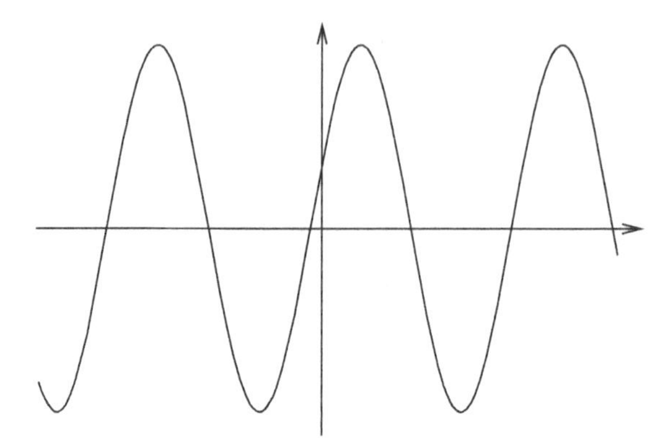

Figure 2. The graph of $A\cos(ct - \varphi)$

the most basic trigonometric functions $\cos t$ and $\sin t$. It will be important in what follows to recall the connection between these functions and complex numbers, as given in Euler's identity $e^{it} = \cos t + i \sin t$. The second observation is that simple harmonic motion is determined as a function of time by two initial conditions, one determining the position, and the other the velocity (specified, for example, at time $t = 0$). This property is shared by more general oscillatory systems, as we shall see below.

Standing and traveling waves

As it turns out, the vibrating string can be viewed in terms of one-dimensional wave motions. Here we want to describe two kinds of motions that lend themselves to simple graphic representations.

- First, we consider **standing waves**. These are wavelike motions described by the graphs $y = u(x,t)$ developing in time t as shown in Figure 3.

  In other words, there is an initial profile $y = \varphi(x)$ representing the wave at time $t = 0$, and an amplifying factor $\psi(t)$, depending on t, so that $y = u(x,t)$ with

  $$u(x,t) = \varphi(x)\psi(t).$$

  The nature of standing waves suggests the mathematical idea of "separation of variables," to which we will return later.

- A second type of wave motion that is often observed in nature is that of a **traveling wave**. Its description is particularly simple:

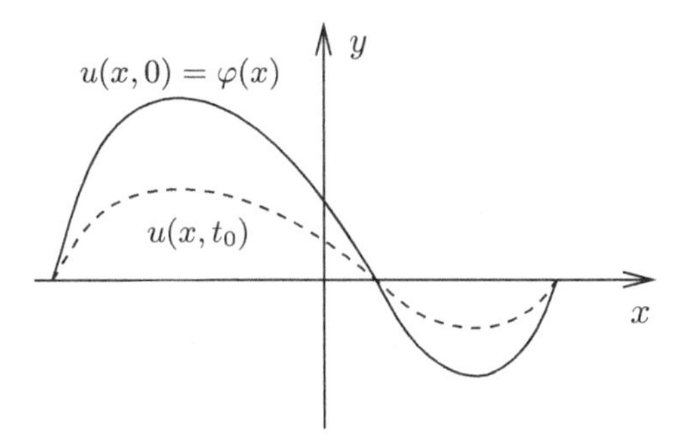

Figure 3. A standing wave at different moments in time: $t = 0$ and $t = t_0$

there is an initial profile $F(x)$ so that $u(x,t)$ equals $F(x)$ when $t = 0$. As t evolves, this profile is displaced to the right by ct units, where c is a positive constant, namely

$$u(x,t) = F(x - ct).$$

Graphically, the situation is depicted in Figure 4.

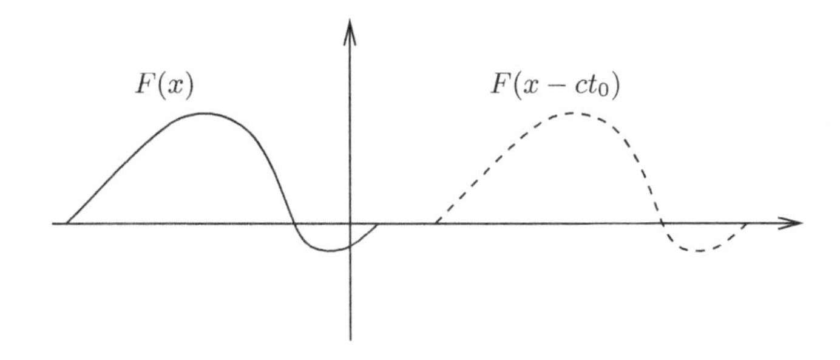

Figure 4. A traveling wave at two different moments in time: $t = 0$ and $t = t_0$

Since the movement in t is at the rate c, that constant represents the **velocity** of the wave. The function $F(x - ct)$ is a one-dimensional traveling wave moving to the right. Similarly, $u(x,t) = F(x + ct)$ is a one-dimensional traveling wave moving to the left.

Harmonics and superposition of tones

The final physical observation we want to mention (without going into any details now) is one that musicians have been aware of since time immemorial. It is the existence of harmonics, or overtones. The **pure tones** are accompanied by combinations of **overtones** which are primarily responsible for the timbre (or tone color) of the instrument. The idea of combination or superposition of tones is implemented mathematically by the basic concept of linearity, as we shall see below.

We now turn our attention to our main problem, that of describing the motion of a vibrating string. First, we derive the wave equation, that is, the partial differential equation that governs the motion of the string.

1.1 Derivation of the wave equation

Imagine a homogeneous string placed in the (x, y)-plane, and stretched along the x-axis between $x = 0$ and $x = L$. If it is set to vibrate, its displacement $y = u(x, t)$ is then a function of x and t, and the goal is to derive the differential equation which governs this function.

For this purpose, we consider the string as being subdivided into a large number N of masses (which we think of as individual particles) distributed uniformly along the x-axis, so that the nth particle has its x-coordinate at $x_n = nL/N$. We shall therefore conceive of the vibrating string as a complex system of N particles, each oscillating in the *vertical direction only*; however, unlike the simple harmonic oscillator we considered previously, each particle will have its oscillation linked to its immediate neighbor by the tension of the string.

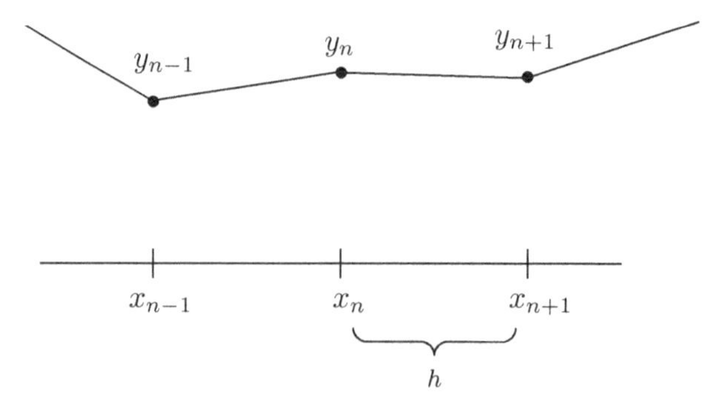

Figure 5. A vibrating string as a discrete system of masses

We then set $y_n(t) = u(x_n, t)$, and note that $x_{n+1} - x_n = h$, with $h = L/N$. If we assume that the string has constant density $\rho > 0$, it is reasonable to assign mass equal to ρh to each particle. By Newton's law, $\rho h y_n''(t)$ equals the force acting on the nth particle. We now make the simple assumption that this force is due to the effect of the two nearby particles, the ones with x-coordinates at x_{n-1} and x_{n+1} (see Figure 5). We further assume that the force (or tension) coming from the right of the nth particle is proportional to $(y_{n+1} - y_n)/h$, where h is the distance between x_{n+1} and x_n; hence we can write the tension as

$$\left(\frac{\tau}{h}\right)(y_{n+1} - y_n),$$

where $\tau > 0$ is a constant equal to the coefficient of tension of the string. There is a similar force coming from the left, and it is

$$\left(\frac{\tau}{h}\right)(y_{n-1} - y_n).$$

Altogether, adding these forces gives us the desired relation between the oscillators $y_n(t)$, namely

$$\rho h y_n''(t) = \frac{\tau}{h}\left\{y_{n+1}(t) + y_{n-1}(t) - 2y_n(t)\right\}. \tag{2}$$

On the one hand, with the notation chosen above, we see that

$$y_{n+1}(t) + y_{n-1}(t) - 2y_n(t) = u(x_n + h, t) + u(x_n - h, t) - 2u(x_n, t).$$

On the other hand, for any reasonable function $F(x)$ (that is, one that has continuous second derivatives) we have

$$\frac{F(x+h) + F(x-h) - 2F(x)}{h^2} \to F''(x) \quad \text{as } h \to 0.$$

Thus we may conclude, after dividing by h in (2) and letting h tend to zero (that is, N goes to infinity), that

$$\rho \frac{\partial^2 u}{\partial t^2} = \tau \frac{\partial^2 u}{\partial x^2},$$

or

$$\frac{1}{c^2}\frac{\partial^2 u}{\partial t^2} = \frac{\partial^2 u}{\partial x^2}, \qquad \text{with } c = \sqrt{\tau/\rho}.$$

This relation is known as the **one-dimensional wave equation**, or more simply as the **wave equation**. For reasons that will be apparent later, the coefficient $c > 0$ is called the **velocity** of the motion.

In connection with this partial differential equation, we make an important simplifying mathematical remark. This has to do with **scaling**, or in the language of physics, a "change of units." That is, we can think of the coordinate x as $x = aX$ where a is an appropriate positive constant. Now, in terms of the new coordinate X, the interval $0 \leq x \leq L$ becomes $0 \leq X \leq L/a$. Similarly, we can replace the time coordinate t by $t = bT$, where b is another positive constant. If we set $U(X,T) = u(x,t)$, then

$$\frac{\partial U}{\partial X} = a\frac{\partial u}{\partial x}, \qquad \frac{\partial^2 U}{\partial X^2} = a^2\frac{\partial^2 u}{\partial x^2},$$

and similarly for the derivatives in t. So if we choose a and b appropriately, we can transform the one-dimensional wave equation into

$$\frac{\partial^2 U}{\partial T^2} = \frac{\partial^2 U}{\partial X^2},$$

which has the effect of setting the velocity c equal to 1. Moreover, we have the freedom to transform the interval $0 \leq x \leq L$ to $0 \leq X \leq \pi$. (We shall see that the choice of π is convenient in many circumstances.) All this is accomplished by taking $a = L/\pi$ and $b = L/(c\pi)$. Once we solve the new equation, we can of course return to the original equation by making the inverse change of variables. Hence, we do not sacrifice generality by thinking of the wave equation as given on the interval $[0, \pi]$ with velocity $c = 1$.

1.2 Solution to the wave equation

Having derived the equation for the vibrating string, we now explain two methods to solve it:

- using traveling waves,
- using the superposition of standing waves.

While the first approach is very simple and elegant, it does not directly give full insight into the problem; the second method accomplishes that, and moreover is of wide applicability. It was first believed that the second method applied only in the simple cases where the initial position and velocity of the string were themselves given as a superposition of standing waves. However, as a consequence of Fourier's ideas, it became clear that the problem could be worked either way for all initial conditions.

Traveling waves

To simplify matters as before, we assume that $c = 1$ and $L = \pi$, so that the equation we wish to solve becomes

$$\frac{\partial^2 u}{\partial t^2} = \frac{\partial^2 u}{\partial x^2} \quad \text{on } 0 \leq x \leq \pi.$$

The crucial observation is the following: if F is any twice differentiable function, then $u(x,t) = F(x+t)$ and $u(x,t) = F(x-t)$ solve the wave equation. The verification of this is a simple exercise in differentiation. Note that the graph of $u(x,t) = F(x-t)$ at time $t = 0$ is simply the graph of F, and that at time $t = 1$ it becomes the graph of F translated to the right by 1. Therefore, we recognize that $F(x-t)$ is a traveling wave which travels to the right with speed 1. Similarly, $u(x,t) = F(x+t)$ is a wave traveling to the left with speed 1. These motions are depicted in Figure 6.

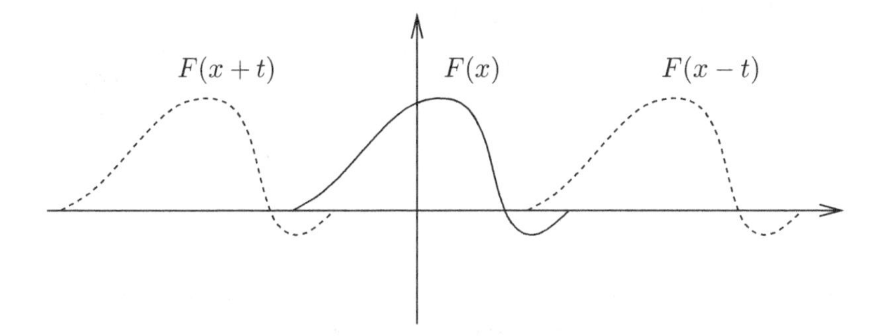

Figure 6. Waves traveling in both directions

Our discussion of tones and their combinations leads us to observe that the wave equation is **linear**. This means that if $u(x,t)$ and $v(x,t)$ are particular solutions, then so is $\alpha u(x,t) + \beta v(x,t)$, where α and β are any constants. Therefore, we may superpose two waves traveling in opposite directions to find that whenever F and G are twice differentiable functions, then

$$u(x,t) = F(x+t) + G(x-t)$$

is a solution of the wave equation. In fact, we now show that all solutions take this form.

We drop for the moment the assumption that $0 \leq x \leq \pi$, and suppose that u is a twice differentiable function which solves the wave equation for all real x and t. Consider the following new set of variables $\xi = x + t$, $\eta = x - t$, and define $v(\xi, \eta) = u(x,t)$. The change of variables formula shows that v satisfies

$$\frac{\partial^2 v}{\partial \xi \partial \eta} = 0.$$

Integrating this relation twice gives $v(\xi, \eta) = F(\xi) + G(\eta)$, which then implies

$$u(x,t) = F(x+t) + G(x-t),$$

for some functions F and G.

We must now connect this result with our original problem, that is, the physical motion of a string. There, we imposed the restrictions $0 \leq x \leq \pi$, the initial shape of the string $u(x, 0) = f(x)$, and also the fact that the string has fixed end points, namely $u(0,t) = u(\pi, t) = 0$ for all t. To use the simple observation above, we first extend f to all of $\mathbb{R}$ by making it odd[1] on $[-\pi, \pi]$, and then periodic[2] in x of period 2π, and similarly for $u(x,t)$, the solution of our problem. Then the extension u solves the wave equation on all of $\mathbb{R}$, and $u(x,0) = f(x)$ for all $x \in \mathbb{R}$. Therefore, $u(x,t) = F(x+t) + G(x-t)$, and setting $t = 0$ we find that

$$F(x) + G(x) = f(x).$$

Since many choices of F and G will satisfy this identity, this suggests imposing another initial condition on u (similar to the two initial conditions in the case of simple harmonic motion), namely the initial velocity of the string which we denote by $g(x)$:

$$\frac{\partial u}{\partial t}(x,0) = g(x),$$

where of course $g(0) = g(\pi) = 0$. Again, we extend g to $\mathbb{R}$ first by making it odd over $[-\pi, \pi]$, and then periodic of period 2π. The two initial conditions of position and velocity now translate into the following system:

$$\begin{cases} F(x) + G(x) = f(x), \\ F'(x) - G'(x) = g(x). \end{cases}$$

[1] A function f defined on a set U is **odd** if $-x \in U$ whenever $x \in U$ and $f(-x) = -f(x)$, and **even** if $f(-x) = f(x)$.

[2] A function f on $\mathbb{R}$ is **periodic** of period ω if $f(x + \omega) = f(x)$ for all x.

Differentiating the first equation and adding it to the second, we obtain

$$2F'(x) = f'(x) + g(x).$$

Similarly

$$2G'(x) = f'(x) - g(x),$$

and hence there are constants C_1 and C_2 so that

$$F(x) = \frac{1}{2}\left[f(x) + \int_0^x g(y)\,dy\right] + C_1$$

and

$$G(x) = \frac{1}{2}\left[f(x) - \int_0^x g(y)\,dy\right] + C_2.$$

Since $F(x) + G(x) = f(x)$ we conclude that $C_1 + C_2 = 0$, and therefore, our final solution of the wave equation with the given initial conditions takes the form

$$u(x,t) = \frac{1}{2}\left[f(x+t) + f(x-t)\right] + \frac{1}{2}\int_{x-t}^{x+t} g(y)\,dy.$$

The form of this solution is known as **d'Alembert's formula**. Observe that the extensions we chose for f and g guarantee that the string always has fixed ends, that is, $u(0,t) = u(\pi,t) = 0$ for all t.

A final remark is in order. The passage from $t \geq 0$ to $t \in \mathbb{R}$, and then back to $t \geq 0$, which was made above, exhibits the time reversal property of the wave equation. In other words, a solution u to the wave equation for $t \geq 0$, leads to a solution u^- defined for negative time $t < 0$ simply by setting $u^-(x,t) = u(x,-t)$, a fact which follows from the invariance of the wave equation under the transformation $t \mapsto -t$. The situation is quite different in the case of the heat equation.

Superposition of standing waves

We turn to the second method of solving the wave equation, which is based on two fundamental conclusions from our previous physical observations. By our considerations of standing waves, we are led to look for special solutions to the wave equation which are of the form $\varphi(x)\psi(t)$. This procedure, which works equally well in other contexts (in the case of the heat equation, for instance), is called **separation of variables** and constructs solutions that are called pure tones. Then by the linearity of the wave equation, we can expect to combine these pure tones into a more complex combination of sound. Pushing this idea further, we can hope ultimately to express the general solution of the wave equation in terms of sums of these particular solutions.

Note that one side of the wave equation involves only differentiation in x, while the other, only differentiation in t. This observation provides another reason to look for solutions of the equation in the form $u(x,t) = \varphi(x)\psi(t)$ (that is, to "separate variables"), the hope being to reduce a difficult partial differential equation into a system of simpler ordinary differential equations. In the case of the wave equation, with u of the above form, we get

$$\varphi(x)\psi''(t) = \varphi''(x)\psi(t),$$

and therefore

$$\frac{\psi''(t)}{\psi(t)} = \frac{\varphi''(x)}{\varphi(x)}.$$

The key observation here is that the left-hand side depends only on t, and the right-hand side only on x. This can happen only if both sides are equal to a constant, say λ. Therefore, the wave equation reduces to the following

$$\begin{cases} \psi''(t) - \lambda\psi(t) = 0 \\ \varphi''(x) - \lambda\varphi(x) = 0. \end{cases} \tag{3}$$

We focus our attention on the first equation in the above system. At this point, the reader will recognize the equation we obtained in the study of simple harmonic motion. Note that we need to consider only the case when $\lambda < 0$, since when $\lambda \geq 0$ the solution ψ will not oscillate as time varies. Therefore, we may write $\lambda = -m^2$, and the solution of the equation is then given by

$$\psi(t) = A\cos mt + B\sin mt.$$

Similarly, we find that the solution of the second equation in (3) is

$$\varphi(x) = \tilde{A}\cos mx + \tilde{B}\sin mx.$$

Now we take into account that the string is attached at $x = 0$ and $x = \pi$. This translates into $\varphi(0) = \varphi(\pi) = 0$, which in turn gives $\tilde{A} = 0$, and if $\tilde{B} \neq 0$, then m must be an integer. If $m = 0$, the solution vanishes identically, and if $m \leq -1$, we may rename the constants and reduce to the case $m \geq 1$ since the function $\sin y$ is odd and $\cos y$ is even. Finally, we arrive at the guess that for each $m \geq 1$, the function

$$u_m(x,t) = (A_m \cos mt + B_m \sin mt) \sin mx,$$

which we recognize as a **standing wave**, is a solution to the wave equation. Note that in the above argument we divided by φ and ψ, which sometimes vanish, so one must actually check by hand that the standing wave u_m solves the equation. This straightforward calculation is left as an exercise to the reader.

Before proceeding further with the analysis of the wave equation, we pause to discuss standing waves in more detail. The terminology comes from looking at the graph of $u_m(x,t)$ for each fixed t. Suppose first that $m = 1$, and take $u(x,t) = \cos t \sin x$. Then, Figure 7 (a) gives the graph of u for different values of t.

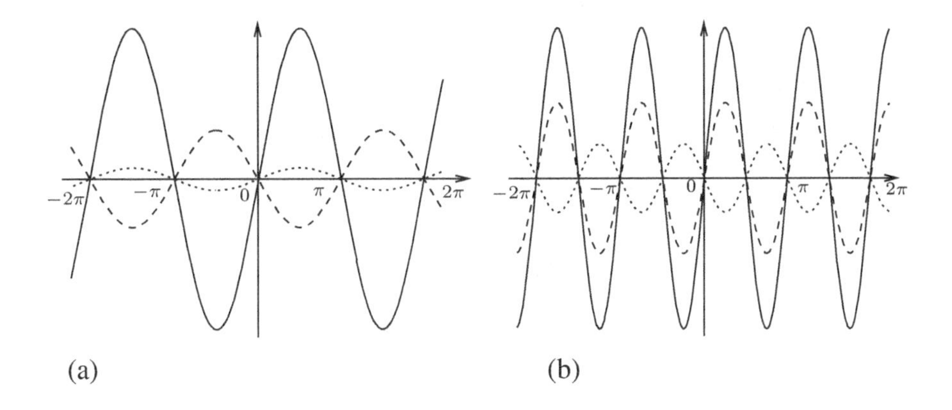

Figure 7. Fundamental tone (a) and overtones (b) at different moments in time

The case $m = 1$ corresponds to the **fundamental tone** or **first harmonic** of the vibrating string.

We now take $m = 2$ and look at $u(x,t) = \cos 2t \sin 2x$. This corresponds to the **first overtone** or **second harmonic**, and this motion is described in Figure 7 (b). Note that $u(\pi/2, t) = 0$ for all t. Such points, which remain motionless in time, are called nodes, while points whose motion has maximum amplitude are named anti-nodes.

For higher values of m we get more overtones or higher harmonics. Note that as m increases, the frequency increases, and the period $2\pi/m$

decreases. Therefore, the fundamental tone has a lower frequency than the overtones.

We now return to the original problem. Recall that the wave equation is linear in the sense that if u and v solve the equation, so does $\alpha u + \beta v$ for any constants α and β. This allows us to construct more solutions by taking linear combinations of the standing waves u_m. This technique, called **superposition**, leads to our final guess for a solution of the wave equation

$$u(x,t) = \sum_{m=1}^{\infty} (A_m \cos mt + B_m \sin mt) \sin mx. \tag{4}$$

Note that the above sum is infinite, so that questions of convergence arise, but since most of our arguments so far are formal, we will not worry about this point now.

Suppose the above expression gave *all* the solutions to the wave equation. If we then require that the initial position of the string at time $t = 0$ is given by the shape of the graph of the function f on $[0, \pi]$, with of course $f(0) = f(\pi) = 0$, we would have $u(x, 0) = f(x)$, hence

$$\sum_{m=1}^{\infty} A_m \sin mx = f(x).$$

Since the initial shape of the string can be any reasonable function f, we must ask the following basic question:

> Given a function f on $[0, \pi]$ (with $f(0) = f(\pi) = 0$), can we find coefficients A_m so that

$$f(x) = \sum_{m=1}^{\infty} A_m \sin mx \ ? \tag{5}$$

This question is stated loosely, but a lot of our effort in the next two chapters of this book will be to formulate the question precisely and attempt to answer it. This was the basic problem that initiated the study of Fourier analysis.

A simple observation allows us to guess a formula giving A_m if the expansion (5) were to hold. Indeed, we multiply both sides by $\sin nx$

and integrate between $[0,\pi]$; working formally, we obtain

$$\begin{aligned}\int_0^\pi f(x)\sin nx\,dx &= \int_0^\pi \left(\sum_{m=1}^\infty A_m \sin mx\right)\sin nx\,dx \\ &= \sum_{m=1}^\infty A_m \int_0^\pi \sin mx\,\sin nx\,dx = A_n\cdot\frac{\pi}{2},\end{aligned}$$

where we have used the fact that

$$\int_0^\pi \sin mx\,\sin nx\,dx = \begin{cases} 0 & \text{if } m\neq n,\\ \pi/2 & \text{if } m=n.\end{cases}$$

Therefore, the guess for A_n, called the nth Fourier sine coefficient of f, is

$$A_n = \frac{2}{\pi}\int_0^\pi f(x)\sin nx\,dx. \tag{6}$$

We shall return to this formula, and other similar ones, later.

One can transform the question about Fourier sine series on $[0,\pi]$ to a more general question on the interval $[-\pi,\pi]$. If we could express f on $[0,\pi]$ in terms of a sine series, then this expansion would also hold on $[-\pi,\pi]$ if we extend f to this interval by making it odd. Similarly, one can ask if an even function $g(x)$ on $[-\pi,\pi]$ can be expressed as a cosine series, namely

$$g(x) = \sum_{m=0}^\infty A'_m \cos mx.$$

More generally, since an arbitrary function F on $[-\pi,\pi]$ can be expressed as $f+g$, where f is odd and g is even,[3] we may ask if F can be written as

$$F(x) = \sum_{m=1}^\infty A_m \sin mx + \sum_{m=0}^\infty A'_m \cos mx,$$

or by applying Euler's identity $e^{ix} = \cos x + i\sin x$, we could hope that F takes the form

$$F(x) = \sum_{m=-\infty}^{\infty} a_m e^{imx}.$$

[3]Take, for example, $f(x) = [F(x) - F(-x)]/2$ and $g(x) = [F(x)+F(-x)]/2$.

By analogy with (6), we can use the fact that

$$\frac{1}{2\pi}\int_{-\pi}^{\pi} e^{imx}e^{-inx}\,dx = \begin{cases} 0 & \text{if } n \neq m \\ 1 & \text{if } n = m, \end{cases}$$

to see that one expects that

$$a_n = \frac{1}{2\pi}\int_{-\pi}^{\pi} F(x)e^{-inx}\,dx.$$

The quantity a_n is called the nth **Fourier coefficient** of F.

We can now reformulate the problem raised above:

Question: Given any reasonable function F on $[-\pi,\pi]$, with Fourier coefficients defined above, is it true that

$$F(x) = \sum_{m=-\infty}^{\infty} a_m e^{imx} \ ? \tag{7}$$

This formulation of the problem, in terms of complex exponentials, is the form we shall use the most in what follows.

Joseph Fourier (1768-1830) was the first to believe that an "arbitrary" function F could be given as a series (7). In other words, his idea was that any function is the linear combination (possibly infinite) of the most basic trigonometric functions $\sin mx$ and $\cos mx$, where m ranges over the integers.[4] Although this idea was implicit in earlier work, Fourier had the conviction that his predecessors lacked, and he used it in his study of heat diffusion; this began the subject of "Fourier analysis." This discipline, which was first developed to solve certain physical problems, has proved to have many applications in mathematics and other fields as well, as we shall see later.

We return to the wave equation. To formulate the problem correctly, we must impose two initial conditions, as our experience with simple harmonic motion and traveling waves indicated. The conditions assign the initial position and velocity of the string. That is, we require that u satisfy the differential equation and the two conditions

$$u(x,0) = f(x) \quad \text{and} \quad \frac{\partial u}{\partial t}(x,0) = g(x),$$

[4]The first proof that a general class of functions can be represented by Fourier series was given later by Dirichlet; see Problem 6, Chapter 4.

where f and g are pre-assigned functions. Note that this is consistent with (4) in that this requires that f and g be expressible as

$$f(x) = \sum_{m=1}^{\infty} A_m \sin mx \quad \text{and} \quad g(x) = \sum_{m=1}^{\infty} mB_m \sin mx.$$

1.3 Example: the plucked string

We now apply our reasoning to the particular problem of the plucked string. For simplicity we choose units so that the string is taken on the interval $[0, \pi]$, and it satisfies the wave equation with $c = 1$. The string is assumed to be plucked to height h at the point p with $0 < p < \pi$; this is the initial position. That is, we take as our initial position the triangular shape given by

$$f(x) = \begin{cases} \dfrac{xh}{p} & \text{for } 0 \leq x \leq p \\ \dfrac{h(\pi - x)}{\pi - p} & \text{for } p \leq x \leq \pi, \end{cases}$$

which is depicted in Figure 8.

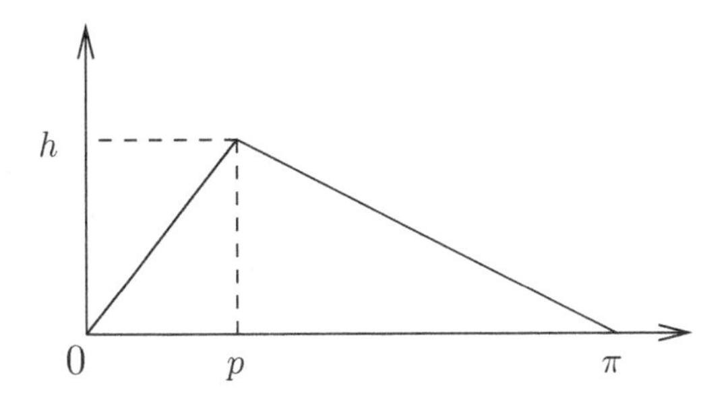

Figure 8. Initial position of a plucked string

We also choose an initial velocity $g(x)$ identically equal to 0. Then, we can compute the Fourier coefficients of f (Exercise 9), and assuming that the answer to the question raised before (5) is positive, we obtain

$$f(x) = \sum_{m=1}^{\infty} A_m \sin mx \quad \text{with} \quad A_m = \frac{2h}{m^2} \frac{\sin mp}{p(\pi - p)}.$$

Thus

$$u(x,t) = \sum_{m=1}^{\infty} A_m \cos mt \sin mx, \tag{8}$$

and note that this series converges absolutely. The solution can also be expressed in terms of traveling waves. In fact

$$u(x,t) = \frac{f(x+t) + f(x-t)}{2}. \tag{9}$$

Here $f(x)$ is defined for all x as follows: first, f is extended to $[-\pi, \pi]$ by making it odd, and then f is extended to the whole real line by making it periodic of period 2π, that is, $f(x + 2\pi k) = f(x)$ for all integers k.

Observe that (8) implies (9) in view of the trigonometric identity

$$\cos v \sin u = \frac{1}{2} [\sin(u+v) + \sin(u-v)].$$

As a final remark, we should note an unsatisfactory aspect of the solution to this problem, which however is in the nature of things. Since the initial data $f(x)$ for the plucked string is not twice continuously differentiable, neither is the function u (given by (9)). Hence u is not truly a solution of the wave equation: while $u(x,t)$ does represent the position of the plucked string, it does not satisfy the partial differential equation we set out to solve! This state of affairs may be understood properly only if we realize that u does solve the equation, but in an appropriate generalized sense. A better understanding of this phenomenon requires ideas relevant to the study of "weak solutions" and the theory of "distributions." These topics we consider only later, in Books III and IV.

2 The heat equation

We now discuss the problem of heat diffusion by following the same framework as for the wave equation. First, we derive the time-dependent heat equation, and then study the steady-state heat equation in the disc, which leads us back to the basic question (7).

2.1 Derivation of the heat equation

Consider an infinite metal plate which we model as the plane $\mathbb{R}^2$, and suppose we are given an initial heat distribution at time $t = 0$. Let the temperature at the point (x, y) at time t be denoted by $u(x, y, t)$.

Consider a small square centered at (x_0, y_0) with sides parallel to the axis and of side length h, as shown in Figure 9. The amount of heat energy in S at time t is given by

$$H(t) = \sigma \iint_S u(x, y, t)\, dx\, dy\,,$$

where $\sigma > 0$ is a constant called the specific heat of the material. Therefore, the heat flow into S is

$$\frac{\partial H}{\partial t} = \sigma \iint_S \frac{\partial u}{\partial t}\, dx\, dy\,,$$

which is approximately equal to

$$\sigma h^2 \frac{\partial u}{\partial t}(x_0, y_0, t),$$

since the area of S is h^2. Now we apply Newton's law of cooling, which states that heat flows from the higher to lower temperature at a rate proportional to the difference, that is, the gradient.

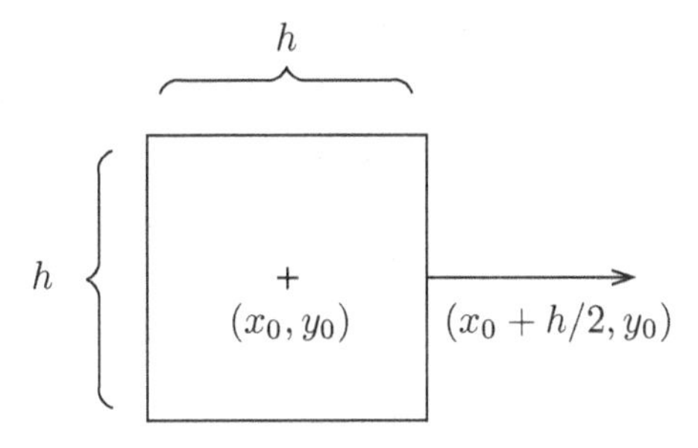

Figure 9. Heat flow through a small square

The heat flow through the vertical side on the right is therefore

$$-\kappa h \frac{\partial u}{\partial x}(x_0 + h/2, y_0, t)\,,$$

where $\kappa > 0$ is the conductivity of the material. A similar argument for the other sides shows that the total heat flow through the square S is given by

$$\kappa h \left[\frac{\partial u}{\partial x}(x_0 + h/2, y_0, t) - \frac{\partial u}{\partial x}(x_0 - h/2, y_0, t) \right.$$
$$\left. + \frac{\partial u}{\partial y}(x_0, y_0 + h/2, t) - \frac{\partial u}{\partial y}(x_0, y_0 - h/2, t) \right].$$

Applying the mean value theorem and letting h tend to zero, we find that

$$\frac{\sigma}{\kappa} \frac{\partial u}{\partial t} = \frac{\partial^2 u}{\partial x^2} + \frac{\partial^2 u}{\partial y^2};$$

this is called the **time-dependent heat equation**, often abbreviated to the heat equation.

2.2 Steady-state heat equation in the disc

After a long period of time, there is no more heat exchange, so that the system reaches thermal equilibrium and $\partial u / \partial t = 0$. In this case, the time-dependent heat equation reduces to the **steady-state heat equation**

$$\frac{\partial^2 u}{\partial x^2} + \frac{\partial^2 u}{\partial y^2} = 0. \tag{10}$$

The operator $\partial^2/\partial x^2 + \partial^2/\partial y^2$ is of such importance in mathematics and physics that it is often abbreviated as $\triangle$ and given a name: the Laplace operator or **Laplacian**. So the steady-state heat equation is written as

$$\triangle u = 0,$$

and solutions to this equation are called **harmonic functions**.

Consider the unit disc in the plane

$$D = \{(x, y) \in \mathbb{R}^2 : x^2 + y^2 < 1\},$$

whose boundary is the unit circle C. In polar coordinates (r, θ), with $0 \leq r$ and $0 \leq \theta < 2\pi$, we have

$$D = \{(r, \theta) : 0 \leq r < 1\} \quad \text{and} \quad C = \{(r, \theta) : r = 1\}.$$

The problem, often called the **Dirichlet problem** (for the Laplacian on the unit disc), is to solve the steady-state heat equation in the unit disc subject to the boundary condition $u = f$ on C. This corresponds to fixing a predetermined temperature distribution on the circle, waiting a long time, and then looking at the temperature distribution inside the disc.

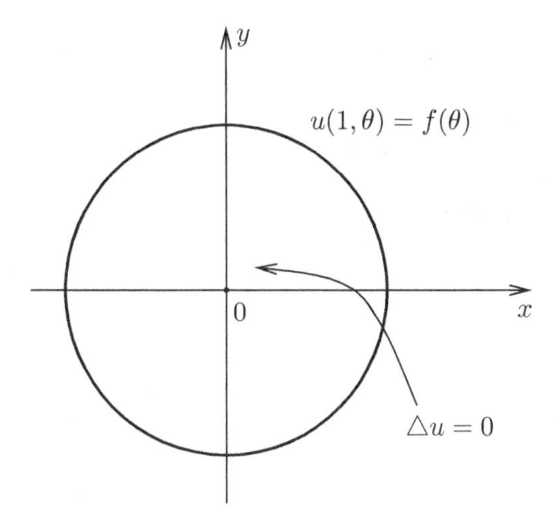

Figure 10. The Dirichlet problem for the disc

While the method of separation of variables will turn out to be useful for equation (10), a difficulty comes from the fact that the boundary condition is not easily expressed in terms of rectangular coordinates. Since this boundary condition is best described by the coordinates (r, θ), namely $u(1, \theta) = f(\theta)$, we rewrite the Laplacian in polar coordinates. An application of the chain rule gives (Exercise 10):

$$\triangle u = \frac{\partial^2 u}{\partial r^2} + \frac{1}{r}\frac{\partial u}{\partial r} + \frac{1}{r^2}\frac{\partial^2 u}{\partial \theta^2}.$$

We now multiply both sides by r^2, and since $\triangle u = 0$, we get

$$r^2\frac{\partial^2 u}{\partial r^2} + r\frac{\partial u}{\partial r} = -\frac{\partial^2 u}{\partial \theta^2}.$$

Separating these variables, and looking for a solution of the form $u(r, \theta) = F(r)G(\theta)$, we find

$$\frac{r^2 F''(r) + rF'(r)}{F(r)} = -\frac{G''(\theta)}{G(\theta)}.$$

Since the two sides depend on different variables, they must both be constant, say equal to λ. We therefore get the following equations:

$$\begin{cases} G''(\theta) + \lambda G(\theta) = 0\,, \\ r^2 F''(r) + rF'(r) - \lambda F(r) = 0. \end{cases}$$

Since G must be periodic of period 2π, this implies that $\lambda \geq 0$ and (as we have seen before) that $\lambda = m^2$ where m is an integer; hence

$$G(\theta) = \tilde{A} \cos m\theta + \tilde{B} \sin m\theta.$$

An application of Euler's identity, $e^{ix} = \cos x + i \sin x$, allows one to rewrite G in terms of complex exponentials,

$$G(\theta) = Ae^{im\theta} + Be^{-im\theta}.$$

With $\lambda = m^2$ and $m \neq 0$, two simple solutions of the equation in F are $F(r) = r^m$ and $F(r) = r^{-m}$ (Exercise 11 gives further information about these solutions). If $m = 0$, then $F(r) = 1$ and $F(r) = \log r$ are two solutions. If $m > 0$, we note that r^{-m} grows unboundedly large as r tends to zero, so $F(r)G(\theta)$ is unbounded at the origin; the same occurs when $m = 0$ and $F(r) = \log r$. We reject these solutions as contrary to our intuition. Therefore, we are left with the following special functions:

$$u_m(r, \theta) = r^{|m|} e^{im\theta}, \qquad m \in \mathbb{Z}.$$

We now make the important observation that (10) is *linear*, and so as in the case of the vibrating string, we may superpose the above special solutions to obtain the presumed general solution:

$$u(r, \theta) = \sum_{m=-\infty}^{\infty} a_m r^{|m|} e^{im\theta}.$$

If this expression gave all the solutions to the steady-state heat equation, then for a reasonable f we should have

$$u(1, \theta) = \sum_{m=-\infty}^{\infty} a_m e^{im\theta} = f(\theta).$$

We therefore ask again in this context: given any reasonable function f on $[0, 2\pi]$ with $f(0) = f(2\pi)$, can we find coefficients a_m so that

$$f(\theta) = \sum_{m=-\infty}^{\infty} a_m e^{im\theta} \; ?$$

Historical Note: D'Alembert (in 1747) first solved the equation of the vibrating string using the method of traveling waves. This solution was elaborated by Euler a year later. In 1753, D. Bernoulli proposed the solution which for all intents and purposes is the Fourier series given by (4), but Euler was not entirely convinced of its full generality, since this could hold only if an "arbitrary" function could be expanded in Fourier series. D'Alembert and other mathematicians also had doubts. This viewpoint was changed by Fourier (in 1807) in his study of the heat equation, where his conviction and work eventually led others to a complete proof that a general function could be represented as a Fourier series.

3 Exercises

1. If $z = x + iy$ is a complex number with $x, y \in \mathbb{R}$, we define

$$|z| = (x^2 + y^2)^{1/2}$$

and call this quantity the **modulus** or **absolute value** of z.

(a) What is the geometric interpretation of $|z|$?

(b) Show that if $|z| = 0$, then $z = 0$.

(c) Show that if $\lambda \in \mathbb{R}$, then $|\lambda z| = |\lambda||z|$, where $|\lambda|$ denotes the standard absolute value of a real number.

(d) If z_1 and z_2 are two complex numbers, prove that

$$|z_1 z_2| = |z_1||z_2| \quad \text{and} \quad |z_1 + z_2| \leq |z_1| + |z_2|.$$

(e) Show that if $z \neq 0$, then $|1/z| = 1/|z|$.

2. If $z = x + iy$ is a complex number with $x, y \in \mathbb{R}$, we define the **complex conjugate** of z by

$$\overline{z} = x - iy.$$

(a) What is the geometric interpretation of $\overline{z}$?

(b) Show that $|z|^2 = z\overline{z}$.

(c) Prove that if z belongs to the unit circle, then $1/z = \overline{z}$.

3. A sequence of complex numbers $\{w_n\}_{n=1}^{\infty}$ is said to converge if there exists $w \in \mathbb{C}$ such that

$$\lim_{n\to\infty} |w_n - w| = 0,$$

and we say that w is a limit of the sequence.

(a) Show that a converging sequence of complex numbers has a unique limit.

The sequence $\{w_n\}_{n=1}^{\infty}$ is said to be a **Cauchy sequence** if for every $\epsilon > 0$ there exists a positive integer N such that

$$|w_n - w_m| < \epsilon \qquad \text{whenever } n, m > N.$$

(b) Prove that a sequence of complex numbers converges if and only if it is a Cauchy sequence. [Hint: A similar theorem exists for the convergence of a sequence of real numbers. Why does it carry over to sequences of complex numbers?]

A series $\sum_{n=1}^{\infty} z_n$ of complex numbers is said to converge if the sequence formed by the partial sums

$$S_N = \sum_{n=1}^{N} z_n$$

converges. Let $\{a_n\}_{n=1}^{\infty}$ be a sequence of non-negative real numbers such that the series $\sum_n a_n$ converges.

(c) Show that if $\{z_n\}_{n=1}^{\infty}$ is a sequence of complex numbers satisfying $|z_n| \leq a_n$ for all n, then the series $\sum_n z_n$ converges. [Hint: Use the Cauchy criterion.]

4. For $z \in \mathbb{C}$, we define the **complex exponential** by

$$e^z = \sum_{n=0}^{\infty} \frac{z^n}{n!}.$$

(a) Prove that the above definition makes sense, by showing that the series converges for every complex number z. Moreover, show that the convergence is uniform[5] on every bounded subset of $\mathbb{C}$.

(b) If z_1, z_2 are two complex numbers, prove that $e^{z_1}e^{z_2} = e^{z_1+z_2}$. [Hint: Use the binomial theorem to expand $(z_1 + z_2)^n$, as well as the formula for the binomial coefficients.]

---

[5] A sequence of functions $\{f_n(z)\}_{n=1}^{\infty}$ is said to be uniformly convergent on a set S if there exists a function f on S so that for every $\epsilon > 0$ there is an integer N such that $|f_n(z) - f(z)| < \epsilon$ whenever $n > N$ and $z \in S$.

(c) Show that if z is purely imaginary, that is, $z = iy$ with $y \in \mathbb{R}$, then

$$e^{iy} = \cos y + i \sin y.$$

This is Euler's identity. [Hint: Use power series.]

(d) More generally,

$$e^{x+iy} = e^x (\cos y + i \sin y)$$

whenever $x, y \in \mathbb{R}$, and show that

$$|e^{x+iy}| = e^x.$$

(e) Prove that $e^z = 1$ if and only if $z = 2\pi k i$ for some integer k.

(f) Show that every complex number $z = x + iy$ can be written in the form

$$z = re^{i\theta},$$

where r is unique and in the range $0 \leq r < \infty$, and $\theta \in \mathbb{R}$ is unique up to an integer multiple of 2π. Check that

$$r = |z| \quad \text{and} \quad \theta = \arctan(y/x)$$

whenever these formulas make sense.

(g) In particular, $i = e^{i\pi/2}$. What is the geometric meaning of multiplying a complex number by i? Or by $e^{i\theta}$ for any $\theta \in \mathbb{R}$?

(h) Given $\theta \in \mathbb{R}$, show that

$$\cos\theta = \frac{e^{i\theta} + e^{-i\theta}}{2} \quad \text{and} \quad \sin\theta = \frac{e^{i\theta} - e^{-i\theta}}{2i}.$$

These are also called Euler's identities.

(i) Use the complex exponential to derive trigonometric identities such as

$$\cos(\theta + \vartheta) = \cos\theta\cos\vartheta - \sin\theta\sin\vartheta,$$

and then show that

$$\begin{aligned} 2\sin\theta\sin\varphi &= \cos(\theta - \varphi) - \cos(\theta + \varphi), \\ 2\sin\theta\cos\varphi &= \sin(\theta + \varphi) + \sin(\theta - \varphi). \end{aligned}$$

This calculation connects the solution given by d'Alembert in terms of traveling waves and the solution in terms of superposition of standing waves.

5. Verify that $f(x) = e^{inx}$ is periodic with period 2π and that

$$\frac{1}{2\pi}\int_{-\pi}^{\pi} e^{inx}\,dx = \begin{cases} 1 & \text{if } n = 0, \\ 0 & \text{if } n \neq 0. \end{cases}$$

Use this fact to prove that if $n, m \geq 1$ we have

$$\frac{1}{\pi}\int_{-\pi}^{\pi} \cos nx \, \cos mx \, dx = \begin{cases} 0 & \text{if } n \neq m, \\ 1 & n = m, \end{cases}$$

and similarly

$$\frac{1}{\pi}\int_{-\pi}^{\pi} \sin nx \, \sin mx \, dx = \begin{cases} 0 & \text{if } n \neq m, \\ 1 & n = m. \end{cases}$$

Finally, show that

$$\int_{-\pi}^{\pi} \sin nx \, \cos mx \, dx = 0 \qquad \text{for any } n, m.$$

[Hint: Calculate $e^{inx}e^{-imx} + e^{inx}e^{imx}$ and $e^{inx}e^{-imx} - e^{inx}e^{imx}$.]

6. Prove that if f is a twice continuously differentiable function on $\mathbb{R}$ which is a solution of the equation

$$f''(t) + c^2 f(t) = 0,$$

then there exist constants a and b such that

$$f(t) = a\cos ct + b\sin ct.$$

This can be done by differentiating the two functions $g(t) = f(t)\cos ct - c^{-1}f'(t)\sin c$ and $h(t) = f(t)\sin ct + c^{-1}f'(t)\cos ct$.

7. Show that if a and b are real, then one can write

$$a\cos ct + b\sin ct = A\cos(ct - \varphi),$$

where $A = \sqrt{a^2 + b^2}$, and φ is chosen so that

$$\cos\varphi = \frac{a}{\sqrt{a^2+b^2}} \quad \text{and} \quad \sin\varphi = \frac{b}{\sqrt{a^2+b^2}}.$$

8. Suppose F is a function on (a, b) with two continuous derivatives. Show that whenever x and $x + h$ belong to (a, b), one may write

$$F(x+h) = F(x) + hF'(x) + \frac{h^2}{2}F''(x) + h^2\varphi(h),$$

where $\varphi(h) \to 0$ as $h \to 0$.

Deduce that

$$\frac{F(x+h) + F(x-h) - 2F(x)}{h^2} \to F''(x) \qquad \text{as } h \to 0.$$

[Hint: This is simply a Taylor expansion. It may be obtained by noting that

$$F(x+h) - F(x) = \int_x^{x+h} F'(y)\,dy,$$

and then writing $F'(y) = F'(x) + (y-x)F''(x) + (y-x)\psi(y-x)$, where $\psi(h) \to 0$ as $h \to 0$.]

9. In the case of the plucked string, use the formula for the Fourier sine coefficients to show that

$$A_m = \frac{2h}{m^2} \frac{\sin mp}{p(\pi - p)}.$$

For what position of p are the second, fourth, ... harmonics missing? For what position of p are the third, sixth, ... harmonics missing?

10. Show that the expression of the Laplacian

$$\triangle = \frac{\partial^2}{\partial x^2} + \frac{\partial^2}{\partial y^2}$$

is given in polar coordinates by the formula

$$\triangle = \frac{\partial^2}{\partial r^2} + \frac{1}{r}\frac{\partial}{\partial r} + \frac{1}{r^2}\frac{\partial^2}{\partial \theta^2}.$$

Also, prove that

$$\left|\frac{\partial u}{\partial x}\right|^2 + \left|\frac{\partial u}{\partial y}\right|^2 = \left|\frac{\partial u}{\partial r}\right|^2 + \frac{1}{r^2}\left|\frac{\partial u}{\partial \theta}\right|^2.$$

11. Show that if $n \in \mathbb{Z}$ the only solutions of the differential equation

$$r^2 F''(r) + rF'(r) - n^2 F(r) = 0,$$

which are twice differentiable when $r > 0$, are given by linear combinations of r^n and r^{-n} when $n \neq 0$, and 1 and $\log r$ when $n = 0$.

[Hint: If F solves the equation, write $F(r) = g(r)r^n$, find the equation satisfied by g, and conclude that $rg'(r) + 2ng(r) = c$ where c is a constant.]

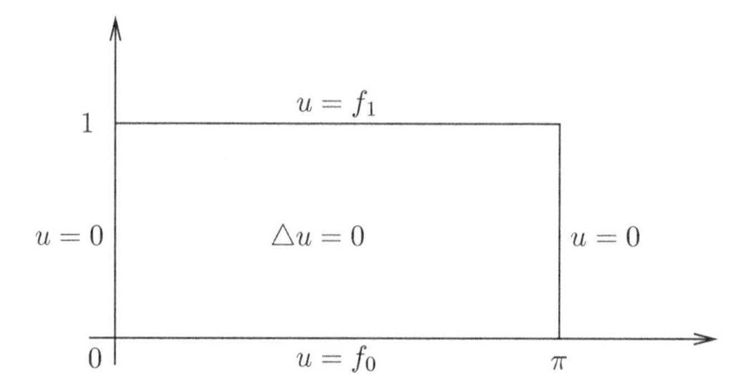

Figure 11. Dirichlet problem in a rectangle

4 Problem

1. Consider the Dirichlet problem illustrated in Figure 11.

More precisely, we look for a solution of the steady-state heat equation $\Delta u = 0$ in the rectangle $R = \{(x, y) : 0 \leq x \leq \pi, \ \ 0 \leq y \leq 1\}$ that vanishes on the vertical sides of R, and so that

$$u(x, 0) = f_0(x) \qquad \text{and} \qquad u(x, 1) = f_1(x),$$

where f_0 and f_1 are initial data which fix the temperature distribution on the horizontal sides of the rectangle.

Use separation of variables to show that if f_0 and f_1 have Fourier expansions

$$f_0(x) = \sum_{k=1}^{\infty} A_k \sin kx \qquad \text{and} \qquad f_1(x) = \sum_{k=1}^{\infty} B_k \sin kx,$$

then

$$u(x, y) = \sum_{k=1}^{\infty} \left(\frac{\sinh k(1-y)}{\sinh k} A_k + \frac{\sinh ky}{\sinh k} B_k \right) \sin kx.$$

We recall the definitions of the hyperbolic sine and cosine functions:

$$\sinh x = \frac{e^x - e^{-x}}{2} \qquad \text{and} \qquad \cosh x = \frac{e^x + e^{-x}}{2}.$$

Compare this result with the solution of the Dirichlet problem in the strip obtained in Problem 3, Chapter 5.

2 Basic Properties of Fourier Series

> Nearly fifty years had passed without any progress on the question of analytic representation of an arbitrary function, when an assertion of Fourier threw new light on the subject. Thus a new era began for the development of this part of Mathematics and this was heralded in a stunning way by major developments in mathematical Physics.
>
> *B. Riemann*, 1854

In this chapter, we begin our rigorous study of Fourier series. We set the stage by introducing the main objects in the subject, and then formulate some basic problems which we have already touched upon earlier.

Our first result disposes of the question of uniqueness: Are two functions with the same Fourier coefficients necessarily equal? Indeed, a simple argument shows that if both functions are continuous, then in fact they must agree.

Next, we take a closer look at the partial sums of a Fourier series. Using the formula for the Fourier coefficients (which involves an integration), we make the key observation that these sums can be written conveniently as integrals:

$$\frac{1}{2\pi}\int D_N(x-y)f(y)\,dy,$$

where $\{D_N\}$ is a family of functions called the Dirichlet kernels. The above expression is the convolution of f with the function D_N. Convolutions will play a critical role in our analysis. In general, given a family of functions $\{K_n\}$, we are led to investigate the limiting properties as n tends to infinity of the convolutions

$$\frac{1}{2\pi}\int K_n(x-y)f(y)\,dy.$$

We find that if the family $\{K_n\}$ satisfies the three important properties of "good kernels," then the convolutions above tend to $f(x)$ as $n \to \infty$ (at least when f is continuous). In this sense, the family $\{K_n\}$ is an "approximation to the identity." Unfortunately, the Dirichlet kernels D_N do not belong to the category of good kernels, which indicates that the question of convergence of Fourier series is subtle.

Instead of pursuing at this stage the problem of convergence, we consider various other methods of summing the Fourier series of a function. The first method, which involves averages of partial sums, leads to convolutions with good kernels, and yields an important theorem of Fejér. From this, we deduce the fact that a continuous function on the circle can be approximated uniformly by trigonometric polynomials. Second, we may also sum the Fourier series in the sense of Abel and again encounter a family of good kernels. In this case, the results about convolutions and good kernels lead to a solution of the Dirichlet problem for the steady-state heat equation in the disc, considered at the end of the previous chapter.

1 Examples and formulation of the problem

We commence with a brief description of the types of functions with which we shall be concerned. Since the Fourier coefficients of f are defined by

$$a_n = \frac{1}{L}\int_0^L f(x)e^{-2\pi inx/L}\,dx, \qquad \text{for } n \in \mathbb{Z},$$

where f is complex-valued on $[0, L]$, it will be necessary to place some integrability conditions on f. We shall therefore assume for the remainder of this book that all functions are at least Riemann integrable.[1] Sometimes it will be illuminating to focus our attention on functions that are more "regular," that is, functions that possess certain continuity or differentiability properties. Below, we list several classes of functions in increasing order of generality. We emphasize that we will not generally restrict our attention to real-valued functions, contrary to what the following pictures may suggest; we will almost always allow functions that take values in the complex numbers $\mathbb{C}$. Furthermore, we sometimes think of our functions as being defined on the circle rather than an interval. We elaborate upon this below.

[1] Limiting ourselves to Riemann integrable functions is natural at this elementary stage of study of the subject. The more advanced notion of Lebesgue integrability will be taken up in Book III.

Everywhere continuous functions

These are the complex-valued functions f which are continuous at every point of the segment $[0, L]$. A typical continuous function is sketched in Figure 1 (a). We shall note later that continuous functions on the circle satisfy the additional condition $f(0) = f(L)$.

Piecewise continuous functions

These are bounded functions on $[0, L]$ which have only finitely many discontinuities. An example of such a function with simple discontinuities is pictured in Figure 1 (b).

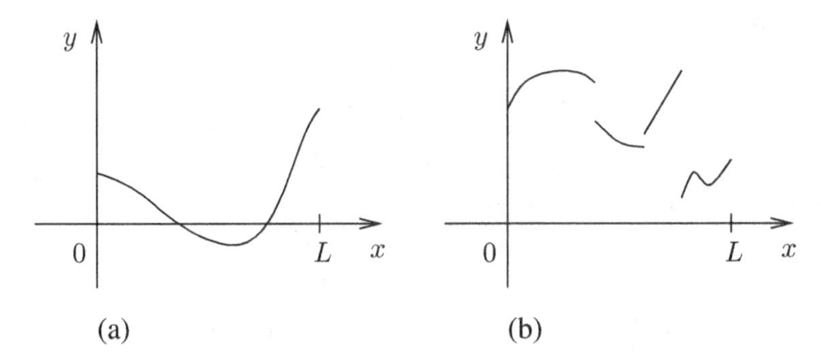

Figure 1. Functions on $[0, L]$: continuous and piecewise continuous

This class of functions is wide enough to illustrate many of the theorems in the next few chapters. However, for logical completeness we consider also the more general class of Riemann integrable functions. This more extended setting is natural since the formula for the Fourier coefficients involves integration.

Riemann integrable functions

This is the most general class of functions we will be concerned with. Such functions are bounded, but may have infinitely many discontinuities. We recall the definition of integrability. A real-valued function f defined on $[0, L]$ is **Riemann integrable** (which we abbreviate as **integrable**[2]) if it is *bounded*, and if for every $\epsilon > 0$, there is a subdivision $0 = x_0 < x_1 < \cdots < x_{N-1} < x_N = L$ of the interval $[0, L]$, so that if $\mathcal{U}$

[2]Starting in Book III, the term "integrable" will be used in the broader sense of Lebesgue theory.

and $\mathcal{L}$ are, respectively, the upper and lower sums of f for this subdivision, namely

$$\mathcal{U} = \sum_{j=1}^{N} [\sup_{x_{j-1} \le x \le x_j} f(x)](x_j - x_{j-1})$$

and

$$\mathcal{L} = \sum_{j=1}^{N} [\inf_{x_{j-1} \le x \le x_j} f(x)](x_j - x_{j-1}),$$

then we have $\mathcal{U} - \mathcal{L} < \epsilon$. Finally, we say that a complex-valued function is integrable if its real and imaginary parts are integrable. It is worthwhile to remember at this point that the sum and product of two integrable functions are integrable.

A simple example of an integrable function on $[0,1]$ with infinitely many discontinuities is given by

$$f(x) = \begin{cases} 1 & \text{if } 1/(n+1) < x \le 1/n \text{ and } n \text{ is odd,} \\ 0 & \text{if } 1/(n+1) < x \le 1/n \text{ and } n \text{ is even,} \\ 0 & \text{if } x = 0. \end{cases}$$

This example is illustrated in Figure 2. Note that f is discontinuous when $x = 1/n$ and at $x = 0$.

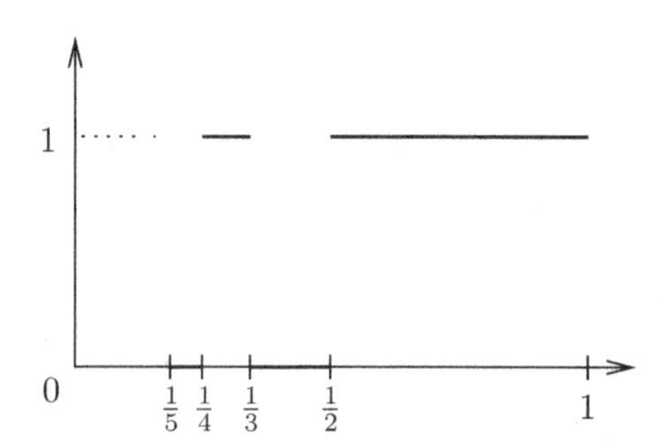

Figure 2. A Riemann integrable function

More elaborate examples of integrable functions whose discontinuities are dense in the interval $[0,1]$ are described in Problem 1. In general, while integrable functions may have infinitely many discontinuities, these functions are actually characterized by the fact that, in a precise sense, their discontinuities are not too numerous: they are "negligible," that is, the set of points where an integrable function is discontinuous has "measure 0." The reader will find further details about Riemann integration in the appendix.

From now on, we shall always assume that our functions are integrable, even if we do not state this requirement explicitly.

Functions on the circle

There is a natural connection between 2π-periodic functions on $\mathbb{R}$ like the exponentials $e^{in\theta}$, functions on an interval of length 2π, and functions on the unit circle. This connection arises as follows.

A point on the unit circle takes the form $e^{i\theta}$, where θ is a real number that is unique up to integer multiples of 2π. If F is a function on the circle, then we may define for each real number θ

$$f(\theta) = F(e^{i\theta}),$$

and observe that with this definition, the function f is periodic on $\mathbb{R}$ of period 2π, that is, $f(\theta + 2\pi) = f(\theta)$ for all θ. The integrability, continuity and other smoothness properties of F are determined by those of f. For instance, we say that F is integrable on the circle if f is integrable on every interval of length 2π. Also, F is continuous on the circle if f is continuous on $\mathbb{R}$, which is the same as saying that f is continuous on any interval of length 2π. Moreover, F is continuously differentiable if f has a continuous derivative, and so forth.

Since f has period 2π, we may restrict it to any interval of length 2π, say $[0, 2\pi]$ or $[-\pi, \pi]$, and still capture the initial function F on the circle. We note that f must take the same value at the end-points of the interval since they correspond to the same point on the circle. Conversely, any function on $[0, 2\pi]$ for which $f(0) = f(2\pi)$ can be extended to a periodic function on $\mathbb{R}$ which can then be identified as a function on the circle. In particular, a continuous function f on the interval $[0, 2\pi]$ gives rise to a continuous function on the circle if and only if $f(0) = f(2\pi)$.

In conclusion, functions on $\mathbb{R}$ that 2π-periodic, and functions on an interval of length 2π that take on the same value at its end-points, are two equivalent descriptions of the same mathematical objects, namely, functions on the circle.

In this connection, we mention an item of notational usage. When our functions are defined on an interval on the line, we often use x as the independent variable; however, when we consider these as functions on the circle, we usually replace the variable x by θ. As the reader will note, we are not strictly bound by this rule since this practice is mostly a matter of convenience.

1.1 Main definitions and some examples

We now begin our study of Fourier analysis with the precise definition of the Fourier series of a function. Here, it is important to pin down where our function is originally defined. If f is an integrable function given on an interval $[a, b]$ of length L (that is, $b - a = L$), then the nth **Fourier coefficient** of f is defined by

$$\hat{f}(n) = \frac{1}{L} \int_a^b f(x) e^{-2\pi i n x/L}\, dx, \qquad n \in \mathbb{Z}.$$

The **Fourier series** of f is given formally[3] by

$$\sum_{n=-\infty}^{\infty} \hat{f}(n) e^{2\pi i n x/L}.$$

We shall sometimes write a_n for the Fourier coefficients of f, and use the notation

$$f(x) \sim \sum_{n=-\infty}^{\infty} a_n e^{2\pi i n x/L}$$

to indicate that the series on the right-hand side is the Fourier series of f.

For instance, if f is an integrable function on the interval $[-\pi, \pi]$, then the nth Fourier coefficient of f is

$$\hat{f}(n) = a_n = \frac{1}{2\pi} \int_{-\pi}^{\pi} f(\theta) e^{-in\theta}\, d\theta, \qquad n \in \mathbb{Z},$$

and the Fourier series of f is

$$f(\theta) \sim \sum_{n=-\infty}^{\infty} a_n e^{in\theta}.$$

Here we use θ as a variable since we think of it as an angle ranging from $-\pi$ to π.

[3]At this point, we do not say anything about the convergence of the series.

Also, if f is defined on $[0, 2\pi]$, then the formulas are the same as above, except that we integrate from 0 to 2π in the definition of the Fourier coefficients.

We may also consider the Fourier coefficients and Fourier series for a function defined on the circle. By our previous discussion, we may think of a function on the circle as a function f on $\mathbb{R}$ which is 2π-periodic. We may restrict the function f to any interval of length 2π, for instance $[0, 2\pi]$ or $[-\pi, \pi]$, and compute its Fourier coefficients. Fortunately, f is *periodic* and Exercise 1 shows that the resulting integrals are independent of the chosen interval. Thus the Fourier coefficients of a function on the circle are well defined.

Finally, we shall sometimes consider a function g given on $[0, 1]$. Then

$$\hat{g}(n) = a_n = \int_0^1 g(x)e^{-2\pi inx}\, dx \quad \text{and} \quad g(x) \sim \sum_{n=-\infty}^{\infty} a_n e^{2\pi inx}.$$

Here we use x for a variable ranging from 0 to 1.

Of course, if f is initially given on $[0, 2\pi]$, then $g(x) = f(2\pi x)$ is defined on $[0, 1]$ and a change of variables shows that the nth Fourier coefficient of f equals the nth Fourier coefficient of g.

Fourier series are part of a larger family called the **trigonometric series** which, by definition, are expressions of the form $\sum_{n=-\infty}^{\infty} c_n e^{2\pi inx/L}$ where $c_n \in \mathbb{C}$. If a trigonometric series involves only finitely many non-zero terms, that is, $c_n = 0$ for all large $|n|$, it is called a **trigonometric polynomial**; its **degree** is the largest value of $|n|$ for which $c_n \neq 0$.

The Nth **partial sum** of the Fourier series of f, for N a positive integer, is a particular example of a trigonometric polynomial. It is given by

$$S_N(f)(x) = \sum_{n=-N}^{N} \hat{f}(n)e^{2\pi inx/L}.$$

Note that by definition, the above sum is *symmetric* since n ranges from $-N$ to N, a choice that is natural because of the resulting decomposition of the Fourier series as sine and cosine series. As a consequence, the convergence of Fourier series will be understood (in this book) as the "limit" as N tends to infinity of these symmetric sums.

In fact, using the partial sums of the Fourier series, we can reformulate the basic question raised in Chapter 1 as follows:

Problem: In what sense does $S_N(f)$ converge to f as $N \to \infty$?

Before proceeding further with this question, we turn to some simple examples of Fourier series.

EXAMPLE 1. Let $f(\theta) = \theta$ for $-\pi \leq \theta \leq \pi$. The calculation of the Fourier coefficients requires a simple integration by parts. First, if $n \neq 0$, then

$$\begin{aligned}\hat{f}(n) &= \frac{1}{2\pi}\int_{-\pi}^{\pi} \theta e^{-in\theta}\, d\theta \\ &= \frac{1}{2\pi}\left[-\frac{\theta}{in}e^{-in\theta}\right]_{-\pi}^{\pi} + \frac{1}{2\pi in}\int_{-\pi}^{\pi} e^{-in\theta}\, d\theta \\ &= \frac{(-1)^{n+1}}{in},\end{aligned}$$

and if $n = 0$ we clearly have

$$\hat{f}(0) = \frac{1}{2\pi}\int_{-\pi}^{\pi} \theta\, d\theta = 0.$$

Hence, the Fourier series of f is given by

$$f(\theta) \sim \sum_{n\neq 0} \frac{(-1)^{n+1}}{in} e^{in\theta} = 2\sum_{n=1}^{\infty} (-1)^{n+1} \frac{\sin n\theta}{n}.$$

The first sum is over all non-zero integers, and the second is obtained by an application of Euler's identities. It is possible to prove by elementary means that the above series converges for every θ, but it is not obvious that it converges to $f(\theta)$. This will be proved later (Exercises 8 and 9 deal with a similar situation).

EXAMPLE 2. Define $f(\theta) = (\pi - \theta)^2/4$ for $0 \leq \theta \leq 2\pi$. Then successive integration by parts similar to that performed in the previous example yield

$$f(\theta) \sim \frac{\pi^2}{12} + \sum_{n=1}^{\infty} \frac{\cos n\theta}{n^2}.$$

EXAMPLE 3. The Fourier series of the function

$$f(\theta) = \frac{\pi}{\sin \pi\alpha} e^{i(\pi-\theta)\alpha}$$

on $[0, 2\pi]$ is

$$f(\theta) \sim \sum_{n=-\infty}^{\infty} \frac{e^{in\theta}}{n+\alpha},$$

whenever α is not an integer.

EXAMPLE 4. The trigonometric polynomial defined for $x \in [-\pi, \pi]$ by

$$D_N(x) = \sum_{n=-N}^{N} e^{inx}$$

is called the Nth **Dirichlet kernel** and is of fundamental importance in the theory (as we shall see later). Notice that its Fourier coefficients a_n have the property that $a_n = 1$ if $|n| \leq N$ and $a_n = 0$ otherwise. A closed form formula for the Dirichlet kernel is

$$D_N(x) = \frac{\sin((N + \frac{1}{2})x)}{\sin(x/2)}.$$

This can be seen by summing the geometric progressions

$$\sum_{n=0}^{N} \omega^n \quad \text{and} \quad \sum_{n=-N}^{-1} \omega^n$$

with $\omega = e^{ix}$. These sums are, respectively, equal to

$$\frac{1 - \omega^{N+1}}{1 - \omega} \quad \text{and} \quad \frac{\omega^{-N} - 1}{1 - \omega}.$$

Their sum is then

$$\frac{\omega^{-N} - \omega^{N+1}}{1 - \omega} = \frac{\omega^{-N-1/2} - \omega^{N+1/2}}{\omega^{-1/2} - \omega^{1/2}} = \frac{\sin((N + \frac{1}{2})x)}{\sin(x/2)},$$

giving the desired result.

EXAMPLE 5. The function $P_r(\theta)$, called the **Poisson kernel**, is defined for $\theta \in [-\pi, \pi]$ and $0 \leq r < 1$ by the absolutely and uniformly convergent series

$$P_r(\theta) = \sum_{n=-\infty}^{\infty} r^{|n|} e^{in\theta}.$$

This function arose implicitly in the solution of the steady-state heat equation on the unit disc discussed in Chapter 1. Note that in calculating the Fourier coefficients of $P_r(\theta)$ we can interchange the order of integration and summation since the sum converges uniformly in θ for each fixed r, and obtain that the nth Fourier coefficient equals $r^{|n|}$. One can also sum the series for $P_r(\theta)$ and see that

$$P_r(\theta) = \frac{1-r^2}{1-2r\cos\theta + r^2}.$$

In fact,

$$P_r(\theta) = \sum_{n=0}^{\infty} \omega^n + \sum_{n=1}^{\infty} \overline{\omega}^n \quad \text{with } \omega = re^{i\theta},$$

where both series converge absolutely. The first sum (an infinite geometric progression) equals $1/(1-\omega)$, and likewise, the second is $\overline{\omega}/(1-\overline{\omega})$. Together, they combine to give

$$\frac{1-\overline{\omega}+(1-\omega)\overline{\omega}}{(1-\omega)(1-\overline{\omega})} = \frac{1-|\omega|^2}{|1-\omega|^2} = \frac{1-r^2}{1-2r\cos\theta+r^2},$$

as claimed. The Poisson kernel will reappear later in the context of Abel summability of the Fourier series of a function.

Let us return to the problem formulated earlier. The definition of the Fourier series of f is purely formal, and it is not obvious whether it converges to f. In fact, the solution of this problem can be very hard, or relatively easy, depending on the sense in which we expect the series to converge, or on what additional restrictions we place on f.

Let us be more precise. Suppose, for the sake of this discussion, that the function f (which is always assumed to be Riemann integrable) is defined on $[-\pi,\pi]$. The first question one might ask is whether the partial sums of the Fourier series of f converge to f pointwise. That is, do we have

$$\lim_{N\to\infty} S_N(f)(\theta) = f(\theta) \quad \text{for every } \theta? \tag{1}$$

We see quite easily that in general we cannot expect this result to be true at every θ, since we can always change an integrable function at one point without changing its Fourier coefficients. As a result, we might ask the same question assuming that f is continuous and periodic. For a long time it was believed that under these additional assumptions the answer would be "yes." It was a surprise when Du Bois-Reymond showed that there exists a continuous function whose Fourier series diverges at a point. We will give such an example in the next chapter. Despite this negative result, we might ask what happens if we add more smoothness conditions on f: for example, we might assume that f is continuously differentiable, or twice continuously differentiable. We will see that then the Fourier series of f converges to f uniformly.

We will also interpret the limit (1) by showing that the Fourier series sums, in the sense of Cesàro or Abel, to the function f at all of its points of continuity. This approach involves appropriate averages of the partial sums of the Fourier series of f.

Finally, we can also define the limit (1) in the mean square sense. In the next chapter, we will show that if f is merely integrable, then

$$\frac{1}{2\pi}\int_{-\pi}^{\pi} |S_N(f)(\theta) - f(\theta)|^2 \, d\theta \to 0 \quad \text{as } N \to \infty.$$

It is of interest to know that the problem of pointwise convergence of Fourier series was settled in 1966 by L. Carleson, who showed, among other things, that if f is integrable in our sense,[4] then the Fourier series of f converges to f except possibly on a set of "measure 0." The proof of this theorem is difficult and beyond the scope of this book.

2 Uniqueness of Fourier series

If we were to assume that the Fourier series of functions f converge to f in an appropriate sense, then we could infer that a function is uniquely determined by its Fourier coefficients. This would lead to the following statement: if f and g have the same Fourier coefficients, then f and g are necessarily equal. By taking the difference $f - g$, this proposition can be reformulated as: if $\hat{f}(n) = 0$ for all $n \in \mathbb{Z}$, then $f = 0$. As stated, this assertion cannot be correct without reservation, since calculating Fourier coefficients requires integration, and we see that, for example, any two functions which differ at finitely many points have the same Fourier series. However, we do have the following positive result.

Theorem 2.1 *Suppose that f is an integrable function on the circle with $\hat{f}(n) = 0$ for all $n \in \mathbb{Z}$. Then $f(\theta_0) = 0$ whenever f is continuous at the point θ_0.*

Thus, in terms of what we know about the set of discontinuities of integrable functions,[5] we can conclude that f vanishes for "most" values of θ.

Proof. We suppose first that f is real-valued, and argue by contradiction. Assume, without loss of generality, that f is defined on

[4] Carleson's proof actually holds for the wider class of functions which are square integrable in the Lebesgue sense.

[5] See the appendix.

$[-\pi, \pi]$, that $\theta_0 = 0$, and $f(0) > 0$. The idea now is to construct a family of trigonometric polynomials $\{p_k\}$ that "peak" at 0, and so that $\int p_k(\theta) f(\theta)\, d\theta \to \infty$ as $k \to \infty$. This will be our desired contradiction since these integrals are equal to zero by assumption.

Since f is continuous at 0, we can choose $0 < \delta \leq \pi/2$, so that $f(\theta) > f(0)/2$ whenever $|\theta| < \delta$. Let

$$p(\theta) = \epsilon + \cos\theta,$$

where $\epsilon > 0$ is chosen so small that $|p(\theta)| < 1 - \epsilon/2$, whenever $\delta \leq |\theta| \leq \pi$. Then, choose a positive η with $\eta < \delta$, so that $p(\theta) \geq 1 + \epsilon/2$, for $|\theta| < \eta$. Finally, let

$$p_k(\theta) = [p(\theta)]^k,$$

and select B so that $|f(\theta)| \leq B$ for all θ. This is possible since f is integrable, hence bounded. Figure 3 illustrates the family $\{p_k\}$. By

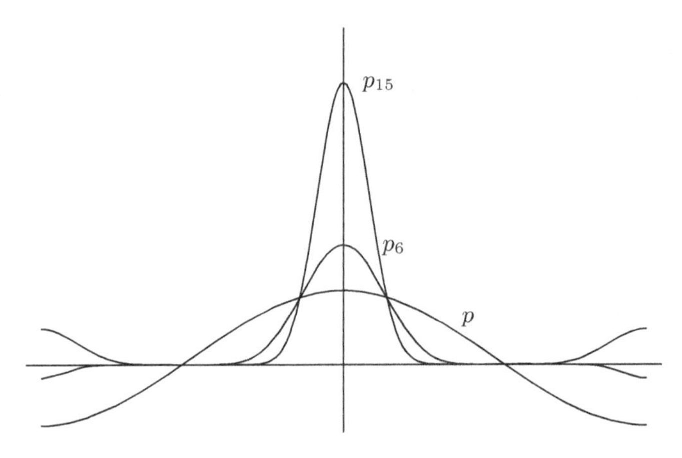

Figure 3. The functions p, p_6, and p_{15} when $\epsilon = 0.1$

construction, each p_k is a trigonometric polynomial, and since $\hat{f}(n) = 0$ for all n, we must have

$$\int_{-\pi}^{\pi} f(\theta) p_k(\theta)\, d\theta = 0 \quad \text{for all } k.$$

However, we have the estimate

$$\left| \int_{\delta \leq |\theta|} f(\theta) p_k(\theta)\, d\theta \right| \leq 2\pi B(1 - \epsilon/2)^k.$$

Also, our choice of δ guarantees that $p(\theta)$ and $f(\theta)$ are non-negative whenever $|\theta| < \delta$, thus

$$\int_{\eta \leq |\theta| < \delta} f(\theta) p_k(\theta)\, d\theta \geq 0.$$

Finally,

$$\int_{|\theta| < \eta} f(\theta) p_k(\theta)\, d\theta \geq 2\eta \frac{f(0)}{2}(1 + \epsilon/2)^k.$$

Therefore, $\int p_k(\theta) f(\theta)\, d\theta \to \infty$ as $k \to \infty$, and this concludes the proof when f is real-valued. In general, write $f(\theta) = u(\theta) + iv(\theta)$, where u and v are real-valued. If we define $\overline{f}(\theta) = \overline{f(\theta)}$, then

$$u(\theta) = \frac{f(\theta) + \overline{f}(\theta)}{2} \quad \text{and} \quad v(\theta) = \frac{f(\theta) - \overline{f}(\theta)}{2i},$$

and since $\hat{\overline{f}}(n) = \overline{\hat{f}(-n)}$, we conclude that the Fourier coefficients of u and v all vanish, hence $f = 0$ at its points of continuity. The idea of constructing a family of functions (trigonometric polynomials in this case) which peak at the origin, together with other nice properties, will play an important role in this book. Such families of functions will be taken up later in Section 4 in connection with the notion of convolution. For now, note that the above theorem implies the following.

Corollary 2.2 *If f is continuous on the circle and $\hat{f}(n) = 0$ for all $n \in \mathbb{Z}$, then $f = 0$.*

The next corollary shows that the problem (1) formulated earlier has a simple positive answer under the assumption that the series of Fourier coefficients converges absolutely.

Corollary 2.3 *Suppose that f is a continuous function on the circle and that the Fourier series of f is absolutely convergent, $\sum_{n=-\infty}^{\infty} |\hat{f}(n)| < \infty$. Then, the Fourier series converges uniformly to f, that is,*

$$\lim_{N \to \infty} S_N(f)(\theta) = f(\theta) \quad \textit{uniformly in } \theta.$$

Proof. Recall that if a sequence of continuous functions converges uniformly, then the limit is also continuous. Now observe that the assumption $\sum |\hat{f}(n)| < \infty$ implies that the partial sums of the Fourier series of f converge absolutely and uniformly, and therefore the function g defined by

$$g(\theta) = \sum_{n=-\infty}^{\infty} \hat{f}(n)e^{in\theta} = \lim_{N\to\infty} \sum_{n=-N}^{N} \hat{f}(n)e^{in\theta}$$

is continuous on the circle. Moreover, the Fourier coefficients of g are precisely $\hat{f}(n)$ since we can interchange the infinite sum with the integral (a consequence of the uniform convergence of the series). Therefore, the previous corollary applied to the function $f - g$ yields $f = g$, as desired.

What conditions on f would guarantee the absolute convergence of its Fourier series? As it turns out, the smoothness of f is directly related to the decay of the Fourier coefficients, and in general, the smoother the function, the faster this decay. As a result, we can expect that relatively smooth functions equal their Fourier series. This is in fact the case, as we now show.

In order to state the result concisely we introduce the standard ***"O" notation***, which we will use freely in the rest of this book. For example, the statement $\hat{f}(n) = O(1/|n|^2)$ as $|n| \to \infty$, means that the left-hand side is bounded by a constant multiple of the right-hand side; that is, there exists $C > 0$ with $|\hat{f}(n)| \leq C/|n|^2$ for all large $|n|$. More generally, $f(x) = O(g(x))$ as $x \to a$ means that for some constant C, $|f(x)| \leq C|g(x)|$ as x approaches a. In particular, $f(x) = O(1)$ means that f is bounded.

Corollary 2.4 *Suppose that f is a twice continuously differentiable function on the circle. Then*

$$\hat{f}(n) = O(1/|n|^2) \quad \text{as } |n| \to \infty,$$

so that the Fourier series of f converges absolutely and uniformly to f.

Proof. The estimate on the Fourier coefficients is proved by integrating by parts twice for $n \neq 0$. We obtain

$$\begin{aligned}
2\pi\hat{f}(n) &= \int_0^{2\pi} f(\theta)e^{-in\theta}\,d\theta \\
&= \left[f(\theta)\cdot\frac{-e^{-in\theta}}{in}\right]_0^{2\pi} + \frac{1}{in}\int_0^{2\pi} f'(\theta)e^{-in\theta}\,d\theta \\
&= \frac{1}{in}\int_0^{2\pi} f'(\theta)e^{-in\theta}\,d\theta \\
&= \frac{1}{in}\left[f'(\theta)\cdot\frac{-e^{-in\theta}}{in}\right]_0^{2\pi} + \frac{1}{(in)^2}\int_0^{2\pi} f''(\theta)e^{-in\theta}\,d\theta \\
&= \frac{-1}{n^2}\int_0^{2\pi} f''(\theta)e^{-in\theta}\,d\theta.
\end{aligned}$$

The quantities in brackets vanish since f and f' are periodic. Therefore

$$2\pi|n|^2|\hat{f}(n)| \leq \left|\int_0^{2\pi} f''(\theta)e^{-in\theta}\,d\theta\right| \leq \int_0^{2\pi} |f''(\theta)|\,d\theta \leq C,$$

where the constant C is independent of n. (We can take $C = 2\pi B$ where B is a bound for f''.) Since $\sum 1/n^2$ converges, the proof of the corollary is complete.

Incidentally, we have also established the following important identity:

$$\widehat{f'}(n) = in\hat{f}(n), \qquad \text{for all } n \in \mathbb{Z}.$$

If $n \neq 0$ the proof is given above, and if $n = 0$ it is left as an exercise to the reader. So if f is differentiable and $f \sim \sum a_n e^{in\theta}$, then $f' \sim \sum a_n ine^{in\theta}$. Also, if f is twice continuously differentiable, then $f'' \sim \sum a_n(in)^2e^{in\theta}$, and so on. Further smoothness conditions on f imply even better decay of the Fourier coefficients (Exercise 10).

There are also stronger versions of Corollary 2.4. It can be shown, for example, that the Fourier series of f converges absolutely, assuming only that f has one continuous derivative. Even more generally, the Fourier series of f converges absolutely (and hence uniformly to f) if f satisfies a **Hölder condition** of order α, with $\alpha > 1/2$, that is,

$$\sup_\theta |f(\theta + t) - f(\theta)| \leq A|t|^\alpha \qquad \text{for all } t.$$

For more on these matters, see the exercises at the end of Chapter 3.

At this point it is worthwhile to introduce a common notation: we say that f belongs to the **class** C^k if f is k times continuously differentiable. Belonging to the class C^k or satisfying a Hölder condition are two possible ways to describe the *smoothness* of a function.

3 Convolutions

The notion of convolution of two functions plays a fundamental role in Fourier analysis; it appears naturally in the context of Fourier series but also serves more generally in the analysis of functions in other settings.

Given two 2π-periodic integrable functions f and g on $\mathbb{R}$, we define their **convolution** $f * g$ on $[-\pi, \pi]$ by

$$(f * g)(x) = \frac{1}{2\pi}\int_{-\pi}^{\pi} f(y)g(x-y)\,dy. \tag{2}$$

The above integral makes sense for each x, since the product of two integrable functions is again integrable. Also, since the functions are periodic, we can change variables to see that

$$(f * g)(x) = \frac{1}{2\pi}\int_{-\pi}^{\pi} f(x-y)g(y)\,dy.$$

Loosely speaking, convolutions correspond to "weighted averages." For instance, if $g = 1$ in (2), then $f * g$ is constant and equal to $\frac{1}{2\pi}\int_{-\pi}^{\pi} f(y)\,dy$, which we may interpret as the average value of f on the circle. Also, the convolution $(f * g)(x)$ plays a role similar to, and in some sense replaces, the pointwise product $f(x)g(x)$ of the two functions f and g.

In the context of this chapter, our interest in convolutions originates from the fact that the partial sums of the Fourier series of f can be expressed as follows:

$$\begin{aligned}
S_N(f)(x) &= \sum_{n=-N}^{N} \hat{f}(n)e^{inx} \\
&= \sum_{n=-N}^{N} \left(\frac{1}{2\pi}\int_{-\pi}^{\pi} f(y)e^{-iny}\,dy\right) e^{inx} \\
&= \frac{1}{2\pi}\int_{-\pi}^{\pi} f(y)\left(\sum_{n=-N}^{N} e^{in(x-y)}\right) dy \\
&= (f * D_N)(x),
\end{aligned}$$

where D_N is the Nth Dirichlet kernel (see Example 4) given by

$$D_N(x) = \sum_{n=-N}^{N} e^{inx}.$$

So we observe that the problem of understanding $S_N(f)$ reduces to the understanding of the convolution $f * D_N$.

We begin by gathering some of the main properties of convolutions.

Proposition 3.1 *Suppose that f, g, and h are 2π-periodic integrable functions. Then:*

(i) $f * (g + h) = (f * g) + (f * h)$.

(ii) $(cf) * g = c(f * g) = f * (cg)$ *for any* $c \in \mathbb{C}$.

(iii) $f * g = g * f$.

(iv) $(f * g) * h = f * (g * h)$.

(v) $f * g$ *is continuous.*

(vi) $\widehat{f * g}(n) = \hat{f}(n)\hat{g}(n)$.

The first four points describe the algebraic properties of convolutions: linearity, commutativity, and associativity. Property (v) exhibits an important principle: the convolution of $f * g$ is "more regular" than f or g. Here, $f * g$ is continuous while f and g are merely (Riemann) integrable. Finally, (vi) is key in the study of Fourier series. In general, the Fourier coefficients of the product fg are not the product of the Fourier coefficients of f and g. However, (vi) says that this relation holds if we replace the product of the two functions f and g by their convolution $f * g$.

Proof. Properties (i) and (ii) follow at once from the linearity of the integral.

The other properties are easily deduced if we assume also that f and g are continuous. In this case, we may freely interchange the order of integration. For instance, to establish (vi) we write

$$\begin{aligned}
\widehat{f * g}(n) &= \frac{1}{2\pi}\int_{-\pi}^{\pi}(f*g)(x)e^{-inx}\,dx \\
&= \frac{1}{2\pi}\int_{-\pi}^{\pi}\frac{1}{2\pi}\left(\int_{-\pi}^{\pi} f(y)g(x-y)\,dy\right)e^{-inx}\,dx \\
&= \frac{1}{2\pi}\int_{-\pi}^{\pi} f(y)e^{-iny}\left(\frac{1}{2\pi}\int_{-\pi}^{\pi} g(x-y)e^{-in(x-y)}\,dx\right)dy \\
&= \frac{1}{2\pi}\int_{-\pi}^{\pi} f(y)e^{-iny}\left(\frac{1}{2\pi}\int_{-\pi}^{\pi} g(x)e^{-inx}\,dx\right)dy \\
&= \hat{f}(n)\hat{g}(n).
\end{aligned}$$

To prove (iii), one first notes that if F is continuous and 2π-periodic, then

$$\int_{-\pi}^{\pi} F(y)\,dy = \int_{-\pi}^{\pi} F(x-y)\,dy \qquad \text{for any } x \in \mathbb{R}.$$

The verification of this identity consists of a change of variables $y \mapsto -y$, followed by a translation $y \mapsto y - x$. Then, one takes $F(y) = f(y)g(x-y)$.

Also, (iv) follows by interchanging two integral signs, and an appropriate change of variables.

Finally, we show that if f and g are continuous, then $f * g$ is continuous. First, we may write

$$(f*g)(x_1) - (f*g)(x_2) = \frac{1}{2\pi}\int_{-\pi}^{\pi} f(y)\left[g(x_1-y) - g(x_2-y)\right]\,dy.$$

Since g is continuous it must be uniformly continuous on any closed and bounded interval. But g is also periodic, so it must be uniformly continuous on all of $\mathbb{R}$; given $\epsilon > 0$ there exists $\delta > 0$ so that $|g(s) - g(t)| < \epsilon$ whenever $|s - t| < \delta$. Then, $|x_1 - x_2| < \delta$ implies $|(x_1 - y) - (x_2 - y)| < \delta$ for any y, hence

$$\begin{aligned}
|(f*g)(x_1) - (f*g)(x_2)| &\leq \frac{1}{2\pi}\left|\int_{-\pi}^{\pi} f(y)\left[g(x_1-y) - g(x_2-y)\right]\,dy\right| \\
&\leq \frac{1}{2\pi}\int_{-\pi}^{\pi} |f(y)|\,|g(x_1-y) - g(x_2-y)|\,dy \\
&\leq \frac{\epsilon}{2\pi}\int_{-\pi}^{\pi} |f(y)|\,dy \\
&\leq \frac{\epsilon}{2\pi}\,2\pi\,B\,,
\end{aligned}$$

where B is chosen so that $|f(x)| \leq B$ for all x. As a result, we conclude that $f * g$ is continuous, and the proposition is proved, at least when f and g are continuous.

In general, when f and g are merely integrable, we may use the results established so far (when f and g are continuous), together with the following approximation lemma, whose proof may be found in the appendix.

Lemma 3.2 *Suppose f is integrable on the circle and bounded by B. Then there exists a sequence $\{f_k\}_{k=1}^{\infty}$ of continuous functions on the circle so that*

$$\sup_{x \in [-\pi,\pi]} |f_k(x)| \leq B \quad \text{for all } k = 1, 2, \ldots,$$

and

$$\int_{-\pi}^{\pi} |f(x) - f_k(x)|\, dx \to 0 \quad \text{as } k \to \infty.$$

Using this result, we may complete the proof of the proposition as follows. Apply Lemma 3.2 to f and g to obtain sequences $\{f_k\}$ and $\{g_k\}$ of approximating continuous functions. Then

$$f * g - f_k * g_k = (f - f_k) * g + f_k * (g - g_k).$$

By the properties of the sequence $\{f_k\}$,

$$\begin{aligned}
|(f - f_k) * g(x)| &\leq \frac{1}{2\pi} \int_{-\pi}^{\pi} |f(x-y) - f_k(x-y)|\, |g(y)|\, dy \\
&\leq \frac{1}{2\pi} \sup_y |g(y)| \int_{-\pi}^{\pi} |f(y) - f_k(y)|\, dy \\
&\to 0 \quad \text{as } k \to \infty.
\end{aligned}$$

Hence $(f - f_k) * g \to 0$ uniformly in x. Similarly, $f_k * (g - g_k) \to 0$ uniformly, and therefore $f_k * g_k$ tends uniformly to $f * g$. Since each $f_k * g_k$ is continuous, it follows that $f * g$ is also continuous, and we have (v).

Next, we establish (vi). For each fixed integer n we must have $\widehat{f_k * g_k}(n) \to \widehat{f * g}(n)$ as k tends to infinity since $f_k * g_k$ converges uniformly to $f * g$. However, we found earlier that $\widehat{f_k}(n)\widehat{g_k}(n) = \widehat{f_k * g_k}(n)$ because both f_k and g_k are continuous. Hence

$$\begin{aligned}
|\hat{f}(n) - \hat{f}_k(n)| &= \frac{1}{2\pi} \left| \int_{-\pi}^{\pi} (f(x) - f_k(x)) e^{-inx}\, dx \right| \\
&\leq \frac{1}{2\pi} \int_{-\pi}^{\pi} |f(x) - f_k(x)|\, dx,
\end{aligned}$$

and as a result we find that $\widehat{f_k}(n) \to \hat{f}(n)$ as k goes to infinity. Similarly $\widehat{g_k}(n) \to \hat{g}(n)$, and the desired property is established once we let k tend to infinity. Finally, properties (iii) and (iv) follow from the same kind of arguments.

4 Good kernels

In the proof of Theorem 2.1 we constructed a sequence of trigonometric polynomials $\{p_k\}$ with the property that the functions p_k peaked at the origin. As a result, we could isolate the behavior of f at the origin. In this section, we return to such families of functions, but this time in a more general setting. First, we define the notion of good kernel, and discuss the characteristic properties of such functions. Then, by the use of convolutions, we show how these kernels can be used to recover a given function.

A family of kernels $\{K_n(x)\}_{n=1}^{\infty}$ on the circle is said to be a family of **good kernels** if it satisfies the following properties:

(a) For all $n \geq 1$,

$$\frac{1}{2\pi} \int_{-\pi}^{\pi} K_n(x)\, dx = 1.$$

(b) There exists $M > 0$ such that for all $n \geq 1$,

$$\int_{-\pi}^{\pi} |K_n(x)|\, dx \leq M.$$

(c) For every $\delta > 0$,

$$\int_{\delta \leq |x| \leq \pi} |K_n(x)|\, dx \to 0, \qquad \text{as } n \to \infty.$$

In practice we shall encounter families where $K_n(x) \geq 0$, in which case (b) is a consequence of (a). We may interpret the kernels $K_n(x)$ as weight distributions on the circle: property (a) says that K_n assigns unit mass to the whole circle $[-\pi, \pi]$, and (c) that this mass concentrates near the origin as n becomes large.[6] Figure 4 (a) illustrates the typical character of a family of good kernels.

The importance of good kernels is highlighted by their use in connection with convolutions.

[6]In the limit, a family of good kernels represents the "Dirac delta function." This terminology comes from physics.

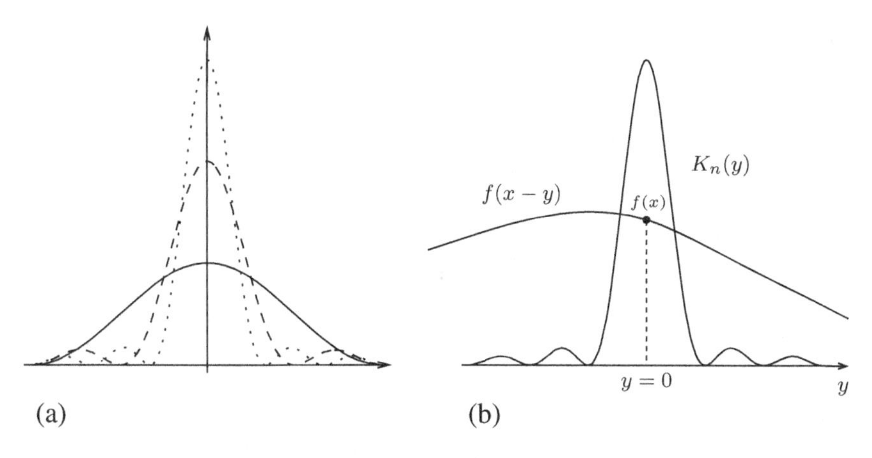

Figure 4. Good kernels

Theorem 4.1 *Let $\{K_n\}_{n=1}^{\infty}$ be a family of good kernels, and f an integrable function on the circle. Then*

$$\lim_{n\to\infty}(f * K_n)(x) = f(x)$$

whenever f is continuous at x. If f is continuous everywhere, then the above limit is uniform.

Because of this result, the family $\{K_n\}$ is sometimes referred to as an **approximation to the identity**.

We have previously interpreted convolutions as weighted averages. In this context, the convolution

$$(f * K_n)(x) = \frac{1}{2\pi}\int_{-\pi}^{\pi} f(x-y)K_n(y)\,dy$$

is the average of $f(x-y)$, where the weights are given by $K_n(y)$. However, the weight distribution K_n concentrates its mass at $y = 0$ as n becomes large. Hence in the integral, the value $f(x)$ is assigned the full mass as $n \to \infty$. Figure 4 (b) illustrates this point.

Proof of Theorem 4.1. If $\epsilon > 0$ and f is continuous at x, choose δ so that $|y| < \delta$ implies $|f(x-y) - f(x)| < \epsilon$. Then, by the first property of good kernels, we can write

$$\begin{aligned}(f * K_n)(x) - f(x) &= \frac{1}{2\pi}\int_{-\pi}^{\pi} K_n(y)f(x-y)\,dy - f(x)\\ &= \frac{1}{2\pi}\int_{-\pi}^{\pi} K_n(y)[f(x-y) - f(x)]\,dy.\end{aligned}$$

Hence,

$$\begin{aligned}|(f*K_n)(x)-f(x)| &= \left|\frac{1}{2\pi}\int_{-\pi}^{\pi}K_n(y)[f(x-y)-f(x)]\,dy\right| \\ &\leq \frac{1}{2\pi}\int_{|y|<\delta}|K_n(y)|\,|f(x-y)-f(x)|\,dy \\ &\qquad + \frac{1}{2\pi}\int_{\delta\leq|y|\leq\pi}|K_n(y)|\,|f(x-y)-f(x)|\,dy \\ &\leq \frac{\epsilon}{2\pi}\int_{-\pi}^{\pi}|K_n(y)|\,dy + \frac{2B}{2\pi}\int_{\delta\leq|y|\leq\pi}|K_n(y)|\,dy,\end{aligned}$$

where B is a bound for f. The first term is bounded by $\epsilon M/2\pi$ because of the second property of good kernels. By the third property we see that for all large n, the second term will be less than ϵ. Therefore, for some constant $C>0$ and all large n we have

$$|(f*K_n)(x)-f(x)|\leq C\epsilon,$$

thereby proving the first assertion in the theorem. If f is continuous everywhere, then it is uniformly continuous, and δ can be chosen independent of x. This provides the desired conclusion that $f*K_n\to f$ uniformly.

Recall from the beginning of Section 3 that

$$S_N(f)(x)=(f*D_N)(x)\,,$$

where $D_N(x)=\sum_{n=-N}^{N}e^{inx}$ is the Dirichlet kernel. It is natural now for us to ask whether D_N is a good kernel, since if this were true, Theorem 4.1 would imply that the Fourier series of f converges to $f(x)$ whenever f is continuous at x. Unfortunately, this is not the case. Indeed, an estimate shows that D_N violates the second property; more precisely, one has (see Problem 2)

$$\int_{-\pi}^{\pi}|D_N(x)|\,dx\geq c\log N, \qquad \text{as } N\to\infty.$$

However, we should note that the formula for D_N as a sum of exponentials immediately gives

$$\frac{1}{2\pi}\int_{-\pi}^{\pi}D_N(x)\,dx=1,$$

so the first property of good kernels is actually verified. The fact that the mean value of D_N is 1, while the integral of its absolute value is large, is a result of cancellations. Indeed, Figure 5 shows that the function $D_N(x)$ takes on positive and negative values and oscillates very rapidly as N gets large.

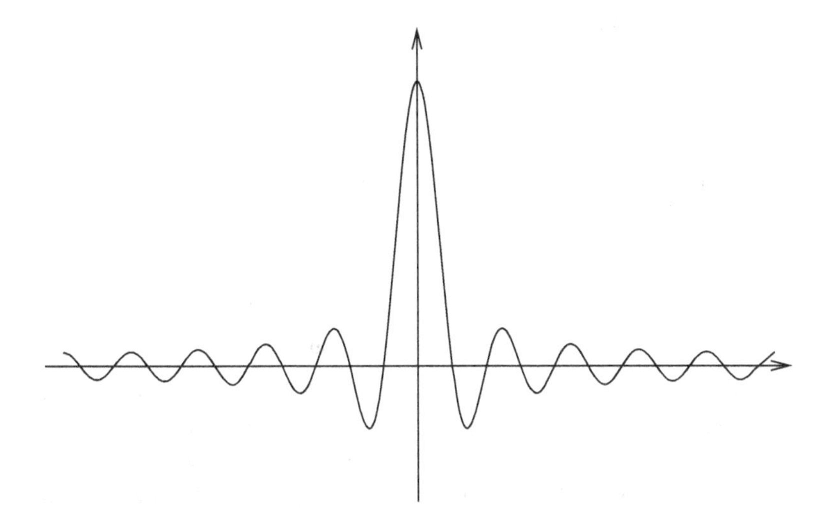

Figure 5. The Dirichlet kernel for large N

This observation suggests that the pointwise convergence of Fourier series is intricate, and may even fail at points of continuity. This is indeed the case, as we will see in the next chapter.

5 Cesàro and Abel summability: applications to Fourier series

Since a Fourier series may fail to converge at individual points, we are led to try to overcome this failure by interpreting the limit

$$\lim_{N\to\infty} S_N(f) = f$$

in a different sense.

5.1 Cesàro means and summation

We begin by taking ordinary averages of the partial sums, a technique which we now describe in more detail.

Suppose we are given a series of complex numbers

$$c_0 + c_1 + c_2 + \cdots = \sum_{k=0}^{\infty} c_k.$$

We define the nth partial sum s_n by

$$s_n = \sum_{k=0}^{n} c_k,$$

and say that the series converges to s if $\lim_{n\to\infty} s_n = s$. This is the most natural and most commonly used type of "summability." Consider, however, the example of the series

$$1 - 1 + 1 - 1 + \cdots = \sum_{k=0}^{\infty} (-1)^k. \tag{3}$$

Its partial sums form the sequence $\{1, 0, 1, 0, \ldots\}$ which has no limit. Because these partial sums alternate evenly between 1 and 0, one might therefore suggest that $1/2$ is the "limit" of the sequence, and hence $1/2$ equals the "sum" of that particular series. We give a precise meaning to this by defining the average of the first N partial sums by

$$\sigma_N = \frac{s_0 + s_1 + \cdots + s_{N-1}}{N}.$$

The quantity σ_N is called the Nth **Cesàro mean**[7] of the sequence $\{s_k\}$ or the Nth **Cesàro sum** of the series $\sum_{k=0}^{\infty} c_k$.

If σ_N converges to a limit σ as N tends to infinity, we say that the series $\sum c_n$ is **Cesàro summable** to σ. In the case of series of functions, we shall understand the limit in the sense of either pointwise or uniform convergence, depending on the situation.

The reader will have no difficulty checking that in the above example (3), the series is Cesàro summable to $1/2$. Moreover, one can show that Cesàro summation is a more inclusive process than convergence. In fact, if a series is convergent to s, then it is also Cesàro summable to the same limit s (Exercise 12).

5.2 Fejér's theorem

An interesting application of Cesàro summability appears in the context of Fourier series.

---

[7]Note that if the series $\sum_{k=1}^{\infty} c_k$ begins with the term $k = 1$, then it is common practice to define $\sigma_N = (s_1 + \cdots + s_N)/N$. This change of notation has little effect on what follows.

We mentioned earlier that the Dirichlet kernels fail to belong to the family of good kernels. Quite surprisingly, their averages are very well behaved functions, in the sense that they do form a family of good kernels.

To see this, we form the Nth Cesàro mean of the Fourier series, which by definition is

$$\sigma_N(f)(x) = \frac{S_0(f)(x) + \cdots + S_{N-1}(f)(x)}{N}.$$

Since $S_n(f) = f * D_n$, we find that

$$\sigma_N(f)(x) = (f * F_N)(x),$$

where $F_N(x)$ is the N-th **Fejér kernel** given by

$$F_N(x) = \frac{D_0(x) + \cdots + D_{N-1}(x)}{N}.$$

Lemma 5.1 *We have*

$$F_N(x) = \frac{1}{N}\frac{\sin^2(Nx/2)}{\sin^2(x/2)},$$

and the Fejér kernel is a good kernel.

The proof of the formula for F_N (a simple application of trigonometric identities) is outlined in Exercise 15. To prove the rest of the lemma, note that F_N is positive and $\frac{1}{2\pi}\int_{-\pi}^{\pi} F_N(x)\,dx = 1$, in view of the fact that a similar identity holds for the Dirichlet kernels D_n. However, $\sin^2(x/2) \geq c_\delta > 0$, if $\delta \leq |x| \leq \pi$, hence $F_N(x) \leq 1/(Nc_\delta)$, from which it follows that

$$\int_{\delta \leq |x| \leq \pi} |F_N(x)|\,dx \to 0 \quad \text{as } N \to \infty.$$

Applying Theorem 4.1 to this new family of good kernels yields the following important result.

Theorem 5.2 *If f is integrable on the circle, then the Fourier series of f is Cesàro summable to f at every point of continuity of f.*

Moreover, if f is continuous on the circle, then the Fourier series of f is uniformly Cesàro summable to f.

We may now state two corollaries. The first is a result that we have already established. The second is new, and of fundamental importance.

Corollary 5.3 *If f is integrable on the circle and $\hat{f}(n) = 0$ for all n, then $f = 0$ at all points of continuity of f.*

The proof is immediate since all the partial sums are 0, hence all the Cesàro means are 0.

Corollary 5.4 *Continuous functions on the circle can be uniformly approximated by trigonometric polynomials.*

This means that if f is continuous on $[-\pi, \pi]$ with $f(-\pi) = f(\pi)$ and $\epsilon > 0$, then there exists a trigonometric polynomial P such that

$$|f(x) - P(x)| < \epsilon \quad \text{for all } -\pi \leq x \leq \pi.$$

This follows immediately from the theorem since the partial sums, hence the Cesàro means, are trigonometric polynomials. Corollary 5.4 is the periodic analogue of the Weierstrass approximation theorem for polynomials which can be found in Exercise 16.

5.3 Abel means and summation

Another method of summation was first considered by Abel and actually predates the Cesàro method.

A series of complex numbers $\sum_{k=0}^{\infty} c_k$ is said to be **Abel summable** to s if for every $0 \leq r < 1$, the series

$$A(r) = \sum_{k=0}^{\infty} c_k r^k$$

converges, and

$$\lim_{r \to 1} A(r) = s.$$

The quantities $A(r)$ are called the **Abel means** of the series. One can prove that if the series converges to s, then it is Abel summable to s. Moreover, the method of Abel summability is even more powerful than the Cesàro method: when the series is Cesàro summable, it is always Abel summable to the same sum. However, if we consider the series

$$1 - 2 + 3 - 4 + 5 - \cdots = \sum_{k=0}^{\infty} (-1)^k (k+1),$$

then one can show that it is Abel summable to $1/4$ since

$$A(r) = \sum_{k=0}^{\infty} (-1)^k (k+1) r^k = \frac{1}{(1+r)^2},$$

but this series is not Cesàro summable; see Exercise 13.

5.4 The Poisson kernel and Dirichlet's problem in the unit disc

To adapt Abel summability to the context of Fourier series, we define the Abel means of the function $f(\theta) \sim \sum_{n=-\infty}^{\infty} a_n e^{in\theta}$ by

$$A_r(f)(\theta) = \sum_{n=-\infty}^{\infty} r^{|n|} a_n e^{in\theta}.$$

Since the index n takes positive and negative values, it is natural to write $c_0 = a_0$, and $c_n = a_n e^{in\theta} + a_{-n} e^{-in\theta}$ for $n > 0$, so that the Abel means of the Fourier series correspond to the definition given in the previous section for numerical series.

We note that since f is integrable, $|a_n|$ is uniformly bounded in n, so that $A_r(f)$ converges absolutely and uniformly for each $0 \leq r < 1$. Just as in the case of Cesàro means, the key fact is that these Abel means can be written as convolutions

$$A_r(f)(\theta) = (f * P_r)(\theta),$$

where $P_r(\theta)$ is the **Poisson kernel** given by

$$P_r(\theta) = \sum_{n=-\infty}^{\infty} r^{|n|} e^{in\theta}. \tag{4}$$

In fact,

$$\begin{aligned} A_r(f)(\theta) &= \sum_{n=-\infty}^{\infty} r^{|n|} a_n e^{in\theta} \\ &= \sum_{n=-\infty}^{\infty} r^{|n|} \left(\frac{1}{2\pi} \int_{-\pi}^{\pi} f(\varphi) e^{-in\varphi}\, d\varphi \right) e^{in\theta} \\ &= \frac{1}{2\pi} \int_{-\pi}^{\pi} f(\varphi) \left(\sum_{n=-\infty}^{\infty} r^{|n|} e^{-in(\varphi - \theta)} \right) d\varphi, \end{aligned}$$

where the interchange of the integral and infinite sum is justified by the uniform convergence of the series.

Lemma 5.5 *If* $0 \leq r < 1$, *then*

$$P_r(\theta) = \frac{1 - r^2}{1 - 2r\cos\theta + r^2}.$$

The Poisson kernel is a good kernel,[8] *as r tends to 1 from below.*

Proof. The identity $P_r(\theta) = \frac{1-r^2}{1-2r\cos\theta+r^2}$ has already been derived in Section 1.1. Note that

$$1 - 2r\cos\theta + r^2 = (1-r)^2 + 2r(1-\cos\theta).$$

Hence if $1/2 \leq r \leq 1$ and $\delta \leq |\theta| \leq \pi$, then

$$1 - 2r\cos\theta + r^2 \geq c_\delta > 0.$$

Thus $P_r(\theta) \leq (1-r^2)/c_\delta$ when $\delta \leq |\theta| \leq \pi$, and the third property of good kernels is verified. Clearly $P_r(\theta) \geq 0$, and integrating the expression (4) term by term (which is justified by the absolute convergence of the series) yields

$$\frac{1}{2\pi}\int_{-\pi}^{\pi} P_r(\theta)\, d\theta = 1,$$

thereby concluding the proof that P_r is a good kernel.

Combining this lemma with Theorem 4.1, we obtain our next result.

Theorem 5.6 *The Fourier series of an integrable function on the circle is Abel summable to f at every point of continuity. Moreover, if f is continuous on the circle, then the Fourier series of f is uniformly Abel summable to f.*

We now return to a problem discussed in Chapter 1, where we sketched the solution of the steady-state heat equation $\triangle u = 0$ in the unit disc with boundary condition $u = f$ on the circle. We expressed the Laplacian in terms of polar coordinates, separated variables, and expected that a solution was given by

$$u(r,\theta) = \sum_{m=-\infty}^{\infty} a_m r^{|m|} e^{im\theta}, \tag{5}$$

where a_m was the mth Fourier coefficient of f. In other words, we were led to take

$$u(r,\theta) = A_r(f)(\theta) = \frac{1}{2\pi}\int_{-\pi}^{\pi} f(\varphi)P_r(\theta-\varphi)\, d\varphi.$$

We are now in a position to show that this is indeed the case.

---

[8] In this case, the family of kernels is indexed by a continuous parameter $0 \leq r < 1$, rather than the discrete n considered previously. In the definition of good kernels, we simply replace n by r and take the limit in property (c) appropriately, for example $r \to 1$ in this case.

Theorem 5.7 *Let f be an integrable function defined on the unit circle. Then the function u defined in the unit disc by the Poisson integral*

$$u(r,\theta) = (f * P_r)(\theta) \tag{6}$$

has the following properties:

(i) *u has two continuous derivatives in the unit disc and satisfies* $\triangle u = 0$.

(ii) *If θ is any point of continuity of f, then*

$$\lim_{r \to 1} u(r,\theta) = f(\theta).$$

If f is continuous everywhere, then this limit is uniform.

(iii) *If f is continuous, then $u(r,\theta)$ is the unique solution to the steady-state heat equation in the disc which satisfies conditions* (i) *and* (ii).

Proof. For (i), we recall that the function u is given by the series (5). Fix $\rho < 1$; inside each disc of radius $r < \rho < 1$ centered at the origin, the series for u can be differentiated term by term, and the differentiated series is uniformly and absolutely convergent. Thus u can be differentiated twice (in fact infinitely many times), and since this holds for all $\rho < 1$, we conclude that u is twice differentiable inside the unit disc. Moreover, in polar coordinates,

$$\triangle u = \frac{\partial^2 u}{\partial r^2} + \frac{1}{r}\frac{\partial u}{\partial r} + \frac{1}{r^2}\frac{\partial^2 u}{\partial \theta^2},$$

so term by term differentiation shows that $\triangle u = 0$.

The proof of (ii) is a simple application of the previous theorem. To prove (iii) we argue as follows. Suppose v solves the steady-state heat equation in the disc and converges to f uniformly as r tends to 1 from below. For each fixed r with $0 < r < 1$, the function $v(r,\theta)$ has a Fourier series

$$\sum_{n=-\infty}^{\infty} a_n(r) e^{in\theta} \quad \text{where} \quad a_n(r) = \frac{1}{2\pi}\int_{-\pi}^{\pi} v(r,\theta) e^{-in\theta}\, d\theta.$$

Taking into account that $v(r,\theta)$ solves the equation

$$\frac{\partial^2 v}{\partial r^2} + \frac{1}{r}\frac{\partial v}{\partial r} + \frac{1}{r^2}\frac{\partial^2 v}{\partial \theta^2} = 0, \tag{7}$$

we find that

$$a_n''(r) + \frac{1}{r}a_n'(r) - \frac{n^2}{r^2}a_n(r) = 0. \tag{8}$$

Indeed, we may first multiply (7) by $e^{-in\theta}$ and integrate in θ. Then, since v is periodic, two integrations by parts give

$$\frac{1}{2\pi}\int_{-\pi}^{\pi} \frac{\partial^2 v}{\partial \theta^2}(r,\theta)e^{-in\theta}\,d\theta = -n^2 a_n(r).$$

Finally, we may interchange the order of differentiation and integration, which is permissible since v has two continuous derivatives; this yields (8).

Therefore, we must have $a_n(r) = A_n r^n + B_n r^{-n}$ for some constants A_n and B_n, when $n \neq 0$ (see Exercise 11 in Chapter 1). To evaluate the constants, we first observe that each term $a_n(r)$ is bounded because v is bounded, therefore $B_n = 0$. To find A_n we let $r \to 1$. Since v converges uniformly to f as $r \to 1$ we find that

$$A_n = \frac{1}{2\pi}\int_{-\pi}^{\pi} f(\theta)e^{-in\theta}\,d\theta.$$

By a similar argument, this formula also holds when $n = 0$. Our conclusion is that for each $0 < r < 1$, the Fourier series of v is given by the series of $u(r,\theta)$, so by the uniqueness of Fourier series for continuous functions, we must have $u = v$.

Remark. By part (iii) of the theorem, we may conclude that if u solves $\Delta u = 0$ in the disc, and converges to 0 uniformly as $r \to 1$, then u must be identically 0. However, if uniform convergence is replaced by pointwise convergence, this conclusion may fail; see Exercise 18.

6 Exercises

1. Suppose f is 2π-periodic and integrable on any finite interval. Prove that if $a, b \in \mathbb{R}$, then

$$\int_a^b f(x)\,dx = \int_{a+2\pi}^{b+2\pi} f(x)\,dx = \int_{a-2\pi}^{b-2\pi} f(x)\,dx.$$

Also prove that

$$\int_{-\pi}^{\pi} f(x+a)\,dx = \int_{-\pi}^{\pi} f(x)\,dx = \int_{-\pi+a}^{\pi+a} f(x)\,dx.$$

2. In this exercise we show how the symmetries of a function imply certain properties of its Fourier coefficients. Let f be a 2π-periodic Riemann integrable function defined on $\mathbb{R}$.

(a) Show that the Fourier series of the function f can be written as

$$f(\theta) \sim \hat{f}(0) + \sum_{n\geq 1} [\hat{f}(n) + \hat{f}(-n)] \cos n\theta + i[\hat{f}(n) - \hat{f}(-n)] \sin n\theta.$$

(b) Prove that if f is even, then $\hat{f}(n) = \hat{f}(-n)$, and we get a cosine series.

(c) Prove that if f is odd, then $\hat{f}(n) = -\hat{f}(-n)$, and we get a sine series.

(d) Suppose that $f(\theta + \pi) = f(\theta)$ for all $\theta \in \mathbb{R}$. Show that $\hat{f}(n) = 0$ for all odd n.

(e) Show that f is real-valued if and only if $\overline{\hat{f}(n)} = \hat{f}(-n)$ for all n.

3. We return to the problem of the plucked string discussed in Chapter 1. Show that the initial condition f is *equal* to its Fourier sine series

$$f(x) = \sum_{m=1}^{\infty} A_m \sin mx \quad \text{with} \quad A_m = \frac{2h}{m^2} \frac{\sin mp}{p(\pi - p)}.$$

[Hint: Note that $|A_m| \leq C/m^2$.]

4. Consider the 2π-periodic odd function defined on $[0, \pi]$ by $f(\theta) = \theta(\pi - \theta)$.

(a) Draw the graph of f.

(b) Compute the Fourier coefficients of f, and show that

$$f(\theta) = \frac{8}{\pi} \sum_{k \text{ odd } \geq 1} \frac{\sin k\theta}{k^3}.$$

5. On the interval $[-\pi, \pi]$ consider the function

$$f(\theta) = \begin{cases} 0 & \text{if } |\theta| > \delta, \\ 1 - |\theta|/\delta & \text{if } |\theta| \leq \delta. \end{cases}$$

Thus the graph of f has the shape of a triangular tent. Show that

$$f(\theta) = \frac{\delta}{2\pi} + 2\sum_{n=1}^{\infty} \frac{1 - \cos n\delta}{n^2 \pi \delta} \cos n\theta.$$

6. Let f be the function defined on $[-\pi, \pi]$ by $f(\theta) = |\theta|$.

(a) Draw the graph of f.

(b) Calculate the Fourier coefficients of f, and show that

$$\hat{f}(n) = \begin{cases} \dfrac{\pi}{2} & \text{if } n = 0, \\ \dfrac{-1+(-1)^n}{\pi n^2} & \text{if } n \neq 0. \end{cases}$$

(c) What is the Fourier series of f in terms of sines and cosines?

(d) Taking $\theta = 0$, prove that

$$\sum_{n \text{ odd } \geq 1} \frac{1}{n^2} = \frac{\pi^2}{8} \quad \text{and} \quad \sum_{n=1}^{\infty} \frac{1}{n^2} = \frac{\pi^2}{6}.$$

See also Example 2 in Section 1.1.

7. Suppose $\{a_n\}_{n=1}^N$ and $\{b_n\}_{n=1}^N$ are two finite sequences of complex numbers. Let $B_k = \sum_{n=1}^k b_n$ denote the partial sums of the series $\sum b_n$ with the convention $B_0 = 0$.

(a) Prove the **summation by parts** formula

$$\sum_{n=M}^{N} a_n b_n = a_N B_N - a_M B_{M-1} - \sum_{n=M}^{N-1} (a_{n+1} - a_n) B_n.$$

(b) Deduce from this formula Dirichlet's test for convergence of a series: if the partial sums of the series $\sum b_n$ are bounded, and $\{a_n\}$ is a sequence of real numbers that decreases monotonically to 0, then $\sum a_n b_n$ converges.

8. Verify that $\dfrac{1}{2i} \sum_{n \neq 0} \dfrac{e^{inx}}{n}$ is the Fourier series of the 2π-periodic **sawtooth** function illustrated in Figure 6, defined by $f(0) = 0$, and

$$f(x) = \begin{cases} -\dfrac{\pi}{2} - \dfrac{x}{2} & \text{if } -\pi < x < 0, \\ \dfrac{\pi}{2} - \dfrac{x}{2} & \text{if } 0 < x < \pi. \end{cases}$$

Note that this function is not continuous. Show that nevertheless, the series converges for every x (by which we mean, as usual, that the symmetric partial sums of the series converge). In particular, the value of the series at the origin, namely 0, is the average of the values of $f(x)$ as x approaches the origin from the left and the right.

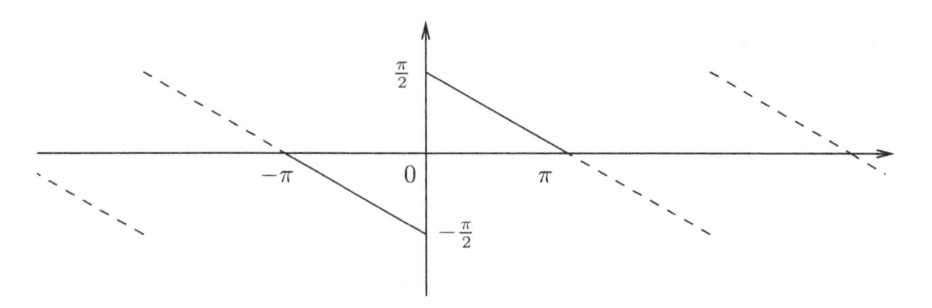

Figure 6. The sawtooth function

[Hint: Use Dirichlet's test for convergence of a series $\sum a_n b_n$.]

9. Let $f(x) = \chi_{[a,b]}(x)$ be the characteristic function of the interval $[a, b] \subset [-\pi, \pi]$, that is,

$$\chi_{[a,b]}(x) = \begin{cases} 1 & \text{if } x \in [a, b], \\ 0 & \text{otherwise.} \end{cases}$$

(a) Show that the Fourier series of f is given by

$$f(x) \sim \frac{b-a}{2\pi} + \sum_{n \neq 0} \frac{e^{-ina} - e^{-inb}}{2\pi in} e^{inx}.$$

The sum extends over all positive and negative integers excluding 0.

(b) Show that if $a \neq -\pi$ or $b \neq \pi$ and $a \neq b$, then the Fourier series does not converge absolutely for any x. [Hint: It suffices to prove that for many values of n one has $|\sin n\theta_0| \geq c > 0$ where $\theta_0 = (b - a)/2$.]

(c) However, prove that the Fourier series converges at every point x. What happens if $a = -\pi$ and $b = \pi$?

10. Suppose f is a periodic function of period 2π which belongs to the class C^k. Show that

$$\hat{f}(n) = O(1/|n|^k) \qquad \text{as } |n| \to \infty.$$

This notation means that there exists a constant C such $|\hat{f}(n)| \leq C/|n|^k$. We could also write this as $|n|^k \hat{f}(n) = O(1)$, where $O(1)$ means bounded.
[Hint: Integrate by parts.]

11. Suppose that $\{f_k\}_{k=1}^{\infty}$ is a sequence of Riemann integrable functions on the interval $[0, 1]$ such that

$$\int_0^1 |f_k(x) - f(x)|\, dx \to 0 \qquad \text{as } k \to \infty.$$

Show that $\hat{f}_k(n) \to \hat{f}(n)$ uniformly in n as $k \to \infty$.

12. Prove that if a series of complex numbers $\sum c_n$ converges to s, then $\sum c_n$ is Cesàro summable to s.

[Hint: Assume $s_n \to 0$ as $n \to \infty$.]

13. The purpose of this exercise is to prove that Abel summability is stronger than the standard or Cesàro methods of summation.

(a) Show that if the series $\sum_{n=1}^{\infty} c_n$ of complex numbers converges to a finite limit s, then the series is Abel summable to s. [Hint: Why is it enough to prove the theorem when $s = 0$? Assuming $s = 0$, show that if $s_N = c_1 + \cdots + c_N$, then $\sum_{n=1}^{N} c_n r^n = (1-r)\sum_{n=1}^{N} s_n r^n + s_N r^{N+1}$. Let $N \to \infty$ to show that

$$\sum c_n r^n = (1-r)\sum s_n r^n.$$

Finally, prove that the right-hand side converges to 0 as $r \to 1$.]

(b) However, show that there exist series which are Abel summable, but that do not converge. [Hint: Try $c_n = (-1)^n$. What is the Abel limit of $\sum c_n$?]

(c) Argue similarly to prove that if a series $\sum_{n=1}^{\infty} c_n$ is Cesàro summable to σ, then it is Abel summable to σ. [Hint: Note that

$$\sum_{n=1}^{\infty} c_n r^n = (1-r)^2 \sum_{n=1}^{\infty} n\sigma_n r^n,$$

and assume $\sigma = 0$.]

(d) Give an example of a series that is Abel summable but not Cesàro summable. [Hint: Try $c_n = (-1)^{n-1} n$. Note that if $\sum c_n$ is Cesàro summable, then c_n/n tends to 0.]

The results above can be summarized by the following implications about series:

$$\text{convergent} \Longrightarrow \text{Cesàro summable} \Longrightarrow \text{Abel summable},$$

and the fact that none of the arrows can be reversed.

14. This exercise deals with a theorem of Tauber which says that under an additional condition on the coefficients c_n, the above arrows can be reversed.

(a) If $\sum c_n$ is Cesàro summable to σ and $c_n = o(1/n)$ (that is, $nc_n \to 0$), then $\sum c_n$ converges to σ. [Hint: $s_n - \sigma_n = [(n-1)c_n + \cdots + c_2]/n$.]

(b) The above statement holds if we replace Cesàro summable by Abel summable. [Hint: Estimate the difference between $\sum_{n=1}^{N} c_n$ and $\sum_{n=1}^{N} c_n r^n$ where $r = 1 - 1/N$.]

15. Prove that the Fejér kernel is given by

$$F_N(x) = \frac{1}{N} \frac{\sin^2(Nx/2)}{\sin^2(x/2)}.$$

[Hint: Remember that $NF_N(x) = D_0(x) + \cdots + D_{N-1}(x)$ where $D_n(x)$ is the Dirichlet kernel. Therefore, if $\omega = e^{ix}$ we have

$$NF_N(x) = \sum_{n=0}^{N-1} \frac{\omega^{-n} - \omega^{n+1}}{1-\omega}.]$$

16. The Weierstrass approximation theorem states: Let f be a continuous function on the closed and bounded interval $[a,b] \subset \mathbb{R}$. Then, for any $\epsilon > 0$, there exists a polynomial P such that

$$\sup_{x\in[a,b]} |f(x) - P(x)| < \epsilon.$$

Prove this by applying Corollary 5.4 of Fejér's theorem and using the fact that the exponential function e^{ix} can be approximated by polynomials uniformly on any interval.

17. In Section 5.4 we proved that the Abel means of f converge to f at all points of continuity, that is,

$$\lim_{r\to 1} A_r(f)(\theta) = \lim_{r\to 1}(P_r * f)(\theta) = f(\theta), \qquad \text{with } 0 < r < 1,$$

whenever f is continuous at θ. In this exercise, we will study the behavior of $A_r(f)(\theta)$ at certain points of discontinuity.

An integrable function is said to have a **jump discontinuity** at θ if the two limits

$$\lim_{\substack{h\to 0\\ h>0}} f(\theta+h) = f(\theta^+) \qquad \text{and} \qquad \lim_{\substack{h\to 0\\ h>0}} f(\theta-h) = f(\theta^-)$$

exist.

(a) Prove that if f has a jump discontinuity at θ, then

$$\lim_{r\to 1} A_r(f)(\theta) = \frac{f(\theta^+) + f(\theta^-)}{2}, \qquad \text{with } 0 \le r < 1.$$

[Hint: Explain why $\frac{1}{2\pi}\int_{-\pi}^{0} P_r(\theta)\,d\theta = \frac{1}{2\pi}\int_0^{\pi} P_r(\theta)\,d\theta = \frac{1}{2}$, then modify the proof given in the text.]

(b) Using a similar argument, show that if f has a jump discontinuity at θ, the Fourier series of f at θ is Cesàro summable to $\frac{f(\theta^+)+f(\theta^-)}{2}$.

18. If $P_r(\theta)$ denotes the Poisson kernel, show that the function

$$u(r,\theta)=\frac{\partial P_r}{\partial\theta},$$

defined for $0\leq r<1$ and $\theta\in\mathbb{R}$, satisfies:

(i) $\triangle u=0$ in the disc.

(ii) $\lim_{r\to 1}u(r,\theta)=0$ for each θ.

However, u is not identically zero.

19. Solve Laplace's equation $\triangle u=0$ in the semi infinite strip

$$S=\{(x,y):0<x<1,\ 0<y\},$$

subject to the following boundary conditions

$$\begin{cases} u(0,y)=0 & \text{when } 0\leq y,\\ u(1,y)=0 & \text{when } 0\leq y,\\ u(x,0)=f(x) & \text{when } 0\leq x\leq 1\end{cases}$$

where f is a given function, with of course $f(0)=f(1)=0$. Write

$$f(x)=\sum_{n=1}^{\infty}a_n\sin(n\pi x)$$

and expand the general solution in terms of the special solutions given by

$$u_n(x,y)=e^{-n\pi y}\sin(n\pi x).$$

Express u as an integral involving f, analogous to the Poisson integral formula (6).

20. Consider the Dirichlet problem in the annulus defined by $\{(r,\theta):\rho<r<1\}$, where $0<\rho<1$ is the inner radius. The problem is to solve

$$\frac{\partial^2 u}{\partial r^2}+\frac{1}{r}\frac{\partial u}{\partial r}+\frac{1}{r^2}\frac{\partial^2 u}{\partial\theta^2}=0$$

subject to the boundary conditions

$$\begin{cases} u(1,\theta)=f(\theta),\\ u(\rho,\theta)=g(\theta),\end{cases}$$

where f and g are given continuous functions.

Arguing as we have previously for the Dirichlet problem in the disc, we can hope to write

$$u(r,\theta) = \sum c_n(r) e^{in\theta}$$

with $c_n(r) = A_n r^n + B_n r^{-n}$, $n \neq 0$. Set

$$f(\theta) \sim \sum a_n e^{in\theta} \quad \text{and} \quad g(\theta) \sim \sum b_n e^{in\theta}.$$

We want $c_n(1) = a_n$ and $c_n(\rho) = b_n$. This leads to the solution

$$u(r,\theta) = \sum_{n\neq 0} \left(\frac{1}{\rho^n - \rho^{-n}}\right) \left[\left((\rho/r)^n - (r/\rho)^n \right) a_n + (r^n - r^{-n}) b_n \right] e^{in\theta}$$
$$+ a_0 + (b_0 - a_0)\frac{\log r}{\log \rho}.$$

Show that as a result we have

$$u(r,\theta) - (P_r * f)(\theta) \to 0 \qquad \text{as } r \to 1 \text{ uniformly in } \theta,$$

and

$$u(r,\theta) - (P_{\rho/r} * g)(\theta) \to 0 \qquad \text{as } r \to \rho \text{ uniformly in } \theta.$$

7 Problems

1. One can construct Riemann integrable functions on $[0,1]$ that have a dense set of discontinuities as follows.

(a) Let $f(x) = 0$ when $x < 0$, and $f(x) = 1$ if $x \geq 0$. Choose a countable dense sequence $\{r_n\}$ in $[0,1]$. Then, show that the function

$$F(x) = \sum_{n=1}^{\infty} \frac{1}{n^2} f(x - r_n)$$

is integrable and has discontinuities at all points of the sequence $\{r_n\}$. [Hint: F is monotonic and bounded.]

(b) Consider next

$$F(x) = \sum_{n=1}^{\infty} 3^{-n} g(x - r_n),$$

where $g(x) = \sin 1/x$ when $x \neq 0$, and $g(0) = 0$. Then F is integrable, discontinuous at each $x = r_n$, and fails to be monotonic in any subinterval of $[0,1]$. [Hint: Use the fact that $3^{-k} > \sum_{n>k} 3^{-n}$.]

(c) The original example of Riemann is the function

$$F(x) = \sum_{n=1}^{\infty} \frac{(nx)}{n^2},$$

where $(x) = x$ for $x \in (-1/2, 1/2]$ and (x) is continued to $\mathbb{R}$ by periodicity, that is, $(x+1) = (x)$. It can be shown that F is discontinuous whenever $x = m/2n$, where $m, n \in \mathbb{Z}$ with m odd and $n \neq 0$.

2. Let D_N denote the Dirichlet kernel

$$D_N(\theta) = \sum_{k=-N}^{N} e^{ik\theta} = \frac{\sin((N+1/2)\theta)}{\sin(\theta/2)},$$

and define

$$L_N = \frac{1}{2\pi} \int_{-\pi}^{\pi} |D_N(\theta)|\, d\theta.$$

(a) Prove that

$$L_N \geq c \log N$$

for some constant $c > 0$. [Hint: Show that $|D_N(\theta)| \geq c \frac{\sin((N+1/2)\theta)}{|\theta|}$, change variables, and prove that

$$L_N \geq c \int_{\pi}^{N\pi} \frac{|\sin\theta|}{|\theta|}\, d\theta + O(1).$$

Write the integral as a sum $\sum_{k=1}^{N-1} \int_{k\pi}^{(k+1)\pi}$. To conclude, use the fact that $\sum_{k=1}^{n} 1/k \geq c \log n$.] A more careful estimate gives

$$L_N = \frac{4}{\pi^2} \log N + O(1).$$

(b) Prove the following as a consequence: for each $n \geq 1$, there exists a continuous function f_n such that $|f_n| \leq 1$ and $|S_n(f_n)(0)| \geq c' \log n$. [Hint: The function g_n which is equal to 1 when D_n is positive and -1 when D_n is negative has the desired property but is not continuous. Approximate g_n in the integral norm (in the sense of Lemma 3.2) by continuous functions h_k satisfying $|h_k| \leq 1$.]

3.* Littlewood provided a refinement of Tauber's theorem:

(a) If $\sum c_n$ is Abel summable to s and $c_n = O(1/n)$, then $\sum c_n$ converges to s.

(b) As a consequence, if $\sum c_n$ is Cesàro summable to s and $c_n = O(1/n)$, then $\sum c_n$ converges to s.

These results may be applied to Fourier series. By Exercise 17, they imply that if f is an integrable function that satisfies $\hat{f}(\nu) = O(1/|\nu|)$, then:

(i) If f is continuous at θ, then

$$S_N(f)(\theta) \to f(\theta) \qquad \text{as } N \to \infty.$$

(ii) If f has a jump discontinuity at θ, then

$$S_N(f)(\theta) \to \frac{f(\theta^+) + f(\theta^-)}{2} \qquad \text{as } N \to \infty.$$

(iii) If f is continuous on $[-\pi, \pi]$, then $S_N(f) \to f$ uniformly.

For the simpler assertion (b), hence a proof of (i), (ii), and (iii), see Problem 5 in Chapter 4.

3 Convergence of Fourier Series

> The sine and cosine series, by which one can represent an arbitrary function in a given interval, enjoy among other remarkable properties that of being convergent. This property did not escape the great geometer (Fourier) who began, through the introduction of the representation of functions just mentioned, a new career for the applications of analysis; it was stated in the Memoir which contains his first research on heat. But no one so far, to my knowledge, gave a general proof of it ...
>
> *G. Dirichlet,* 1829

In this chapter, we continue our study of the problem of convergence of Fourier series. We approach the problem from two different points of view.

The first is "global" and concerns the overall behavior of a function f over the entire interval $[0, 2\pi]$. The result we have in mind is "mean-square convergence": if f is integrable on the circle, then

$$\frac{1}{2\pi}\int_0^{2\pi} |f(\theta) - S_N(f)(\theta)|^2 \, d\theta \to 0 \quad \text{as } N \to \infty.$$

At the heart of this result is the fundamental notion of "orthogonality"; this idea is expressed in terms of vector spaces with inner products, and their related infinite dimensional variants, the Hilbert spaces. A connected result is the Parseval identity which equates the mean-square "norm" of the function with a corresponding norm of its Fourier coefficients. Orthogonality is a fundamental mathematical notion which has many applications in analysis.

The second viewpoint is "local" and concerns the behavior of f near a given point. The main question we consider is the problem of pointwise convergence: does the Fourier series of f converge to the value $f(\theta)$ for a given θ? We first show that this convergence does indeed hold whenever f is differentiable at θ. As a corollary, we obtain the Riemann localization principle, which states that the question of whether or not $S_N(f)(\theta) \to f(\theta)$ is completely determined by the behavior of f in an arbitrarily small interval about θ. This is a remarkable result since the Fourier coefficients, hence the Fourier series, of f depend on the values of f on the whole interval $[0, 2\pi]$.

Even though convergence of the Fourier series holds at points where f is differentiable, it may fail if f is merely continuous. The chapter concludes with the presentation of a continuous function whose Fourier series does not converge at a given point, as promised earlier.

1 Mean-square convergence of Fourier series

The aim of this section is the proof of the following theorem.

Theorem 1.1 *Suppose f is integrable on the circle. Then*

$$\frac{1}{2\pi}\int_0^{2\pi} |f(\theta) - S_N(f)(\theta)|^2 \, d\theta \to 0 \qquad \text{as } N \to \infty.$$

As we remarked earlier, the key concept involved is that of orthogonality. The correct setting for orthogonality is in a vector space equipped with an inner product.

1.1 Vector spaces and inner products

We now review the definitions of a vector space over $\mathbb{R}$ or $\mathbb{C}$, an inner product, and its associated norm. In addition to the familiar finite-dimensional vector spaces $\mathbb{R}^d$ and $\mathbb{C}^d$, we also examine two infinite-dimensional examples which play a central role in the proof of Theorem 1.1.

Preliminaries on vector spaces

A vector space V over the real numbers $\mathbb{R}$ is a set whose elements may be "added" together, and "multiplied" by scalars. More precisely, we may associate to any pair $X, Y \in V$ an element in V called their sum and denoted by $X + Y$. We require that this addition respects the usual laws of arithmetic, such as commutativity $X + Y = Y + X$, and associativity $X + (Y + Z) = (X + Y) + Z$, etc. Also, given any $X \in V$ and real number λ, we assign an element $\lambda X \in V$ called the product of X by λ. This scalar multiplication must satisfy the standard properties, for instance $\lambda_1(\lambda_2 X) = (\lambda_1\lambda_2)X$ and $\lambda(X + Y) = \lambda X + \lambda Y$. We may instead allow scalar multiplication by numbers in $\mathbb{C}$; we then say that V is a vector space over the complex numbers.

For example, the set $\mathbb{R}^d$ of d-tuples of real numbers $(x_1, x_2, \ldots, x_d)$ is a vector space over the reals. Addition is defined componentwise by

$$(x_1, \ldots, x_d) + (y_1, \ldots, y_d) = (x_1 + y_1, \ldots, x_d + y_d),$$

and so is multiplication by a scalar $\lambda \in \mathbb{R}$:

$$\lambda(x_1, \ldots, x_d) = (\lambda x_1, \ldots, \lambda x_d).$$

Similarly, the space $\mathbb{C}^d$ (the complex version of the previous example) is the set of d-tuples of complex numbers $(z_1, z_2, \ldots, z_d)$. It is a vector space over $\mathbb{C}$ with addition defined componentwise by

$$(z_1, \ldots, z_d) + (w_1, \ldots, w_d) = (z_1 + w_1, \ldots, z_d + w_d).$$

Multiplication by scalars $\lambda \in \mathbb{C}$ is given by

$$\lambda(z_1, \ldots, z_d) = (\lambda z_1, \ldots, \lambda z_d).$$

An **inner product** on a vector space V over $\mathbb{R}$ associates to any pair X, Y of elements in V a real number which we denote by (X, Y). In particular, the inner product must be symmetric $(X, Y) = (Y, X)$ and linear in both variables; that is,

$$(\alpha X + \beta Y, Z) = \alpha(X, Z) + \beta(Y, Z)$$

whenever $\alpha, \beta \in \mathbb{R}$ and $X, Y, Z \in V$. Also, we require that the inner product be positive-definite, that is, $(X, X) \geq 0$ for all X in V. In particular, given an inner product $(\cdot, \cdot)$ we may define the norm of X by

$$\|X\| = (X, X)^{1/2}.$$

If in addition $\|X\| = 0$ implies $X = 0$, we say that the inner product is strictly positive-definite.

For example, the space $\mathbb{R}^d$ is equipped with a (strictly positive-definite) inner product defined by

$$(X, Y) = x_1 y_1 + \cdots + x_d y_d$$

when $X = (x_1, \ldots, x_d)$ and $Y = (y_1, \ldots, y_d)$. Then

$$\|X\| = (X, X)^{1/2} = \sqrt{x_1^2 + \cdots + x_d^2},$$

which is the usual Euclidean distance. One also uses the notation $|X|$ instead of $\|X\|$.

For vector spaces over the complex numbers, the inner product of two elements is a complex number. Moreover, these inner products are called Hermitian (instead of symmetric) since they must satisfy $(X,Y) = \overline{(Y,X)}$. Hence the inner product is linear in the first variable, but conjugate-linear in the second:

$$(\alpha X + \beta Y, Z) = \alpha(X,Z) + \beta(Y,Z) \quad \text{and}$$

$$(X, \alpha Y + \beta Z) = \overline{\alpha}(X,Y) + \overline{\beta}(X,Z).$$

Also, we must have $(X,X) \geq 0$, and the norm of X is defined by $\|X\| = (X,X)^{1/2}$ as before. Again, the inner product is strictly positive-definite if $\|X\| = 0$ implies $X = 0$.

For example, the inner product of two vectors $Z = (z_1, \ldots, z_d)$ and $W = (w_1, \ldots, w_d)$ in $\mathbb{C}^d$ is defined by

$$(Z, W) = z_1\overline{w_1} + \cdots + z_d\overline{w_d}.$$

The norm of the vector Z is then given by

$$\|Z\| = (Z,Z)^{1/2} = \sqrt{|z_1|^2 + \cdots + |z_d|^2}.$$

The presence of an inner product on a vector space allows one to define the geometric notion of "orthogonality." Let V be a vector space (over $\mathbb{R}$ or $\mathbb{C}$) with inner product $(\cdot,\cdot)$ and associated norm $\|\cdot\|$. Two elements X and Y are **orthogonal** if $(X,Y) = 0$, and we write $X \perp Y$. Three important results can be derived from this notion of orthogonality:

(i) The Pythagorean theorem: if X and Y are orthogonal, then

$$\|X + Y\|^2 = \|X\|^2 + \|Y\|^2.$$

(ii) The Cauchy-Schwarz inequality: for any $X, Y \in V$ we have

$$|(X,Y)| \leq \|X\| \, \|Y\|.$$

(iii) The triangle inequality: for any $X, Y \in V$ we have

$$\|X + Y\| \leq \|X\| + \|Y\|.$$

The proofs of these facts are simple. For (i) it suffices to expand $(X+Y, X+Y)$ and use the assumption that $(X, Y) = 0$.

For (ii), we first dispose of the case when $\|Y\| = 0$ by showing that this implies $(X, Y) = 0$ for all X. Indeed, for all real t we have

$$0 \leq \|X + tY\|^2 = \|X\|^2 + 2t\,\mathrm{Re}(X, Y)$$

and $\mathrm{Re}(X, Y) \neq 0$ contradicts the inequality if we take t to be large and positive (or negative). Similarly, by considering $\|X + itY\|^2$, we find that $\mathrm{Im}(X, Y) = 0$.

If $\|Y\| \neq 0$, we may set $c = (X, Y)/(Y, Y)$; then $X - cY$ is orthogonal to Y, and therefore also to cY. If we write $X = X - cY + cY$ and apply the Pythagorean theorem, we get

$$\|X\|^2 = \|X - cY\|^2 + \|cY\|^2 \geq |c|^2\|Y\|^2.$$

Taking square roots on both sides gives the result. Note that we have equality in the above precisely when $X = cY$.

Finally, for (iii) we first note that

$$\|X + Y\|^2 = (X, X) + (X, Y) + (Y, X) + (Y, Y).$$

But $(X, X) = \|X\|^2$, $(Y, Y) = \|Y\|^2$, and by the Cauchy-Schwarz inequality

$$|(X, Y) + (Y, X)| \leq 2\,\|X\|\,\|Y\|,$$

therefore

$$\|X + Y\|^2 \leq \|X\|^2 + 2\,\|X\|\,\|Y\| + \|Y\|^2 = (\|X\| + \|Y\|)^2.$$

Two important examples

The vector spaces $\mathbb{R}^d$ and $\mathbb{C}^d$ are finite dimensional. In the context of Fourier series, we need to work with two infinite-dimensional vector spaces, which we now describe.

EXAMPLE 1. The vector space $\ell^2(\mathbb{Z})$ over $\mathbb{C}$ is the set of all (two-sided) infinite sequences of complex numbers

$$(\ldots, a_{-n}, \ldots, a_{-1}, a_0, a_1, \ldots, a_n, \ldots)$$

such that

$$\sum_{n\in\mathbb{Z}} |a_n|^2 < \infty;$$

that is, the series converges. Addition is defined componentwise, and so is scalar multiplication. The inner product between the two vectors $A = (\ldots, a_{-1}, a_0, a_1, \ldots)$ and $B = (\ldots, b_{-1}, b_0, b_1, \ldots)$ is defined by the absolutely convergent series

$$(A, B) = \sum_{n\in\mathbb{Z}} a_n \overline{b_n}.$$

The norm of A is then given by

$$\|A\| = (A, A)^{1/2} = \left(\sum_{n\in\mathbb{Z}} |a_n|^2\right)^{1/2}.$$

We must first check that $\ell^2(\mathbb{Z})$ is a vector space. This requires that if A and B are two elements in $\ell^2(\mathbb{Z})$, then so is the vector $A + B$. To see this, for each integer $N > 0$ we let A_N denote the truncated element

$$A_N = (\ldots, 0, 0, a_{-N}, \ldots, a_{-1}, a_0, a_1, \ldots, a_N, 0, 0, \ldots),$$

where we have set $a_n = 0$ whenever $|n| > N$. We define the truncated element B_N similarly. Then, by the triangle inequality which holds in a finite dimensional Euclidean space, we have

$$\|A_N + B_N\| \leq \|A_N\| + \|B_N\| \leq \|A\| + \|B\|.$$

Thus

$$\sum_{|n|\leq N} |a_n + b_n|^2 \leq (\|A\| + \|B\|)^2,$$

and letting N tend to infinity gives $\sum_{n\in\mathbb{Z}} |a_n + b_n|^2 < \infty$. It also follows that $\|A + B\| \leq \|A\| + \|B\|$, which is the triangle inequality. The Cauchy-Schwarz inequality, which states that the sum $\sum_{n\in\mathbb{Z}} a_n \bar{b}_n$ converges absolutely and that $|(A, B)| \leq \|A\| \, \|B\|$, can be deduced in the same way from its finite analogue.

In the three examples $\mathbb{R}^d$, $\mathbb{C}^d$, and $\ell^2(\mathbb{Z})$, the vector spaces with their inner products and norms satisfy two important properties:

(i) The inner product is strictly positive-definite, that is, $\|X\| = 0$ implies $X = 0$.

(ii) The vector space is **complete**, which by definition means that every Cauchy sequence in the norm converges to a limit in the vector space.

An inner product space with these two properties is called a **Hilbert space**. We see that $\mathbb{R}^d$ and $\mathbb{C}^d$ are examples of finite-dimensional Hilbert spaces, while $\ell^2(\mathbb{Z})$ is an example of an infinite-dimensional Hilbert space (see Exercises 1 and 2). If either of the conditions above fail, the space is called a **pre-Hilbert space**.

We now give an important example of a pre-Hilbert space where both conditions (i) and (ii) fail.

EXAMPLE 2. Let $\mathcal{R}$ denote the set of complex-valued Riemann integrable functions on $[0, 2\pi]$ (or equivalently, integrable functions on the circle). This is a vector space over $\mathbb{C}$. Addition is defined pointwise by

$$(f+g)(\theta) = f(\theta) + g(\theta).$$

Naturally, multiplication by a scalar $\lambda \in \mathbb{C}$ is given by

$$(\lambda f)(\theta) = \lambda \cdot f(\theta).$$

An inner product is defined on this vector space by

$$(f, g) = \frac{1}{2\pi}\int_0^{2\pi} f(\theta)\overline{g(\theta)}\, d\theta. \tag{1}$$

The norm of f is then

$$\|f\| = \left(\frac{1}{2\pi}\int_0^{2\pi} |f(\theta)|^2\, d\theta\right)^{1/2}.$$

One needs to check that the analogue of the Cauchy-Schwarz and triangle inequalities hold in this example; that is, $|(f,g)| \leq \|f\|\,\|g\|$ and $\|f+g\| \leq \|f\| + \|g\|$. While these facts can be obtained as consequences of the corresponding inequalities in the previous examples, the argument is a little elaborate and we prefer to proceed differently.

We first observe that $2AB \leq (A^2 + B^2)$ for any two real numbers A and B. If we set $A = \lambda^{1/2}|f(\theta)|$ and $B = \lambda^{-1/2}|g(\theta)|$ with $\lambda > 0$, we get

$$|f(\theta)\overline{g(\theta)}| \leq \frac{1}{2}(\lambda|f(\theta)|^2 + \lambda^{-1}|g(\theta)|^2).$$

We then integrate this in θ to obtain

$$|(f,g)| \leq \frac{1}{2\pi}\int_0^{2\pi} |f(\theta)|\,|\overline{g(\theta)}|\, d\theta \leq \frac{1}{2}(\lambda\|f\|^2 + \lambda^{-1}\|g\|^2).$$

Then, put $\lambda = \|g\|/\|f\|$ to get the Cauchy-Schwarz inequality. The triangle inequality is then a simple consequence, as we have seen above.

Of course, in our choice of λ we must assume that $\|f\| \neq 0$ and $\|g\| \neq 0$, which leads us to the following observation.

In $\mathcal{R}$, condition (i) for a Hilbert space fails, since $\|f\| = 0$ implies only that f vanishes at its points of continuity. This is not a very serious problem since in the appendix we show that an integrable function is continuous except for a "negligible" set, so that $\|f\| = 0$ implies that f vanishes except on a set of "measure zero." One can get around the difficulty that f is not identically zero by adopting the convention that such functions are actually the zero function, since for the purpose of integration, f behaves precisely like the zero function.

A more essential difficulty is that the space $\mathcal{R}$ is not complete. One way to see this is to start with the function

$$f(\theta) = \begin{cases} 0 & \text{for } \theta = 0, \\ \log(1/\theta) & \text{for } 0 < \theta \leq 2\pi. \end{cases}$$

Since f is not bounded, it does not belong to the space $\mathcal{R}$. Moreover, the sequence of truncations f_n defined by

$$f_n(\theta) = \begin{cases} 0 & \text{for } 0 \leq \theta \leq 1/n, \\ f(\theta) & \text{for } 1/n < \theta \leq 2\pi \end{cases}$$

can easily be seen to form a Cauchy sequence in $\mathcal{R}$ (see Exercise 5). However, this sequence cannot converge to an element in $\mathcal{R}$, since that limit, if it existed, would have to be f; for another example, see Exercise 7.

This and more complicated examples motivate the search for the completion of $\mathcal{R}$, the class of Riemann integrable functions on $[0, 2\pi]$. The construction and identification of this completion, the Lebesgue class $L^2([0, 2\pi])$, represents an important turning point in the development of analysis (somewhat akin to the much earlier completion of the rationals, that is, the passage from $\mathbb{Q}$ to $\mathbb{R}$). A further discussion of these fundamental ideas will be postponed until Book III, where we take up the Lebesgue theory of integration.

We now turn to the proof of Theorem 1.1.

1.2 Proof of mean-square convergence

Consider the space $\mathcal{R}$ of integrable functions on the circle with inner product

$$(f, g) = \frac{1}{2\pi} \int_0^{2\pi} f(\theta)\overline{g(\theta)}\, d\theta$$

and norm $\|f\|$ defined by

$$\|f\|^2 = (f, f) = \frac{1}{2\pi} \int_0^{2\pi} |f(\theta)|^2 \, d\theta.$$

With this notation, we must prove that $\|f - S_N(f)\| \to 0$ as N tends to infinity.

For each integer n, let $e_n(\theta) = e^{in\theta}$, and observe that the family $\{e_n\}_{n\in\mathbb{Z}}$ is **orthonormal**; that is,

$$(e_n, e_m) = \begin{cases} 1 & \text{if } n = m \\ 0 & \text{if } n \neq m. \end{cases}$$

Let f be an integrable function on the circle, and let a_n denote its Fourier coefficients. An important observation is that these Fourier coefficients are represented by inner products of f with the elements in the orthonormal set $\{e_n\}_{n\in\mathbb{Z}}$:

$$(f, e_n) = \frac{1}{2\pi} \int_0^{2\pi} f(\theta) e^{-in\theta} \, d\theta = a_n.$$

In particular, $S_N(f) = \sum_{|n|\leq N} a_n e_n$. Then the orthonormal property of the family $\{e_n\}$ and the fact that $a_n = (f, e_n)$ imply that the difference $f - \sum_{|n|\leq N} a_n e_n$ is orthogonal to e_n for all $|n| \leq N$. Therefore, we must have

$$(f - \sum_{|n|\leq N} a_n e_n) \perp \sum_{|n|\leq N} b_n e_n \tag{2}$$

for any complex numbers b_n. We draw two conclusions from this fact.

First, we can apply the Pythagorean theorem to the decomposition

$$f = f - \sum_{|n|\leq N} a_n e_n + \sum_{|n|\leq N} a_n e_n,$$

where we now choose $b_n = a_n$, to obtain

$$\|f\|^2 = \|f - \sum_{|n|\leq N} a_n e_n\|^2 + \|\sum_{|n|\leq N} a_n e_n\|^2.$$

Since the orthonormal property of the family $\{e_n\}_{n\in\mathbb{Z}}$ implies that

$$\|\sum_{|n|\leq N} a_n e_n\|^2 = \sum_{|n|\leq N} |a_n|^2,$$

we deduce that

$$\|f\|^2 = \|f - S_N(f)\|^2 + \sum_{|n|\leq N} |a_n|^2. \tag{3}$$

The second conclusion we may draw from (2) is the following simple lemma.

Lemma 1.2 (Best approximation) *If f is integrable on the circle with Fourier coefficients a_n, then*

$$\|f - S_N(f)\| \leq \|f - \sum_{|n|\leq N} c_n e_n\|$$

for any complex numbers c_n. Moreover, equality holds precisely when $c_n = a_n$ for all $|n| \leq N$.

Proof. This follows immediately by applying the Pythagorean theorem to

$$f - \sum_{|n|\leq N} c_n e_n = f - S_N(f) + \sum_{|n|\leq N} b_n e_n,$$

where $b_n = a_n - c_n$.

This lemma has a clear geometric interpretation. It says that the trigonometric polynomial of degree at most N which is closest to f in the norm $\|\cdot\|$ is the partial sum $S_N(f)$. This geometric property of the partial sums is depicted in Figure 1, where the orthogonal projection of f in the plane spanned by $\{e_{-N}, \ldots, e_0, \ldots, e_N\}$ is simply $S_N(f)$.

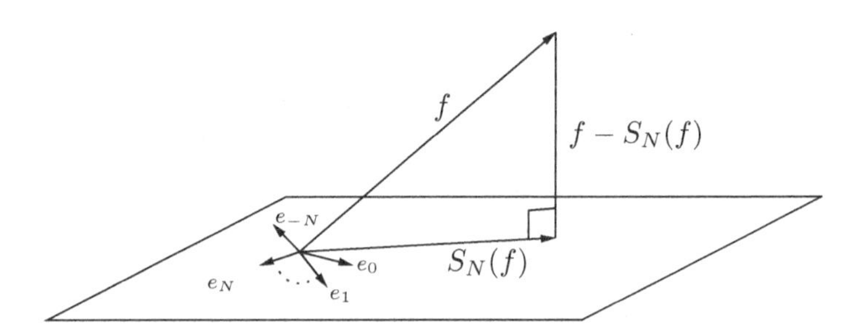

Figure 1. The best approximation lemma

We can now give the proof that $\|S_N(f) - f\| \to 0$ using the best approximation lemma, as well as the important fact that trigonometric polynomials are dense in the space of continuous functions on the circle.

Suppose that f is continuous on the circle. Then, given $\epsilon > 0$, there exists (by Corollary 5.4 in Chapter 2) a trigonometric polynomial P, say of degree M, such that

$$|f(\theta) - P(\theta)| < \epsilon \quad \text{for all } \theta.$$

In particular, taking squares and integrating this inequality yields $\|f - P\| < \epsilon$, and by the best approximation lemma we conclude that

$$\|f - S_N(f)\| < \epsilon \quad \text{whenever } N \geq M.$$

This proves Theorem 1.1 when f is continuous.

If f is merely integrable, we can no longer approximate f uniformly by trigonometric polynomials. Instead, we apply the approximation Lemma 3.2 in Chapter 2 and choose a continuous function g on the circle which satisfies

$$\sup_{\theta \in [0,2\pi]} |g(\theta)| \leq \sup_{\theta \in [0,2\pi]} |f(\theta)| = B,$$

and

$$\int_0^{2\pi} |f(\theta) - g(\theta)|\, d\theta < \epsilon^2.$$

Then we get

$$\begin{aligned} \|f - g\|^2 &= \frac{1}{2\pi} \int_0^{2\pi} |f(\theta) - g(\theta)|^2\, d\theta \\ &= \frac{1}{2\pi} \int_0^{2\pi} |f(\theta) - g(\theta)|\, |f(\theta) - g(\theta)|\, d\theta \\ &\leq \frac{2B}{2\pi} \int_0^{2\pi} |f(\theta) - g(\theta)|\, d\theta \\ &\leq C\epsilon^2. \end{aligned}$$

Now we may approximate g by a trigonometric polynomial P so that $\|g - P\| < \epsilon$. Then $\|f - P\| < C'\epsilon$, and we may again conclude by applying the best approximation lemma. This completes the proof that the partial sums of the Fourier series of f converge to f in the mean square norm $\|\cdot\|$.

Note that this result and the relation (3) imply that if a_n is the nth Fourier coefficient of an integrable function f, then the series $\sum_{n=-\infty}^{\infty} |a_n|^2$ converges, and in fact we have **Parseval's identity**

$$\sum_{n=-\infty}^{\infty} |a_n|^2 = \|f\|^2.$$

This identity provides an important connection between the norms in the two vector spaces $\ell^2(\mathbb{Z})$ and $\mathcal{R}$.

We now summarize the results of this section.

Theorem 1.3 *Let f be an integrable function on the circle with $f \sim \sum_{n=-\infty}^{\infty} a_n e^{in\theta}$. Then we have:*

(i) *Mean-square convergence of the Fourier series*

$$\frac{1}{2\pi} \int_0^{2\pi} |f(\theta) - S_N(f)(\theta)|^2 \, d\theta \to 0 \quad \text{as } N \to \infty.$$

(ii) *Parseval's identity*

$$\sum_{n=-\infty}^{\infty} |a_n|^2 = \frac{1}{2\pi} \int_0^{2\pi} |f(\theta)|^2 \, d\theta.$$

Remark 1. If $\{e_n\}$ is *any* orthonormal family of functions on the circle, and $a_n = (f, e_n)$, then we may deduce from the relation (3) that

$$\sum_{n=-\infty}^{\infty} |a_n|^2 \leq \|f\|^2.$$

This is known as **Bessel's inequality**. Equality holds (as in Parseval's identity) precisely when the family $\{e_n\}$ is also a "basis," in the sense that $\| \sum_{|n| \leq N} a_n e_n - f \| \to 0$ as $N \to \infty$.

Remark 2. We may associate to every integrable function the sequence $\{a_n\}$ formed by its Fourier coefficients. Parseval's identity guarantees that $\{a_n\} \in \ell^2(\mathbb{Z})$. Since $\ell^2(\mathbb{Z})$ is a Hilbert space, the failure of $\mathcal{R}$ to be complete, discussed earlier, may be understood as follows: there exist sequences $\{a_n\}_{n \in \mathbb{Z}}$ such that $\sum_{n \in \mathbb{Z}} |a_n|^2 < \infty$, yet no Riemann integrable function F has n^{th} Fourier coefficient equal to a_n for all n. An example is given in Exercise 6.

Since the terms of a converging series tend to 0, we deduce from Parseval's identity or Bessel's inequality the following result.

Theorem 1.4 (Riemann-Lebesgue lemma) *If f is integrable on the circle, then $\hat{f}(n) \to 0$ as $|n| \to \infty$.*

An equivalent reformulation of this proposition is that if f is integrable on $[0, 2\pi]$, then

$$\int_0^{2\pi} f(\theta) \sin(N\theta) \, d\theta \to 0 \quad \text{as } N \to \infty$$

and

$$\int_0^{2\pi} f(\theta)\cos(N\theta)\,d\theta \to 0 \quad \text{as } N \to \infty.$$

To conclude this section, we give a more general version of the Parseval identity which we will use in the next chapter.

Lemma 1.5 *Suppose F and G are integrable on the circle with*

$$F \sim \sum a_n e^{in\theta} \quad \textit{and} \quad G \sim \sum b_n e^{in\theta}.$$

Then

$$\frac{1}{2\pi}\int_0^{2\pi} F(\theta)\overline{G(\theta)}\,d\theta = \sum_{n=-\infty}^{\infty} a_n\overline{b_n}.$$

Recall from the discussion in Example 1 that the series $\sum_{n=-\infty}^{\infty} a_n\overline{b_n}$ converges absolutely.

Proof. The proof follows from Parseval's identity and the fact that

$$(F,G) = \frac{1}{4}\left[\|F+G\|^2 - \|F-G\|^2 + i\left(\|F+iG\|^2 - \|F-iG\|^2\right)\right]$$

which holds in every Hermitian inner product space. The verification of this fact is left to the reader.

2 Return to pointwise convergence

The mean-square convergence theorem does not provide further insight into the problem of pointwise convergence. Indeed, Theorem 1.1 by itself does not guarantee that the Fourier series converges for any θ. Exercise 3 helps to explain this statement. However, if a function is differentiable at a point θ_0, then its Fourier series converges at θ_0. After proving this result, we give an example of a continuous function with diverging Fourier series at one point. These phenomena are indicative of the intricate nature of the problem of pointwise convergence in the theory of Fourier series.

2.1 A local result

Theorem 2.1 *Let f be an integrable function on the circle which is differentiable at a point θ_0. Then $S_N(f)(\theta_0) \to f(\theta_0)$ as N tends to infinity.*

Proof. Define

$$F(t) = \begin{cases} \dfrac{f(\theta_0 - t) - f(\theta_0)}{t} & \text{if } t \neq 0 \text{ and } |t| < \pi \\ -f'(\theta_0) & \text{if } t = 0. \end{cases}$$

First, F is bounded near 0 since f is differentiable there. Second, for all small δ the function F is integrable on $[-\pi, -\delta] \cup [\delta, \pi]$ because f has this property and $|t| > \delta$ there. As a consequence of Proposition 1.4 in the appendix, the function F is integrable on all of $[-\pi, \pi]$. We know that $S_N(f)(\theta_0) = (f * D_N)(\theta_0)$, where D_N is the Dirichlet kernel. Since $\frac{1}{2\pi} \int D_N = 1$, we find that

$$\begin{aligned} S_N(f)(\theta_0) - f(\theta_0) &= \frac{1}{2\pi} \int_{-\pi}^{\pi} f(\theta_0 - t) D_N(t)\, dt - f(\theta_0) \\ &= \frac{1}{2\pi} \int_{-\pi}^{\pi} [f(\theta_0 - t) - f(\theta_0)] D_N(t)\, dt \\ &= \frac{1}{2\pi} \int_{-\pi}^{\pi} F(t) t D_N(t)\, dt. \end{aligned}$$

We recall that

$$tD_N(t) = \frac{t}{\sin(t/2)} \sin((N + 1/2)t),$$

where the quotient $\frac{t}{\sin(t/2)}$ is continuous in the interval $[-\pi, \pi]$. Since we can write

$$\sin((N + 1/2)t) = \sin(Nt)\cos(t/2) + \cos(Nt)\sin(t/2),$$

we can apply the Riemann-Lebesgue lemma to the Riemann integrable functions $F(t)t\cos(t/2)/\sin(t/2)$ and $F(t)t$ to finish the proof of the theorem.

Observe that the conclusion of the theorem still holds if we only assume that f satisfies a **Lipschitz condition** at θ_0; that is,

$$|f(\theta) - f(\theta_0)| \leq M|\theta - \theta_0|$$

for some $M \geq 0$ and all θ. This is the same as saying that f satisfies a Hölder condition of order $\alpha = 1$.

A striking consequence of this theorem is the localization principle of Riemann. This result states that the convergence of $S_N(f)(\theta_0)$ depends only on the behavior of f near θ_0. This is not clear at first, since forming the Fourier series requires integrating f over the *whole* circle.

Theorem 2.2 *Suppose f and g are two integrable functions defined on the circle, and for some θ_0 there exists an open interval I containing θ_0 such that*

$$f(\theta) = g(\theta) \quad \text{for all } \theta \in I.$$

Then $S_N(f)(\theta_0) - S_N(g)(\theta_0) \to 0$ as N tends to infinity.

Proof. The function $f - g$ is 0 in I, so it is differentiable at θ_0, and we may apply the previous theorem to conclude the proof.

2.2 A continuous function with diverging Fourier series

We now turn our attention to an example of a continuous periodic function whose Fourier series diverges at a point. Thus, Theorem 2.1 fails if the differentiability assumption is replaced by the weaker assumption of continuity. Our counter-example shows that this hypothesis which had appeared plausible, is in fact false; moreover, its construction also illuminates an important principle of the theory.

The principle that is involved here will be referred to as "symmetry-breaking."[1] The symmetry that we have in mind is the symmetry between the frequencies $e^{in\theta}$ and $e^{-in\theta}$ which appear in the Fourier expansion of a function. For example, the partial sum operator S_N is defined in a way that reflects this symmetry. Also, the Dirichlet, Fejèr, and Poisson kernels are symmetric in this sense. When we break the symmetry, that is, when we split the Fourier series $\sum_{n=-\infty}^{\infty} a_n e^{in\theta}$ into the two pieces $\sum_{n\geq 0} a_n e^{in\theta}$ and $\sum_{n<0} a_n e^{in\theta}$, we introduce new and far-reaching phenomena.

We give a simple example. Start with the sawtooth function f which is odd in θ and which equals $i(\pi - \theta)$ when $0 < \theta < \pi$. Then, by Exercise 8 in Chapter 2, we know that

$$f(\theta) \sim \sum_{n\neq 0} \frac{e^{in\theta}}{n}. \tag{4}$$

Consider now the result of breaking the symmetry and the resulting series

$$\sum_{n=-\infty}^{n=-1} \frac{e^{in\theta}}{n}.$$

Then, unlike (4), the above is no longer the Fourier series of a Riemann integrable function. Indeed, suppose it were the Fourier series of an

[1] We have borrowed this terminology from physics, where it is used in a very different context.

integrable function, say $\tilde{f}$, where in particular $\tilde{f}$ is bounded. Using the Abel means, we then have

$$|A_r(\tilde{f})(0)| = \sum_{n=1}^{\infty} \frac{r^n}{n},$$

which tends to infinity as r tends to 1, because $\sum 1/n$ diverges. This gives the desired contradiction since

$$|A_r(\tilde{f})(0)| \leq \frac{1}{2\pi} \int_{-\pi}^{\pi} |\tilde{f}(\theta)| P_r(\theta)\, d\theta \leq \sup_{\theta} |\tilde{f}(\theta)|,$$

where $P_r(\theta)$ denotes the Poisson kernel discussed in the previous chapter.

The sawtooth function is the object from which we will fashion our counter-example. We proceed as follows. For each $N \geq 1$ we define the following two functions on $[-\pi, \pi]$,

$$f_N(\theta) = \sum_{1 \leq |n| \leq N} \frac{e^{in\theta}}{n} \quad \text{and} \quad \tilde{f}_N(\theta) = \sum_{-N \leq n \leq -1} \frac{e^{in\theta}}{n}.$$

We contend that:

(i) $|\tilde{f}_N(0)| \geq c \log N$.

(ii) $f_N(\theta)$ is uniformly bounded in N and θ.

The first statement is a consequence of the fact that $\sum_{n=1}^{N} 1/n \geq \log N$, which is easily established (see also Figure 2):

$$\sum_{n=1}^{N} \frac{1}{n} \geq \sum_{n=1}^{N-1} \int_n^{n+1} \frac{dx}{x} = \int_1^N \frac{dx}{x} = \log N.$$

To prove (ii), we argue in the same spirit as in the proof of Tauber's theorem, which says that if the series $\sum c_n$ is Abel summable to s and $c_n = o(1/n)$, then $\sum c_n$ actually converges to s (see Exercise 14 in Chapter 2). In fact, the proof of Tauber's theorem is quite similar to that of the lemma below.

Lemma 2.3 *Suppose that the Abel means $A_r = \sum_{n=1}^{\infty} r^n c_n$ of the series $\sum_{n=1}^{\infty} c_n$ are bounded as r tends to 1 (with $r < 1$). If $c_n = O(1/n)$, then the partial sums $S_N = \sum_{n=1}^{N} c_n$ are bounded.*

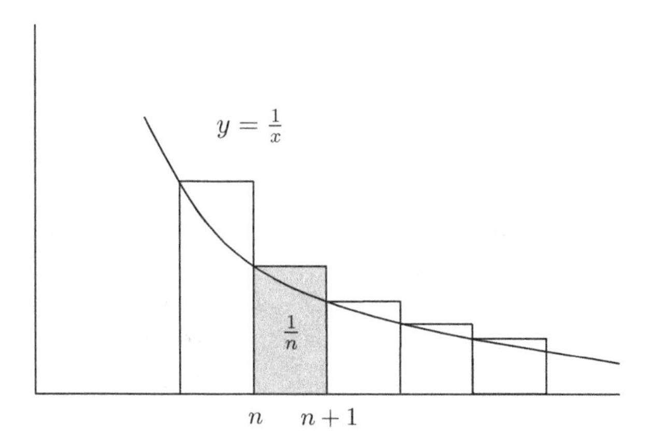

Figure 2. Comparing a sum with an integral

Proof. Let $r = 1 - 1/N$ and choose M so that $n|c_n| \leq M$. We estimate the difference

$$S_N - A_r = \sum_{n=1}^{N}(c_n - r^n c_n) - \sum_{n=N+1}^{\infty} r^n c_n$$

as follows:

$$\begin{aligned} |S_N - A_r| &\leq \sum_{n=1}^{N} |c_n|(1 - r^n) + \sum_{n=N+1}^{\infty} r^n |c_n| \\ &\leq M \sum_{n=1}^{N}(1 - r) + \frac{M}{N} \sum_{n=N+1}^{\infty} r^n \\ &\leq MN(1 - r) + \frac{M}{N}\frac{1}{1 - r} \\ &= 2M, \end{aligned}$$

where we have used the simple observation that

$$1 - r^n = (1 - r)(1 + r + \cdots + r^{n-1}) \leq n(1 - r).$$

So we see that if M satisfies both $|A_r| \leq M$ and $n|c_n| \leq M$, then $|S_N| \leq 3M$.

We apply the lemma to the series

$$\sum_{n \neq 0} \frac{e^{in\theta}}{n},$$

which is the Fourier series of the sawtooth function f used above. Here $c_n = e^{in\theta}/n + e^{-in\theta}/(-n)$ for $n \neq 0$, so clearly $c_n = O(1/|n|)$. Finally, the Abel means of this series are $A_r(f)(\theta) = (f * P_r)(\theta)$. But f is bounded and P_r is a good kernel, so $S_N(f)(\theta)$ is uniformly bounded in N and θ, as was to be shown.

We now come to the heart of the matter. Notice that f_N and $\tilde{f}_N$ are trigonometric polynomials of degree N (that is, they have non-zero Fourier coefficients only when $|n| \leq N$). From these, we form trigonometric polynomials P_N and $\tilde{P}_N$, now of degrees $3N$ and $2N-1$, by displacing the frequencies of f_N and $\tilde{f}_N$ by $2N$ units. In other words, we define $P_N(\theta) = e^{i(2N)\theta} f_N(\theta)$ and $\tilde{P}_N(\theta) = e^{i(2N)\theta} \tilde{f}_N(\theta)$. So while f_N has non-vanishing Fourier coefficients when $0 < |n| \leq N$, now the coefficients of P_N are non-vanishing for $N \leq n \leq 3N$, $n \neq 2N$. Moreover, while $n = 0$ is the center of symmetry of f_N, now $n = 2N$ is the center of symmetry of P_N. We next consider the partial sums S_M.

Lemma 2.4

$$S_M(P_N) = \begin{cases} P_N & \text{if } M \geq 3N, \\ \tilde{P}_N & \text{if } M = 2N, \\ 0 & \text{if } M < N. \end{cases}$$

This is clear from what has been said above and from Figure 3.

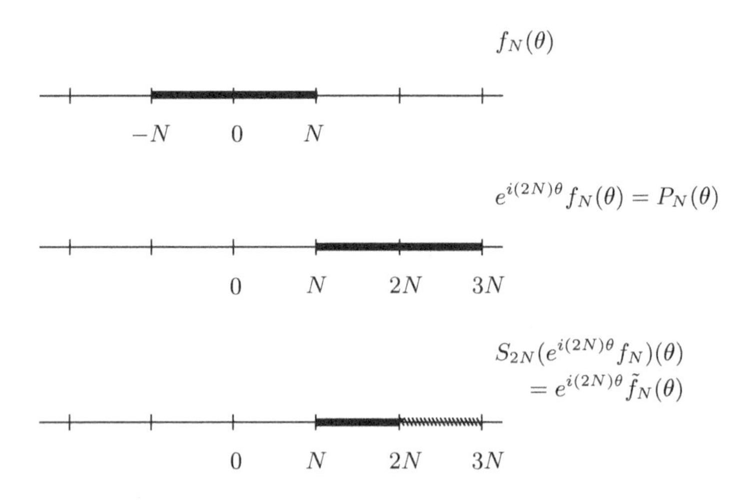

Figure 3. Breaking symmetry in Lemma 2.4

The effect is that when $M = 2N$, the operator S_M breaks the symmetry of P_N, but in the other cases covered in the lemma, the action of S_M

is relatively benign, since then the outcome is either P_N or 0.

Finally, we need to find a convergent series of positive terms $\sum \alpha_k$ and a sequence of integers $\{N_k\}$ which increases rapidly enough so that:

(i) $N_{k+1} > 3N_k$,

(ii) $\alpha_k \log N_k \to \infty$ as $k \to \infty$.

We choose (for example) $\alpha_k = 1/k^2$ and $N_k = 3^{2^k}$ which are easily seen to satisfy the above criteria.

Finally, we can write down our desired function. It is

$$f(\theta) = \sum_{k=1}^{\infty} \alpha_k P_{N_k}(\theta).$$

Due to the uniform boundedness of the P_N (recall that $|P_N(\theta)| = |f_N(\theta)|$), the series above converges uniformly to a continuous periodic function. However, by our lemma we get

$$|S_{2N_m}(f)(0)| \geq c\alpha_m \log N_m + O(1) \to \infty \quad \text{as } m \to \infty.$$

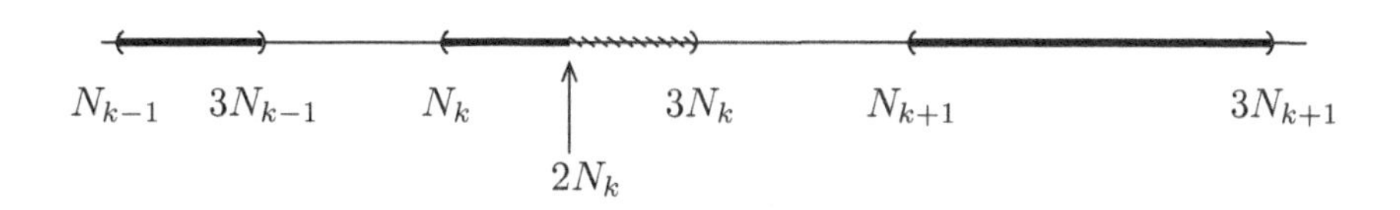

Figure 4. Symmetry broken in the middle interval $(N_k, 3N_k)$

Indeed, the terms that correspond to N_k with $k < m$ or $k > m$ contribute $O(1)$ or 0, respectively (because the P_N's are uniformly bounded), while the term that corresponds to N_m is in absolute value greater than $c\alpha_m \log N_m$ because $|\tilde{P}_N(\theta)| = |\tilde{f}_N(\theta)| \geq c \log N$. So the partial sums of the Fourier series of f at 0 are not bounded, and we are done since this proves the divergence of the Fourier series of f at $\theta = 0$. To produce a function whose series diverges at any other preassigned $\theta = \theta_0$, it suffices to consider the function $f(\theta - \theta_0)$.

3 Exercises

1. Show that the first two examples of inner product spaces, namely $\mathbb{R}^d$ and $\mathbb{C}^d$, are complete.

[Hint: Every Cauchy sequence in $\mathbb{R}$ has a limit.]

2. Prove that the vector space $\ell^2(\mathbb{Z})$ is complete.

[Hint: Suppose $A_k = \{a_{k,n}\}_{n\in\mathbb{Z}}$ with $k = 1, 2, \ldots$ is a Cauchy sequence. Show that for each n, $\{a_{k,n}\}_{k=1}^{\infty}$ is a Cauchy sequence of complex numbers, therefore it converges to a limit, say b_n. By taking partial sums of $\|A_k - A_{k'}\|$ and letting $k' \to \infty$, show that $\|A_k - B\| \to 0$ as $k \to \infty$, where $B = (\ldots, b_{-1}, b_0, b_1, \ldots)$. Finally, prove that $B \in \ell^2(\mathbb{Z})$.]

3. Construct a sequence of integrable functions $\{f_k\}$ on $[0, 2\pi]$ such that

$$\lim_{k\to\infty} \frac{1}{2\pi}\int_0^{2\pi} |f_k(\theta)|^2\, d\theta = 0$$

but $\lim_{k\to\infty} f_k(\theta)$ fails to exist for any θ.

[Hint: Choose a sequence of intervals $I_k \subset [0, 2\pi]$ whose lengths tend to 0, and so that each point belongs to infinitely many of them; then let $f_k = \chi_{I_k}$.]

4. Recall the vector space $\mathcal{R}$ of integrable functions, with its inner product and norm

$$\|f\| = \left(\frac{1}{2\pi}\int_0^{2\pi} |f(x)|^2\, dx\right)^{1/2}.$$

(a) Show that there exist non-zero integrable functions f for which $\|f\| = 0$.

(b) However, show that if $f \in \mathcal{R}$ with $\|f\| = 0$, then $f(x) = 0$ whenever f is continuous at x.

(c) Conversely, show that if $f \in \mathcal{R}$ vanishes at all of its points of continuity, then $\|f\| = 0$.

5. Let

$$f(\theta) = \begin{cases} 0 & \text{for } \theta = 0 \\ \log(1/\theta) & \text{for } 0 < \theta \le 2\pi, \end{cases}$$

and define a sequence of functions in $\mathcal{R}$ by

$$f_n(\theta) = \begin{cases} 0 & \text{for } 0 \le \theta \le 1/n \\ f(\theta) & \text{for } 1/n < \theta \le 2\pi. \end{cases}$$

Prove that $\{f_n\}_{n=1}^{\infty}$ is a Cauchy sequence in $\mathcal{R}$. However, f does not belong to $\mathcal{R}$.

[Hint: Show that $\int_a^b (\log\theta)^2\,d\theta \to 0$ if $0 < a < b$ and $b \to 0$, by using the fact that the derivative of $\theta(\log\theta)^2 - 2\theta\log\theta + 2\theta$ is equal to $(\log\theta)^2$.]

6. Consider the sequence $\{a_k\}_{k=-\infty}^{\infty}$ defined by

$$a_k = \begin{cases} 1/k & \text{if } k \geq 1 \\ 0 & \text{if } k \leq 0. \end{cases}$$

Note that $\{a_k\} \in \ell^2(\mathbb{Z})$, but that no Riemann integrable function has k^{th} Fourier coefficient equal to a_k for all k.

7. Show that the trigonometric series

$$\sum_{n\geq 2} \frac{1}{\log n} \sin nx$$

converges for every x, yet it is not the Fourier series of a Riemann integrable function.

The same is true for $\sum \frac{\sin nx}{n^\alpha}$ for $0 < \alpha < 1$, but the case $1/2 < \alpha < 1$ is more difficult. See Problem 1.

8. Exercise 6 in Chapter 2 dealt with the sums

$$\sum_{n \text{ odd } \geq 1} \frac{1}{n^2} \quad \text{and} \quad \sum_{n=1}^{\infty} \frac{1}{n^2}.$$

Similar sums can be derived using the methods of this chapter.

(a) Let f be the function defined on $[-\pi, \pi]$ by $f(\theta) = |\theta|$. Use Parseval's identity to find the sums of the following two series:

$$\sum_{n=0}^{\infty} \frac{1}{(2n+1)^4} \quad \text{and} \quad \sum_{n=1}^{\infty} \frac{1}{n^4}.$$

In fact, they are $\pi^4/96$ and $\pi^4/90$, respectively.

(b) Consider the 2π-periodic odd function defined on $[0, \pi]$ by $f(\theta) = \theta(\pi - \theta)$. Show that

$$\sum_{n=0}^{\infty} \frac{1}{(2n+1)^6} = \frac{\pi^6}{960} \quad \text{and} \quad \sum_{n=1}^{\infty} \frac{1}{n^6} = \frac{\pi^6}{945}.$$

Remark. The general expression when k is even for $\sum_{n=1}^{\infty} 1/n^k$ in terms of π^k is given in Problem 4. However, finding a formula for the sum $\sum_{n=1}^{\infty} 1/n^3$, or more generally $\sum_{n=1}^{\infty} 1/n^k$ with k odd, is a famous unresolved question.

9. Show that for α not an integer, the Fourier series of

$$\frac{\pi}{\sin \pi\alpha} e^{i(\pi - x)\alpha}$$

on $[0, 2\pi]$ is given by

$$\sum_{n=-\infty}^{\infty} \frac{e^{inx}}{n + \alpha}.$$

Apply Parseval's formula to show that

$$\sum_{n=-\infty}^{\infty} \frac{1}{(n+\alpha)^2} = \frac{\pi^2}{(\sin \pi\alpha)^2}.$$

10. Consider the example of a vibrating string which we analyzed in Chapter 1. The displacement $u(x, t)$ of the string at time t satisfies the wave equation

$$\frac{1}{c^2}\frac{\partial^2 u}{\partial t^2} = \frac{\partial^2 u}{\partial x^2}, \qquad c^2 = \tau/\rho.$$

The string is subject to the initial conditions

$$u(x, 0) = f(x) \quad \text{and} \quad \frac{\partial u}{\partial t}(x, 0) = g(x),$$

where we assume that $f \in C^1$ and g is continuous. We define the total **energy** of the string by

$$E(t) = \frac{1}{2}\rho \int_0^L \left(\frac{\partial u}{\partial t}\right)^2 dx + \frac{1}{2}\tau \int_0^L \left(\frac{\partial u}{\partial x}\right)^2 dx.$$

The first term corresponds to the "kinetic energy" of the string (in analogy with $(1/2)mv^2$, the kinetic energy of a particle of mass m and velocity v), and the second term corresponds to its "potential energy."

Show that the total energy of the string is conserved, in the sense that $E(t)$ is constant. Therefore,

$$E(t) = E(0) = \frac{1}{2}\rho \int_0^L g(x)^2\, dx + \frac{1}{2}\tau \int_0^L f'(x)^2\, dx.$$

11. The inequalities of Wirtinger and Poincaré establish a relationship between the norm of a function and that of its derivative.

(a) If f is T-periodic, continuous, and piecewise C^1 with $\int_0^T f(t)\,dt = 0$, show that

$$\int_0^T |f(t)|^2\,dt \leq \frac{T^2}{4\pi^2}\int_0^T |f'(t)|^2\,dt,$$

with equality if and only if $f(t) = A\sin(2\pi t/T) + B\cos(2\pi t/T)$. [Hint: Apply Parseval's identity.]

(b) If f is as above and g is just C^1 and T-periodic, prove that

$$\left|\int_0^T \overline{f(t)}g(t)\,dt\right|^2 \leq \frac{T^2}{4\pi^2}\int_0^T |f(t)|^2\,dt \int_0^T |g'(t)|^2\,dt.$$

(c) For any compact interval $[a, b]$ and any continuously differentiable function f with $f(a) = f(b) = 0$, show that

$$\int_a^b |f(t)|^2\,dt \leq \frac{(b-a)^2}{\pi^2}\int_a^b |f'(t)|^2\,dt.$$

Discuss the case of equality, and prove that the constant $(b-a)^2/\pi^2$ cannot be improved. [Hint: Extend f to be odd with respect to a and periodic of period $T = 2(b-a)$ so that its integral over an interval of length T is 0. Apply part a) to get the inequality, and conclude that equality holds if and only if $f(t) = A\sin(\pi\frac{t-a}{b-a})$].

12. Prove that $\displaystyle\int_0^\infty \frac{\sin x}{x}\,dx = \frac{\pi}{2}$.

[Hint: Start with the fact that the integral of $D_N(\theta)$ equals 2π, and note that the difference $(1/\sin(\theta/2)) - 2/\theta$ is continuous on $[-\pi, \pi]$. Apply the Riemann-Lebesgue lemma.]

13. Suppose that f is periodic and of class C^k. Show that

$$\hat{f}(n) = o(1/|n|^k),$$

that is, $|n|^k\hat{f}(n)$ goes to 0 as $|n| \to \infty$. This is an improvement over Exercise 10 in Chapter 2.

[Hint: Use the Riemann-Lebesgue lemma.]

14. Prove that the Fourier series of a continuously differentiable function f on the circle is absolutely convergent.

[Hint: Use the Cauchy-Schwarz inequality and Parseval's identity for f'.]

15. Let f be 2π-periodic and Riemann integrable on $[-\pi, \pi]$.

(a) Show that

$$\hat{f}(n) = -\frac{1}{2\pi}\int_{-\pi}^{\pi} f(x+\pi/n)e^{-inx}\,dx$$

hence

$$\hat{f}(n) = \frac{1}{4\pi}\int_{-\pi}^{\pi} [f(x) - f(x+\pi/n)]e^{-inx}\,dx.$$

(b) Now assume that f satisfies a Hölder condition of order α, namely

$$|f(x+h) - f(x)| \leq C|h|^{\alpha}$$

for some $0 < \alpha \leq 1$, some $C > 0$, and all x, h. Use part a) to show that

$$\hat{f}(n) = O(1/|n|^{\alpha}).$$

(c) Prove that the above result cannot be improved by showing that the function

$$f(x) = \sum_{k=0}^{\infty} 2^{-k\alpha} e^{i2^k x},$$

where $0 < \alpha < 1$, satisfies

$$|f(x+h) - f(x)| \leq C|h|^{\alpha},$$

and $\hat{f}(N) = 1/N^{\alpha}$ whenever $N = 2^k$.

[Hint: For (c), break up the sum as follows $f(x+h) - f(x) = \sum_{2^k \leq 1/|h|} + \sum_{2^k > 1/|h|}$. To estimate the first sum use the fact that $|1 - e^{i\theta}| \leq |\theta|$ whenever θ is small. To estimate the second sum, use the obvious inequality $|e^{ix} - e^{iy}| \leq 2$.]

16. Let f be a 2π-periodic function which satisfies a Lipschitz condition with constant K; that is,

$$|f(x) - f(y)| \leq K|x - y| \quad \text{for all } x, y.$$

This is simply the Hölder condition with $\alpha = 1$, so by the previous exercise, we see that $\hat{f}(n) = O(1/|n|)$. Since the harmonic series $\sum 1/n$ diverges, we cannot say anything (yet) about the absolute convergence of the Fourier series of f. The outline below actually proves that the Fourier series of f converges absolutely and uniformly.

(a) For every positive h we define $g_h(x) = f(x+h) - f(x-h)$. Prove that

$$\frac{1}{2\pi}\int_0^{2\pi} |g_h(x)|^2\,dx = \sum_{n=-\infty}^{\infty} 4|\sin nh|^2|\hat{f}(n)|^2,$$

and show that

$$\sum_{n=-\infty}^{\infty} |\sin nh|^2|\hat{f}(n)|^2 \leq K^2h^2.$$

(b) Let p be a positive integer. By choosing $h = \pi/2^{p+1}$, show that

$$\sum_{2^{p-1}<|n|\leq 2^p} |\hat{f}(n)|^2 \leq \frac{K^2\pi^2}{2^{2p+1}}.$$

(c) Estimate $\sum_{2^{p-1}<|n|\leq 2^p} |\hat{f}(n)|$, and conclude that the Fourier series of f converges absolutely, hence uniformly. [Hint: Use the Cauchy-Schwarz inequality to estimate the sum.]

(d) In fact, modify the argument slightly to prove Bernstein's theorem: If f satisfies a Hölder condition of order $\alpha > 1/2$, then the Fourier series of f converges absolutely.

17. If f is a bounded monotonic function on $[-\pi, \pi]$, then

$$\hat{f}(n) = O(1/|n|).$$

[Hint: One may assume that f is increasing, and say $|f| \leq M$. First check that the Fourier coefficients of the characteristic function of $[a, b]$ satisfy $O(1/|n|)$. Now show that a sum of the form

$$\sum_{k=1}^{N} \alpha_k \chi_{[a_k,a_{k+1}]}(x)$$

with $-\pi = a_1 < a_2 < \cdots < a_N < a_{N+1} = \pi$ and $-M \leq \alpha_1 \leq \cdots \leq \alpha_N \leq M$ has Fourier coefficients that are $O(1/|n|)$ uniformly in N. Summing by parts one gets a telescopic sum $\sum(\alpha_{k+1} - \alpha_k)$ which can be bounded by $2M$. Now approximate f by functions of the above type.]

18. Here are a few things we have learned about the decay of Fourier coefficients:

(a) if f is of class C^k, then $\hat{f}(n) = o(1/|n|^k)$;

(b) if f is Lipschitz, then $\hat{f}(n) = O(1/|n|)$;

(c) if f is monotonic, then $\hat{f}(n) = O(1/|n|)$;

(d) if f is satisfies a Hölder condition with exponent α where $0 < \alpha < 1$, then $\hat{f}(n) = O(1/|n|^{\alpha})$;

(e) if f is merely Riemann integrable, then $\sum |\hat{f}(n)|^2 < \infty$ and therefore $\hat{f}(n) = o(1)$.

Nevertheless, show that the Fourier coefficients of a continuous function can tend to 0 arbitrarily slowly by proving that for every sequence of nonnegative real numbers $\{\epsilon_n\}$ converging to 0, there exists a continuous function f such that $|\hat{f}(n)| \geq \epsilon_n$ for infinitely many values of n.

[Hint: Choose a subsequence $\{\epsilon_{n_k}\}$ so that $\sum_k \epsilon_{n_k} < \infty$.]

19. Give another proof that the sum $\sum_{0<|n|\leq N} e^{inx}/n$ is uniformly bounded in N and $x \in [-\pi, \pi]$ by using the fact that

$$\frac{1}{2i} \sum_{0<|n|\leq N} \frac{e^{inx}}{n} = \sum_{n=1}^{N} \frac{\sin nx}{n} = \frac{1}{2} \int_0^x (D_N(t) - 1)\, dt,$$

where D_N is the Dirichlet kernel. Now use the fact that $\int_0^\infty \frac{\sin t}{t}\, dt < \infty$ which was proved in Exercise 12.

20. Let $f(x)$ denote the sawtooth function defined by $f(x) = (\pi - x)/2$ on the interval $(0, 2\pi)$ with $f(0) = 0$ and extended by periodicity to all of $\mathbb{R}$. The Fourier series of f is

$$f(x) \sim \frac{1}{2i} \sum_{|n|\neq 0} \frac{e^{inx}}{n} = \sum_{n=1}^{\infty} \frac{\sin nx}{n},$$

and f has a jump discontinuity at the origin with

$$f(0^+) = \frac{\pi}{2}, \quad f(0^-) = -\frac{\pi}{2}, \quad \text{and hence} \quad f(0^+) - f(0^-) = \pi.$$

Show that

$$\max_{0<x\leq\pi/N} S_N(f)(x) - \frac{\pi}{2} = \int_0^\pi \frac{\sin t}{t}\, dt - \frac{\pi}{2},$$

which is roughly 9% of the jump π. This result is a manifestation of Gibbs's phenomenon which states that near a jump discontinuity, the Fourier series of a function overshoots (or undershoots) it by approximately 9% of the jump.

[Hint: Use the expression for $S_N(f)$ given in Exercise 19.]

4 Problems

1. For each $0 < \alpha < 1$ the series

$$\sum_{n=1}^{\infty} \frac{\sin nx}{n^{\alpha}}$$

converges for every x but is not the Fourier series of a Riemann integrable function.

(a) If the **conjugate Dirichlet kernel** is defined by

$$\tilde{D}_N(x) = \sum_{|n| \leq N} \text{sign}(x)\, e^{inx} \qquad \text{where } \text{sign}(x) = \begin{cases} 1 & \text{if } n > 0 \\ 0 & \text{if } n = 0 \\ -1 & \text{if } n < 0, \end{cases}$$

then show that

$$\tilde{D}_N(x) = \frac{\cos(x/2) - \cos((N+1/2)x)}{\sin(x/2)},$$

and

$$\int_{-\pi}^{\pi} |\tilde{D}_N(x)|\, dx \leq c \log N.$$

(b) As a result, if f is Riemann integrable, then

$$(f * \tilde{D}_N)(0) = O(\log N).$$

(c) In the present case, this leads to

$$\sum_{n=1}^{N} \frac{1}{n^{\alpha}} = O(\log N),$$

which is a contradiction.

2. An important fact we have proved is that the family $\{e^{inx}\}_{n \in \mathbb{Z}}$ is orthonormal in $\mathcal{R}$ and it is also complete, in the sense that the Fourier series of f converges to f in the norm. In this exercise, we consider another family possessing these same properties.

On $[-1, 1]$ define

$$L_n(x) = \frac{d^n}{dx^n}(x^2 - 1)^n, \qquad n = 0, 1, 2, \ldots.$$

Then L_n is a polynomial of degree n which is called the nth **Legendre polynomial**.

(a) Show that if f is indefinitely differentiable on $[-1,1]$, then

$$\int_{-1}^{1} L_n(x)f(x)\,dx = (-1)^n \int_{-1}^{1} (x^2-1)^n f^{(n)}(x)\,dx.$$

In particular, show that L_n is orthogonal to x^m whenever $m < n$. Hence $\{L_n\}_{n=0}^{\infty}$ is an orthogonal family.

(b) Show that

$$\|L_n\|^2 = \int_{-1}^{1} |L_n(x)|^2\,dx = \frac{(n!)^2 2^{2n+1}}{2n+1}.$$

[Hint: First, note that $\|L_n\|^2 = (-1)^n(2n)!\int_{-1}^{1}(x^2-1)^n\,dx$. Write $(x^2-1)^n = (x-1)^n(x+1)^n$ and integrate by parts n times to calculate this last integral.]

(c) Prove that any polynomial of degree n that is orthogonal to $1, x, x^2, \ldots, x^{n-1}$ is a constant multiple of L_n.

(d) Let $\mathcal{L}_n = L_n/\|L_n\|$, which are the normalized Legendre polynomials. Prove that $\{\mathcal{L}_n\}$ is the family obtained by applying the "Gram-Schmidt process" to $\{1, x, \ldots, x^n, \ldots\}$, and conclude that every Riemann integrable function f on $[-1,1]$ has a **Legendre expansion**

$$\sum_{n=0}^{\infty} \langle f, \mathcal{L}_n\rangle \mathcal{L}_n$$

which converges to f in the mean-square sense.

3. Let α be a complex number not equal to an integer.

(a) Calculate the Fourier series of the 2π-periodic function defined on $[-\pi,\pi]$ by $f(x) = \cos(\alpha x)$.

(b) Prove the following formulas due to Euler:

$$\sum_{n=1}^{\infty} \frac{1}{n^2-\alpha^2} = \frac{1}{2\alpha^2} - \frac{\pi}{2\alpha\tan(\alpha\pi)}.$$

For all $u \in \mathbb{C} - \pi\mathbb{Z}$,

$$\cot u = \frac{1}{u} + 2\sum_{n=1}^{\infty} \frac{u}{u^2 - n^2\pi^2}.$$

(c) Show that for all $\alpha \in \mathbb{C} - \mathbb{Z}$ we have

$$\frac{\alpha\pi}{\sin(\alpha\pi)} = 1 + 2\alpha^2 \sum_{n=1}^{\infty} \frac{(-1)^{n-1}}{n^2 - \alpha^2}.$$

(d) For all $0 < \alpha < 1$, show that

$$\int_0^{\infty} \frac{t^{\alpha-1}}{t+1}\, dt = \frac{\pi}{\sin(\alpha\pi)}.$$

[Hint: Split the integral as $\int_0^1 + \int_1^\infty$ and change variables $t = 1/u$ in the second integral. Now both integrals are of the form

$$\int_0^1 \frac{t^{\gamma-1}}{1+t}\, dt, \qquad 0 < \gamma < 1,$$

which one can show is equal to $\sum_{k=0}^{\infty} \frac{(-1)^k}{k+\gamma}$. Use part (c) to conclude the proof.]

4. In this problem, we find the formula for the sum of the series

$$\sum_{n=1}^{\infty} \frac{1}{n^k}$$

where k is any even integer. These sums are expressed in terms of the Bernoulli numbers; the related Bernoulli polynomials are discussed in the next problem.

Define the **Bernoulli numbers** B_n by the formula

$$\frac{z}{e^z - 1} = \sum_{n=0}^{\infty} \frac{B_n}{n!} z^n.$$

(a) Show that $B_0 = 1$, $B_1 = -1/2$, $B_2 = 1/6$, $B_3 = 0$, $B_4 = -1/30$, and $B_5 = 0$.

(b) Show that for $n \geq 1$ we have

$$B_n = -\frac{1}{n+1} \sum_{k=0}^{n-1} \binom{n+1}{k} B_k.$$

(c) By writing

$$\frac{z}{e^z - 1} = 1 - \frac{z}{2} + \sum_{n=2}^{\infty} \frac{B_n}{n!} z^n,$$

show that $B_n = 0$ if n is odd and > 1. Also prove that

$$z \cot z = 1 + \sum_{n=1}^{\infty} (-1)^n \frac{2^{2n} B_{2n}}{(2n)!} z^{2n}.$$

(d) The **zeta function** is defined by

$$\zeta(s) = \sum_{n=1}^{\infty} \frac{1}{n^s}, \qquad \text{for all } s > 1.$$

Deduce from the result in (c), and the expression for the cotangent function obtained in the previous problem, that

$$x \cot x = 1 - 2 \sum_{m=1}^{\infty} \frac{\zeta(2m)}{\pi^{2m}} x^{2m}.$$

(e) Conclude that

$$2\zeta(2m) = (-1)^{m+1} \frac{(2\pi)^{2m}}{(2m)!} B_{2m}.$$

5. Define the **Bernoulli polynomials** $B_n(x)$ by the formula

$$\frac{ze^{xz}}{e^z - 1} = \sum_{n=0}^{\infty} \frac{B_n(x)}{n!} z^n.$$

(a) The functions $B_n(x)$ are polynomials in x and

$$B_n(x) = \sum_{k=0}^{n} \binom{n}{k} B_k x^{n-k}.$$

Show that $B_0(x) = 1$, $B_1(x) = x - 1/2$, $B_2(x) = x^2 - x + 1/6$, and $B_3(x) = x^3 - \frac{3}{2}x^2 + \frac{1}{2}x$.

(b) If $n \geq 1$, then

$$B_n(x+1) - B_n(x) = nx^{n-1},$$

and if $n \geq 2$, then

$$B_n(0) = B_n(1) = B_n.$$

(c) Define $S_m(n) = 1^m + 2^m + \cdots + (n-1)^m$. Show that

$$(m+1)S_m(n) = B_{m+1}(n) - B_{m+1}.$$

(d) Prove that the Bernoulli polynomials are the only polynomials that satisfy

(i) $B_0(x) = 1$,

(ii) $B_n'(x) = nB_{n-1}(x)$ for $n \geq 1$,

(iii) $\int_0^1 B_n(x)\,dx = 0$ for $n \geq 1$, and show that from (b) one obtains

$$\int_x^{x+1} B_n(t)\,dt = x^n.$$

(e) Calculate the Fourier series of $B_1(x)$ to conclude that for $0 < x < 1$ we have

$$B_1(x) = x - 1/2 = \frac{-1}{\pi} \sum_{k=1}^{\infty} \frac{\sin(2\pi kx)}{k}.$$

Integrate and conclude that

$$B_{2n}(x) = (-1)^{n+1} \frac{2(2n)!}{(2\pi)^{2n}} \sum_{k=1}^{\infty} \frac{\cos(2\pi kx)}{k^{2n}},$$

$$B_{2n+1}(x) = (-1)^{n+1} \frac{2(2n+1)!}{(2\pi)^{2n+1}} \sum_{k=1}^{\infty} \frac{\sin(2\pi kx)}{k^{2n+1}}.$$

Finally, show that for $0 < x < 1$,

$$B_n(x) = -\frac{n!}{(2\pi i)^n} \sum_{k \neq 0} \frac{e^{2\pi ikx}}{k^n}.$$

We observe that the Bernoulli polynomials are, up to normalization, successive integrals of the sawtooth function.

4 Some Applications of Fourier Series

> Fourier series and analogous expansions intervene very naturally in the general theory of curves and surfaces. In effect, this theory, conceived from the point of view of analysis, deals obviously with the study of arbitrary functions. I was thus led to use Fourier series in several questions of geometry, and I have obtained in this direction a number of results which will be presented in this work. One notes that my considerations form only a beginning of a principal series of researches, which would without doubt give many new results.
>
> *A. Hurwitz,* 1902

In the previous chapters we introduced some basic facts about Fourier analysis, motivated by problems that arose in physics. The motion of a string and the diffusion of heat were two instances that led naturally to the expansion of a function in terms of a Fourier series. We propose next to give the reader a flavor of the broader impact of Fourier analysis, and illustrate how these ideas reach out to other areas of mathematics. In particular, consider the following three problems:

I. Among all simple closed curves of length ℓ in the plane $\mathbb{R}^2$, which one encloses the largest area?

II. Given an irrational number γ, what can be said about the distribution of the fractional parts of the sequence of numbers $n\gamma$, for $n = 1, 2, 3, \ldots$?

III. Does there exist a continuous function that is nowhere differentiable?

The first problem is clearly geometric in nature, and at first sight, would seem to have little to do with Fourier series. The second question lies on the border between number theory and the study of dynamical systems, and gives us the simplest example of the idea of "ergodicity." The third problem, while analytic in nature, resisted many attempts before the solution was finally discovered. It is remarkable that all three questions can be resolved quite simply and directly by the use of Fourier series.

In the last section of this chapter, we return to a problem that provided our initial motivation. We consider the time-dependent heat equation on the circle. Here our investigation will lead us to the important but enigmatic heat kernel for the circle. However, the mysteries surrounding its basic properties will not be fully understood until we can apply the Poisson summation formula, which we will do in the next chapter.

1 The isoperimetric inequality

Let Γ denote a closed curve in the plane which does not intersect itself. Also, let ℓ denote the length of Γ, and $\mathcal{A}$ the area of the bounded region in $\mathbb{R}^2$ enclosed by Γ. The problem now is to determine for a given ℓ the curve Γ which maximizes $\mathcal{A}$ (if any such curve exists).

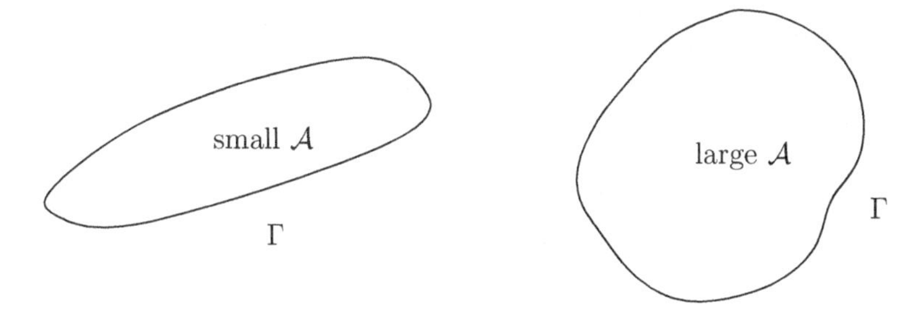

Figure 1. The isoperimetric problem

A little experimentation and reflection suggests that the solution should be a circle. This conclusion can be reached by the following heuristic considerations. The curve can be thought of as a closed piece of string lying flat on a table. If the region enclosed by the string is not convex (for example), one can deform part of the string and increase the area enclosed by it. Also, playing with some simple examples, one can convince oneself that the "flatter" the curve is in some portion, the less efficient it is in enclosing area. Therefore we want to maximize the "roundness" of the curve at each point.

Although the circle is the correct guess, making the above ideas precise is a difficult matter.

The key idea in the solution we give to the isoperimetric problem consists of an application of Parseval's identity for Fourier series. However, before we can attempt a solution to this problem, we must define the notion of a simple closed curve, its length, and what we mean by the area of the region enclosed by it.

Curves, length and area

A parametrized curve γ is a mapping

$$\gamma : [a, b] \to \mathbb{R}^2.$$

The image of γ is a set of points in the plane which we call a **curve** and denote by Γ. The curve Γ is **simple** if it does not intersect itself, and **closed** if its two end-points coincide. In terms of the parametrization above, these two conditions translate into $\gamma(s_1) \neq \gamma(s_2)$ unless $s_1 = a$ and $s_2 = b$, in which case $\gamma(a) = \gamma(b)$. We may extend γ to a periodic function on $\mathbb{R}$ of period $b - a$, and think of γ as a function on the circle. We also always impose some smoothness on our curves by assuming that γ is of class C^1, and that its derivative γ' satisfies $\gamma'(s) \neq 0$. Altogether, these conditions guarantee that Γ has a well-defined tangent at each point, which varies continuously as the point on the curve varies. Moreover, the parametrization γ induces an orientation on Γ as the parameter s travels from a to b.

Any C^1 bijective mapping $s : [c, d] \to [a, b]$ gives rise to another parametrization of Γ by the formula

$$\eta(t) = \gamma(s(t)).$$

Clearly, the conditions that Γ be closed and simple are independent of the chosen parametrization. Also, we say that the two parametrizations γ and η are equivalent if $s'(t) > 0$ for all t; this means that η and γ induce the same orientation on the curve Γ. If, however, $s'(t) < 0$, then η reverses the orientation.

If Γ is parametrized by $\gamma(s) = (x(s), y(s))$, then the **length** of the curve Γ is defined by

$$\ell = \int_a^b |\gamma'(s)|\, ds = \int_a^b \left(x'(s)^2 + y'(s)^2\right)^{1/2} ds.$$

The length of Γ is a notion intrinsic to the curve, and does not depend on its parametrization. To see that this is indeed the case, suppose that $\gamma(s(t)) = \eta(t)$. Then, the change of variables formula and the chain rule imply that

$$\int_a^b |\gamma'(s)|\, ds = \int_c^d |\gamma'(s(t))|\, |s'(t)|\, dt = \int_c^d |\eta'(t)|\, dt,$$

as desired.

In the proof of the theorem below, we shall use a special type of parametrization for Γ. We say that γ is a **parametrization by arc-length** if $|\gamma'(s)| = 1$ for all s. This means that $\gamma(s)$ travels at a constant speed, and as a consequence, the length of Γ is precisely $b - a$. Therefore, after a possible additional translation, a parametrization by arc-length will be defined on $[0, \ell]$. Any curve admits a parametrization by arc-length (Exercise 1).

We now turn to the isoperimetric problem.

The attempt to give a precise formulation of the area $\mathcal{A}$ of the region enclosed by a simple closed curve Γ raises a number of tricky questions. In a variety of simple situations, it is evident that the area is given by the following familiar formula of the calculus:

$$\begin{aligned}
\mathcal{A} &= \frac{1}{2}\left|\int_\Gamma (x\,dy - y\,dx)\right| \\
&= \frac{1}{2}\left|\int_a^b x(s)y'(s) - y(s)x'(s)\,ds\right|;
\end{aligned} \tag{1}$$

see, for example, Exercise 3. Thus in formulating our result we shall adopt the easy expedient of taking (1) as our definition of area. This device allows us to give a quick and neat proof of the isoperimetric inequality. A listing of issues this simplification leaves unresolved can be found after the proof of the theorem.

Statement and proof of the isoperimetric inequality

Theorem 1.1 *Suppose that Γ is a simple closed curve in $\mathbb{R}^2$ of length ℓ, and let $\mathcal{A}$ denote the area of the region enclosed by this curve. Then*

$$\mathcal{A} \le \frac{\ell^2}{4\pi},$$

with equality if and only if Γ is a circle.

The first observation is that we can rescale the problem. This means that we can change the units of measurement by a factor of $\delta > 0$ as follows. Consider the mapping of the plane $\mathbb{R}^2$ to itself, which sends the point (x, y) to $(\delta x, \delta y)$. A look at the formula defining the length of a curve shows that if Γ is of length ℓ, then its image under this mapping has length $\delta\ell$. So this operation magnifies or contracts lengths by a factor of δ depending on whether $\delta \ge 1$ or $\delta \le 1$. Similarly, we see that the mapping magnifies (or contracts) areas by a factor of δ^2. By taking $\delta = 2\pi/\ell$, we see that it suffices to prove that if $\ell = 2\pi$ then $\mathcal{A} \leq \pi$, with equality only if Γ is a circle.

Let $\gamma : [0, 2\pi] \to \mathbb{R}^2$ with $\gamma(s) = (x(s), y(s))$ be a parametrization by arc-length of the curve Γ, that is, $x'(s)^2 + y'(s)^2 = 1$ for all $s \in [0, 2\pi]$. This implies that

$$\frac{1}{2\pi}\int_0^{2\pi} (x'(s)^2 + y'(s)^2)\, ds = 1. \tag{2}$$

Since the curve is closed, the functions $x(s)$ and $y(s)$ are 2π-periodic, so we may consider their Fourier series

$$x(s) \sim \sum a_n e^{ins} \quad \text{and} \quad y(s) \sim \sum b_n e^{ins}.$$

Then, as we remarked in the later part of Section 2 of Chapter 2, we have

$$x'(s) \sim \sum a_n in e^{ins} \quad \text{and} \quad y'(s) \sim \sum b_n in e^{ins}.$$

Parseval's identity applied to (2) gives

$$\sum_{n=-\infty}^{\infty} |n|^2 \left(|a_n|^2 + |b_n|^2\right) = 1. \tag{3}$$

We now apply the bilinear form of Parseval's identity (Lemma 1.5, Chapter 3) to the integral defining $\mathcal{A}$. Since $x(s)$ and $y(s)$ are real-valued, we have $a_n = \overline{a_{-n}}$ and $b_n = \overline{b_{-n}}$, so we find that

$$\mathcal{A} = \frac{1}{2}\left|\int_0^{2\pi} x(s)y'(s) - y(s)x'(s)\, ds\right| = \pi \left|\sum_{n=-\infty}^{\infty} n\left(a_n\overline{b_n} - b_n\overline{a_n}\right)\right|.$$

We observe next that

$$|a_n\overline{b_n} - b_n\overline{a_n}| \leq 2\,|a_n|\,|b_n| \leq |a_n|^2 + |b_n|^2, \tag{4}$$

and since $|n| \leq |n|^2$, we may use (3) to get

$$\begin{aligned} \mathcal{A} &\leq \pi \sum_{n=-\infty}^{\infty} |n|^2 \left(|a_n|^2 + |b_n|^2\right) \\ &\leq \pi, \end{aligned}$$

as desired.

When $\mathcal{A} = \pi$, we see from the above argument that

$$x(s) = a_{-1}e^{-is} + a_0 + a_1 e^{is} \quad \text{and} \quad y(s) = b_{-1}e^{-is} + b_0 + b_1 e^{is}$$

because $|n| < |n|^2$ as soon as $|n| \geq 2$. We know that $x(s)$ and $y(s)$ are real-valued, so $a_{-1} = \overline{a_1}$ and $b_{-1} = \overline{b_1}$. The identity (3) implies that $2\left(|a_1|^2 + |b_1|^2\right) = 1$, and since we have equality in (4) we must have $|a_1| = |b_1| = 1/2$. We write

$$a_1 = \frac{1}{2}\, e^{i\alpha} \quad \text{and} \quad b_1 = \frac{1}{2}\, e^{i\beta}.$$

The fact that $1 = 2|a_1\overline{b_1} - \overline{a_1}b_1|$ implies that $|\sin(\alpha - \beta)| = 1$, hence $\alpha - \beta = k\pi/2$ where k is an odd integer. From this we find that

$$x(s) = a_0 + \cos(\alpha + s) \quad \text{and} \quad y(s) = b_0 \pm \sin(\alpha + s),$$

where the sign in $y(s)$ depends on the parity of $(k-1)/2$. In any case, we see that Γ is a circle, for which the case of equality obviously holds, and the proof of the theorem is complete.

The solution given above (due to Hurwitz in 1901) is indeed very elegant, but clearly leaves some important issues unanswered. We list these as follows. Suppose Γ is a simple closed curve.

(i) How is the "region enclosed by Γ" defined?

(ii) What is the geometric definition of the "area" of this region? Does this definition accord with (1)?

(iii) Can these results be extended to the most general class of simple closed curves relevant to the problem—those curves which are "rectifiable"—that is, those to which we can ascribe a finite length?

It turns out that the clarifications of the problems raised are connected to a number of other significant ideas in analysis. We shall return to these questions in succeeding books of this series.

2 Weyl's equidistribution theorem

We now apply ideas coming from Fourier series to a problem dealing with properties of irrational numbers. We begin with a brief discussion of congruences, a concept needed to understand our main theorem.

The reals modulo the integers

If x is a real number, we let $[x]$ denote the greatest integer less than or equal to x and call the quantity $[x]$ the **integer part** of x. The **fractional part** of x is then defined by $\langle x \rangle = x - [x]$. In particular, $\langle x \rangle \in [0, 1)$ for every $x \in \mathbb{R}$. For example, the integer and fractional parts of 2.7 are 2 and 0.7, respectively, while the integer and fractional parts of -3.4 are -4 and 0.6, respectively.

We may define a relation on $\mathbb{R}$ by saying that the two numbers x and y are equivalent, or congruent, if $x - y \in \mathbb{Z}$. We then write

$$x = y \bmod \mathbb{Z} \quad \text{or} \quad x = y \bmod 1.$$

This means that we identify two real numbers if they differ by an integer. Observe that any real number x is congruent to a unique number in $[0, 1)$ which is precisely $\langle x \rangle$, the fractional part of x. In effect, reducing a real number modulo $\mathbb{Z}$ means looking only at its fractional part and disregarding its integer part.

Now start with a real number $\gamma \neq 0$ and look at the sequence $\gamma, 2\gamma, 3\gamma, \ldots$. An intriguing question is to ask what happens to this sequence if we reduce it modulo $\mathbb{Z}$, that is, if we look at the sequence of fractional parts

$$\langle \gamma \rangle, \ \langle 2\gamma \rangle, \ \langle 3\gamma \rangle, \ \ldots.$$

Here are some simple observations:

(i) If γ is rational, then only finitely many numbers appearing in $\langle n\gamma \rangle$ are distinct.

(ii) If γ is irrational, then the numbers $\langle n\gamma \rangle$ are all distinct.

Indeed, for part (i), note that if $\gamma = p/q$, the first q terms in the sequence are

$$\langle p/q \rangle, \ \langle 2p/q \rangle, \ \ldots, \ \langle (q-1)p/q \rangle, \ \langle qp/q \rangle = 0.$$

The sequence then begins to repeat itself, since

$$\langle (q+1)p/q \rangle = \langle 1 + p/q \rangle = \langle p/q \rangle,$$

and so on. However, see Exercise 6 for a more refined result.

Also, for part (ii) assume that not all numbers are distinct. We therefore have $\langle n_1\gamma \rangle = \langle n_2\gamma \rangle$ for some $n_1 \neq n_2$; then $n_1\gamma - n_2\gamma \in \mathbb{Z}$, hence γ is rational, a contradiction.

In fact, it can be shown that if γ is irrational, then $\langle n\gamma\rangle$ is dense in the interval $[0,1)$, a result originally proved by Kronecker. In other words, the sequence $\langle n\gamma\rangle$ hits every sub-interval of $[0,1)$ (and hence it does so infinitely many times). We will obtain this fact as a corollary of a deeper theorem dealing with the uniform distribution of the sequence $\langle n\gamma\rangle$.

A sequence of numbers $\xi_1, \xi_2, \ldots, \xi_n, \ldots$ in $[0,1)$ is said to be **equidistributed** if for every interval $(a,b) \subset [0,1)$,

$$\lim_{N\to\infty} \frac{\#\{1 \leq n \leq N : \xi_n \in (a,b)\}}{N} = b - a$$

where $\#A$ denotes the cardinality of the finite set A. This means that for large N, the proportion of numbers ξ_n in (a,b) with $n \leq N$ is equal to the ratio of the length of the interval (a,b) to the length of the interval $[0,1)$. In other words, the sequence ξ_n sweeps out the whole interval evenly, and every sub-interval gets its fair share. Clearly, the ordering of the sequence is very important, as the next two examples illustrate.

EXAMPLE 1. The sequence

$$0, \frac{1}{2}, 0, \frac{1}{3}, \frac{2}{3}, 0, \frac{1}{4}, \frac{2}{4}, \frac{3}{4}, 0, \frac{1}{5}, \frac{2}{5}, \cdots$$

appears to be equidistributed since it passes over the interval $[0,1)$ very evenly. Of course this is not a proof, and the reader is invited to give one. For a somewhat related example, see Exercise 8 with $\sigma = 1/2$.

EXAMPLE 2. Let $\{r_n\}_{n=1}^{\infty}$ be *any* enumeration of the rationals in $[0,1)$. Then the sequence defined by

$$\xi_n = \begin{cases} r_{n/2} & \text{if } n \text{ is even,} \\ 0 & \text{if } n \text{ is odd,} \end{cases}$$

is not equidistributed since "half" of the sequence is at 0. Nevertheless, this sequence is obviously dense.

We now arrive at the main theorem of this section.

Theorem 2.1 *If γ is irrational, then the sequence of fractional parts $\langle\gamma\rangle, \langle 2\gamma\rangle, \langle 3\gamma\rangle, \ldots$ is equidistributed in $[0,1)$.*

In particular, $\langle n\gamma\rangle$ is dense in $[0,1)$, and we get Kronecker's theorem as a corollary. In Figure 2 we illustrate the set of points $\langle\gamma\rangle, \langle 2\gamma\rangle, \langle 3\gamma\rangle, \ldots, \langle N\gamma\rangle$ for three different values of N when $\gamma = \sqrt{2}$.

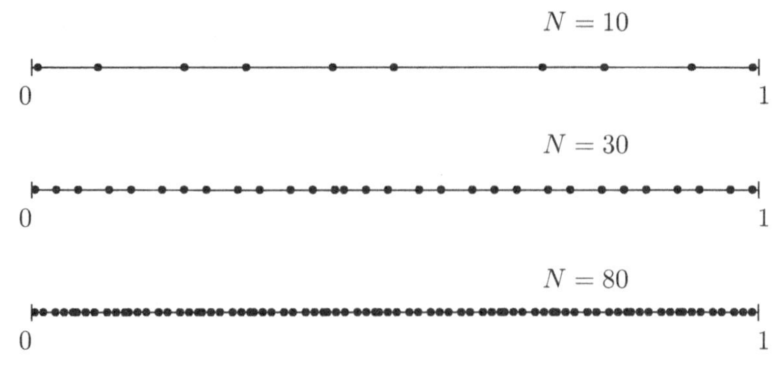

Figure 2. The sequence $\langle\gamma\rangle, \langle 2\gamma\rangle, \langle 3\gamma\rangle, \ldots, \langle N\gamma\rangle$ when $\gamma = \sqrt{2}$

Fix $(a, b) \subset [0, 1)$ and let $\chi_{(a,b)}(x)$ denote the characteristic function of the interval (a, b), that is, the function equal to 1 in (a, b) and 0 in $[0, 1) - (a, b)$. We may extend this function to $\mathbb{R}$ by periodicity (period 1), and still denote this extension by $\chi_{(a,b)}(x)$. Then, as a consequence of the definitions, we find that

$$\#\{1 \leq n \leq N : \langle n\gamma\rangle \in (a, b)\} = \sum_{n=1}^{N} \chi_{(a,b)}(n\gamma),$$

and the theorem can be reformulated as the statement that

$$\frac{1}{N}\sum_{n=1}^{N} \chi_{(a,b)}(n\gamma) \to \int_0^1 \chi_{(a,b)}(x)\, dx, \qquad \text{as } N \to \infty.$$

This step removes the difficulty of working with fractional parts and reduces the number theory to analysis.

The heart of the matter lies in the following result.

Lemma 2.2 *If f is continuous and periodic of period 1, and γ is irrational, then*

$$\frac{1}{N}\sum_{n=1}^{N} f(n\gamma) \to \int_0^1 f(x)\, dx \qquad \text{as } N \to \infty.$$

The proof of the lemma is divided into three steps.

Step 1. We first check the validity of the limit in the case when f is one of the exponentials $1, e^{2\pi ix}, \ldots, e^{2\pi ikx}, \ldots$. If $f = 1$, the limit surely holds. If $f = e^{2\pi ikx}$ with $k \neq 0$, then the integral is 0. Since γ is irrational, we have $e^{2\pi ik\gamma} \neq 1$, therefore

$$\frac{1}{N}\sum_{n=1}^{N} f(n\gamma) = \frac{e^{2\pi ik\gamma}}{N}\frac{1-e^{2\pi ikN\gamma}}{1-e^{2\pi ik\gamma}},$$

which goes to 0 as $N \to \infty$.

Step 2. It is clear that if f and g satisfy the lemma, then so does $Af + Bg$ for any $A, B \in \mathbb{C}$. Therefore, the first step implies that the lemma is true for all trigonometric polynomials.

Step 3. Let $\epsilon > 0$. If f is any continuous periodic function of period 1, choose a trigonometric polynomial P so that $\sup_{x\in\mathbb{R}} |f(x) - P(x)| < \epsilon/3$ (this is possible by Corollary 5.4 in Chapter 2). Then, by step 1, for all large N we have

$$\left|\frac{1}{N}\sum_{n=1}^{N} P(n\gamma) - \int_0^1 P(x)\,dx\right| < \epsilon/3.$$

Therefore

$$\begin{aligned}\left|\frac{1}{N}\sum_{n=1}^{N} f(n\gamma) - \int_0^1 f(x)\,dx\right| &\leq \frac{1}{N}\sum_{n=1}^{N} |f(n\gamma) - P(n\gamma)| + \\ &\quad + \left|\frac{1}{N}\sum_{n=1}^{N} P(n\gamma) - \int_0^1 P(x)\,dx\right| + \\ &\quad + \int_0^1 |P(x) - f(x)|\,dx \\ &< \epsilon,\end{aligned}$$

and the lemma is proved.

Now we can finish the proof of the theorem. Choose two continuous periodic functions f_ϵ^+ and f_ϵ^- of period 1 which approximate $\chi_{(a,b)}(x)$ on $[0, 1)$ from above and below; both f_ϵ^+ and f_ϵ^- are bounded by 1 and agree with $\chi_{(a,b)}(x)$ except in intervals of total length 2ϵ (see Figure 3). In particular, $f_\epsilon^-(x) \leq \chi_{(a,b)}(x) \leq f_\epsilon^+(x)$, and

$$b - a - 2\epsilon \leq \int_0^1 f_\epsilon^-(x)\,dx \quad \text{and} \quad \int_0^1 f_\epsilon^+(x)\,dx \leq b - a + 2\epsilon.$$

If $S_N = \frac{1}{N}\sum_{n=1}^{N} \chi_{(a,b)}(n\gamma)$, then we get

$$\frac{1}{N}\sum_{n=1}^{N} f_\epsilon^-(n\gamma) \leq S_N \leq \frac{1}{N}\sum_{n=1}^{N} f_\epsilon^+(n\gamma).$$

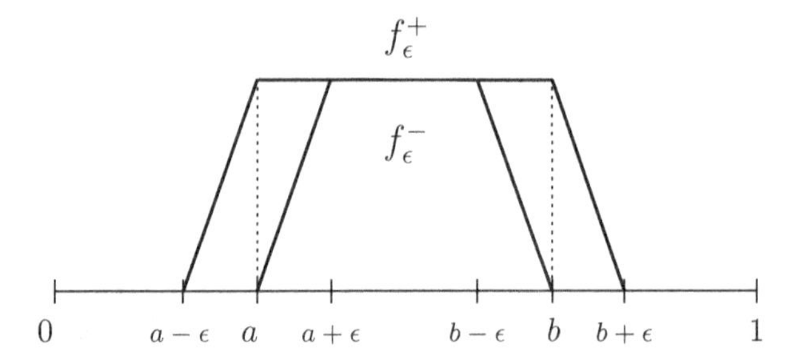

Figure 3. Approximations of $\chi_{(a,b)}(x)$

Therefore

$$b - a - 2\epsilon \leq \liminf_{N\to\infty} S_N \quad \text{and} \quad \limsup_{N\to\infty} S_N \leq b - a + 2\epsilon.$$

Since this is true for every $\epsilon > 0$, the limit $\lim_{N\to\infty} S_N$ exists and must equal $b - a$. This completes the proof of the equidistribution theorem.

This theorem has the following consequence.

Corollary 2.3 *The conclusion of Lemma 2.2 holds for every function f which is Riemann integrable in $[0,1]$, and periodic of period 1.*

Proof. Assume f is real-valued, and consider a partition of the interval $[0,1]$, say $0 = x_0 < x_1 < \cdots < x_N = 1$. Next, define $f_U(x) = \sup_{x_{j-1}\leq y\leq x_j} f(y)$ if $x \in [x_{j-1}, x_j)$ and $f_L(x) = \inf_{x_{j-1}\leq y\leq x_j} f(y)$ for $x \in (x_{j-1}, x_j)$. Then clearly $f_L \leq f \leq f_U$ and

$$\int_0^1 f_L(x)\,dx \leq \int_0^1 f(x)\,dx \leq \int_0^1 f_U(x)\,dx.$$

Moreover, by making the partition sufficiently fine we can guarantee that for a given $\epsilon > 0$,

$$\int_0^1 f_U(x)\,dx - \int_0^1 f_L(x)\,dx \leq \epsilon.$$

However,

$$\frac{1}{N}\sum_{n=1}^{N} f_L(n\gamma) \to \int_0^1 f_L(x)\,dx$$

by the theorem, because each f_L is a finite linear combination of characteristic functions of intervals; similarly we have

$$\frac{1}{N}\sum_{n=1}^{N} f_U(n\gamma) \to \int_0^1 f_U(x)\,dx.$$

From these two assertions we can conclude the proof of the corollary by using the previous approximation argument.

There is an interesting interpretation of the lemma and its corollary, in terms of a simple dynamical system. In this example, the underlying space is the circle parametrized by the angle θ. We also consider a mapping of this space to itself: here, we choose a rotation ρ of the circle by the angle $2\pi\gamma$, that is, the transformation $\rho : \theta \mapsto \theta + 2\pi\gamma$.

We want next to consider how this space, with its underlying action ρ, evolves in time. In other words, we wish to consider the iterates of ρ, namely ρ, ρ^2, ρ^3, ..., ρ^n where

$$\rho^n = \rho \circ \rho \circ \cdots \circ \rho : \theta \mapsto \theta + 2\pi n\gamma,$$

and where we think of the action ρ^n taking place at the time $t = n$.

To each Riemann integrable function f on the circle, we can also associate the corresponding effects of the rotation ρ, and obtain a sequence of functions

$$f(\theta),\ f(\rho(\theta)),\ f(\rho^2(\theta)),\ \ldots,\ f(\rho^n(\theta)),\ \ldots$$

with $f(\rho^n(\theta)) = f(\theta + 2\pi n\gamma)$. In this special context, the **ergodicity** of this system is then the statement that the "time average"

$$\lim_{N\to\infty} \frac{1}{N}\sum_{n=1}^{N} f(\rho^n(\theta))$$

exists for each θ and equals the "space average"

$$\frac{1}{2\pi}\int_0^{2\pi} f(\theta)\,d\theta,$$

whenever γ is irrational. In fact, this assertion is merely a rephrasing of Corollary 2.3, once we make the change of variables $\theta = 2\pi x$.

Returning to the problem of equidistributed sequences, we observe that the proof of Theorem 2.1 gives the following characterization.

Weyl's criterion. *A sequence of real numbers* $\xi_1, \xi_2 \ldots$ *in* $[0, 1)$ *is equidistributed if and only if for all integers* $k \neq 0$ *one has*

$$\frac{1}{N}\sum_{n=1}^{N} e^{2\pi i k \xi_n} \to 0, \qquad \text{as } N \to \infty.$$

One direction of this theorem was in effect proved above, and the converse can be found in Exercise 7. In particular, we find that to understand the equidistributive properties of a sequence ξ_n, it suffices to estimate the size of the corresponding "exponential sum" $\sum_{n=1}^{N} e^{2\pi i k \xi_n}$. For example, it can be shown using Weyl's criterion that the sequence $\langle n^2\gamma\rangle$ is equidistributed whenever γ is irrational. This and other examples can be found in Exercises 8, and 9; also Problems 2, and 3.

As a last remark, we mention a nice geometric interpretation of the distribution properties of $\langle n\gamma\rangle$. Suppose that the sides of a square are reflecting mirrors and that a ray of light leaves a point inside the square. What kind of path will the light trace out?

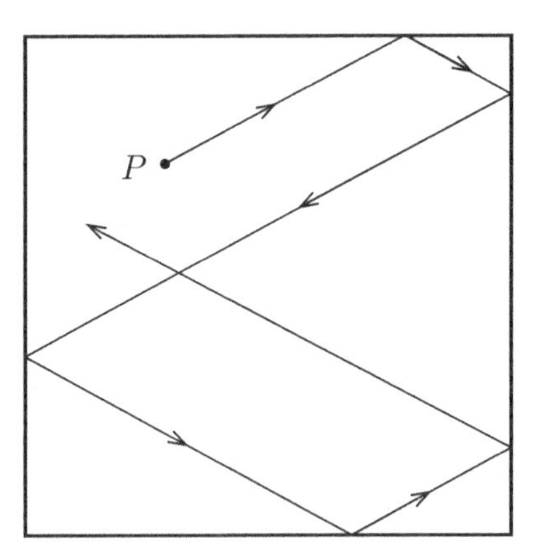

Figure 4. Reflection of a ray of light in a square

To solve this problem, the main idea is to consider the grid of the plane formed by successively reflecting the initial square across its sides. With an appropriate choice of axis, the path traced by the light in the square corresponds to the straight line $P + (t, \gamma t)$ in the plane. As a result, the reader may observe that the path will be either closed and periodic, or it will be dense in the square. The first of these situations will happen if and only if the slope γ of the initial direction of the light (determined with respect to one of the sides of the square) is rational. In the second situation, when γ is irrational, the density follows from Kronecker's theorem. What stronger conclusion does one get from the equidistribution theorem?

3 A continuous but nowhere differentiable function

There are many obvious examples of continuous functions that are not differentiable at one point, say $f(x) = |x|$. It is almost as easy to construct a continuous function that is not differentiable at any given finite set of points, or even at appropriate sets containing countably many points. A more subtle problem is whether there exists a continuous function that is *nowhere* differentiable. In 1861, Riemann guessed that the function defined by

$$R(x) = \sum_{n=1}^{\infty} \frac{\sin(n^2 x)}{n^2} \tag{5}$$

was nowhere differentiable. He was led to consider this function because of its close connection to the theta function which will be introduced in Chapter 5. Riemann never gave a proof, but mentioned this example in one of his lectures. This triggered the interest of Weierstrass who, in an attempt to find a proof, came across the first example of a continuous but nowhere differentiable function. Say $0 < b < 1$ and a is an integer > 1. In 1872 he proved that if $ab > 1 + 3\pi/2$, then the function

$$W(x) = \sum_{n=1}^{\infty} b^n \cos(a^n x)$$

is nowhere differentiable.

But the story is not complete without a final word about Riemann's original function. In 1916 Hardy showed that R is not differentiable at all irrational multiples of π, and also at certain rational multiples of π. However, it was not until much later, in 1969, that Gerver completely settled the problem, first by proving that the function R *is* actually differentiable at all the rational multiples of π of the form $\pi p/q$ with p and q odd integers, and then by showing that R is not differentiable in all of the remaining cases.

In this section, we prove the following theorem.

Theorem 3.1 *If* $0 < \alpha < 1$, *then the function*

$$f_\alpha(x) = f(x) = \sum_{n=0}^{\infty} 2^{-n\alpha} e^{i2^n x}$$

is continuous but nowhere differentiable.

The continuity is clear because of the absolute convergence of the series. The crucial property of f which we need is that it has many vanishing Fourier coefficients. A Fourier series that skips many terms, like the one given above, or like $W(x)$, is called a **lacunary Fourier series.**

The proof of the theorem is really the story of three methods of summing a Fourier series. First, there is the ordinary convergence in terms of the partial sums $S_N(g) = g * D_N$. Next, there is Cesàro summability $\sigma_N(g) = g * F_N$, with F_N the Fejér kernel. A third method, clearly connected with the second, involves the **delayed means** defined by

$$\triangle_N(g) = 2\sigma_{2N}(g) - \sigma_N(g).$$

Hence $\triangle_N(g) = g * [2F_{2N} - F_N]$. These methods can best be visualized as in Figure 5.

Suppose $g(x) \sim \sum a_n e^{inx}$. Then:

- S_N arises by multiplying the term $a_n e^{inx}$ by 1 if $|n| \leq N$, and 0 if $|n| > N$.
- σ_N arises by multiplying $a_n e^{inx}$ by $1 - |n|/N$ for $|n| \leq N$ and 0 for $|n| > N$.
- $\triangle_N$ arises by multiplying $a_n e^{inx}$ by 1 if $|n| \leq N$, by $2(1 - |n|/(2N))$ for $N \leq |n| \leq 2N$, and 0 for $|n| > 2N$.

For example, note that

$$\begin{aligned}
\sigma_N(g)(x) &= \frac{S_0(g)(x) + S_1(g)(x) + \cdots + S_{N-1}(g)(x)}{N} \\
&= \frac{1}{N} \sum_{\ell=0}^{N-1} \sum_{|k| \leq \ell} a_k e^{ikx} \\
&= \frac{1}{N} \sum_{|n| \leq N} (N - |n|) a_n e^{inx} \\
&= \sum_{|n| \leq N} \left(1 - \frac{|n|}{N}\right) a_n e^{inx}.
\end{aligned}$$

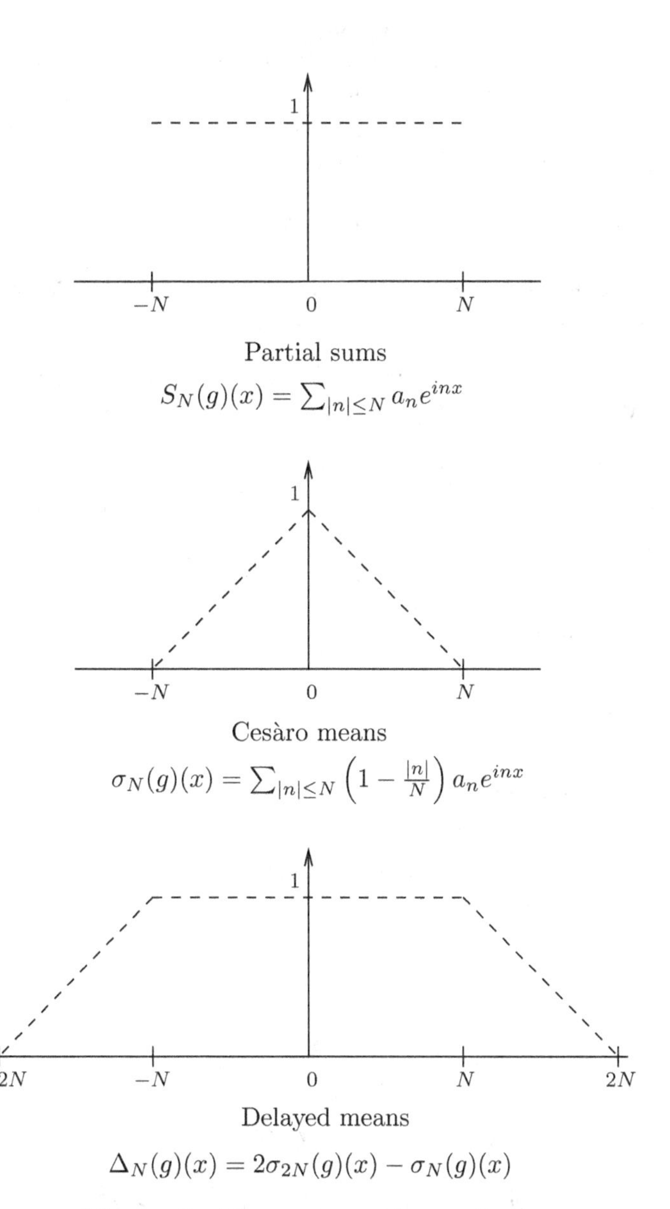

Figure 5. Three summation methods

The proof of the other assertion is similar.

The delayed means have two important features. On the one hand, their properties are closely related to the (good) features of the Cesàro means. On the other hand, for series that have lacunary properties like those of f, the delayed means are essentially equal to the partial sums. In particular, note that for our function $f = f_\alpha$

$$S_N(f) = \triangle_{N'}(f), \tag{6}$$

where N' is the largest integer of the form 2^k with $N' \leq N$. This is clear by examining Figure 5 and the definition of f.

We turn to the proof of the theorem proper and argue by contradiction; that is, we assume that $f'(x_0)$ exists for some x_0.

Lemma 3.2 *Let g be any continuous function that is differentiable at x_0. Then, the Cesàro means satisfy $\sigma_N(g)'(x_0) = O(\log N)$, therefore*

$$\triangle_N(g)'(x_0) = O(\log N).$$

Proof. First we have

$$\sigma_N(g)'(x_0) = \int_{-\pi}^{\pi} F_N'(x_0 - t)g(t)\,dt = \int_{-\pi}^{\pi} F_N'(t)g(x_0 - t)\,dt,$$

where F_N is the Fejér kernel. Since F_N is periodic, we have $\int_{-\pi}^{\pi} F_N'(t)dt=0$ and this implies that

$$\sigma_N(g)'(x_0) = \int_{-\pi}^{\pi} F_N'(t)[g(x_0 - t) - g(x_0)]\,dt.$$

From the assumption that g is differentiable at x_0 we get

$$|\sigma_N(g)'(x_0)| \leq C \int_{-\pi}^{\pi} |F_N'(t)|\,|t|\,dt.$$

Now observe that F_N' satisfies the two estimates

$$|F_N'(t)| \leq AN^2 \quad \text{and} \quad |F_N'(t)| \leq \frac{A}{|t|^2}.$$

For the first inequality, recall that F_N is a trigonometric polynomial of degree N whose coefficients are bounded by 1. Therefore, F_N' is a trigonometric polynomial of degree N whose coefficients are no bigger than N. Hence $|F'(t)| \leq (2N + 1)N \leq AN^2$.

For the second inequality, we recall that

$$F_N(t) = \frac{1}{N}\frac{\sin^2(Nt/2)}{\sin^2(t/2)}.$$

Differentiating this expression, we get two terms:

$$\frac{\sin(Nt/2)\cos(Nt/2)}{\sin^2(t/2)} - \frac{1}{N}\frac{\cos(t/2)\sin^2(Nt/2)}{\sin^3(t/2)}.$$

If we then use the facts that $|\sin(Nt/2)| \leq CN|t|$ and $|\sin(t/2)| \geq c|t|$ (for $|t| \leq \pi$), we get the desired estimates for $F_N'(t)$.

Using all of these estimates we find that

$$\begin{aligned}
|\sigma_N(g)'(x_0)| &\leq C\int_{|t|\geq 1/N} |F_N'(t)|\,|t|\,dt + C\int_{|t|\leq 1/N} |F_N'(t)|\,|t|\,dt \\
&\leq CA\int_{|t|\geq 1/N} \frac{dt}{|t|} + CAN\int_{|t|\leq 1/N} dt \\
&= O(\log N) + O(1) \\
&= O(\log N).
\end{aligned}$$

The proof of the lemma is complete once we invoke the definition of $\triangle_N(g)$.

Lemma 3.3 *If* $2N = 2^n$, *then*

$$\triangle_{2N}(f) - \triangle_N(f) = 2^{-n\alpha}e^{i2^n x}.$$

This follows from our previous observation (6) because $\triangle_{2N}(f) = S_{2N}(f)$ and $\triangle_N(f) = S_N(f)$.

Now, by the first lemma we have

$$\triangle_{2N}(f)'(x_0) - \triangle_N(f)'(x_0) = O(\log N),$$

and the second lemma also implies

$$|\triangle_{2N}(f)'(x_0) - \triangle_N(f)'(x_0)| = 2^{n(1-\alpha)} \geq cN^{1-\alpha}.$$

This is the desired contradiction since $N^{1-\alpha}$ grows faster than $\log N$.

A few additional remarks about our function $f_\alpha(x) = \sum_{n=0}^{\infty} 2^{-n\alpha}e^{i2^n x}$ are in order.

This function is complex-valued as opposed to the examples R and W above, and so the nowhere differentiability of f_α does not imply the same property for its real and imaginary parts. However, a small modification of our proof shows that, in fact, the real part of f_α,

$$\sum_{n=0}^{\infty} 2^{-n\alpha} \cos 2^n x,$$

as well as its imaginary part, are both nowhere differentiable. To see this, observe first that by the same proof, Lemma 3.2 has the following generalization: if g is a continuous function which is differentiable at x_0, then

$$\triangle_N(g)'(x_0 + h) = O(\log N) \quad \text{whenever } |h| \leq c/N.$$

We then proceed with $F(x) = \sum_{n=0}^{\infty} 2^{-n\alpha} \cos 2^n x$, noting as above that $\triangle_{2N}(F) - \triangle_N(F) = 2^{-n\alpha} \cos 2^n x$; as a result, assuming that F is differentiable at x_0, we get that

$$|2^{n(1-\alpha)} \sin(2^n(x_0 + h))| = O(\log N)$$

when $2N = 2^n$, and $|h| \leq c/N$. To get a contradiction, we need only choose h so that $|\sin(2^n(x_0 + h))| = 1$; this is accomplished by setting δ equal to the distance from $2^n x_0$ to the nearest number of the form $(k + 1/2)\pi$, $k \in \mathbb{Z}$ (so $\delta \leq \pi/2$), and taking $h = \pm\delta/2^n$.

Clearly, when $\alpha > 1$ the function f_α is continuously differentiable since the series can be differentiated term by term. Finally, the nowhere differentiability we have proved for $\alpha < 1$ actually extends to $\alpha = 1$ by a suitable refinement of the argument (see Problem 8 in Chapter 5). In fact, using these more elaborate methods one can also show that the Weierstrass function W is nowhere differentiable if $ab \geq 1$.

4 The heat equation on the circle

As a final illustration, we return to the original problem of heat diffusion considered by Fourier.

Suppose we are given an initial temperature distribution at $t = 0$ on a ring and that we are asked to describe the temperature at points on the ring at times $t > 0$.

The ring is modeled by the unit circle. A point on this circle is described by its angle $\theta = 2\pi x$, where the variable x lies between 0 and 1. If $u(x, t)$ denotes the temperature at time t of a point described by the angle θ, then considerations similar to the ones given in Chapter 1 show that u satisfies the differential equation

$$(7) \qquad \frac{\partial u}{\partial t} = c\,\frac{\partial^2 u}{\partial x^2}.$$

The constant c is a positive physical constant which depends on the material of which the ring is made (see Section 2.1 in Chapter 1). After rescaling the time variable, we may assume that $c = 1$. If f is our initial data, we impose the condition

$$u(x, 0) = f(x).$$

To solve the problem, we separate variables and look for special solutions of the form

$$u(x, t) = A(x)B(t).$$

Then inserting this expression for u into the heat equation we get

$$\frac{B'(t)}{B(t)} = \frac{A''(x)}{A(x)}.$$

Both sides are therefore constant, say equal to λ. Since A must be periodic of period 1, we see that the only possibility is $\lambda = -4\pi^2 n^2$, where $n \in \mathbb{Z}$. Then A is a linear combination of the exponentials $e^{2\pi i n x}$ and $e^{-2\pi i n x}$, and $B(t)$ is a multiple of $e^{-4\pi^2 n^2 t}$. By superposing these solutions, we are led to

$$(8) \qquad u(x, t) = \sum_{n=-\infty}^{\infty} a_n e^{-4\pi^2 n^2 t} e^{2\pi i n x},$$

where, setting $t = 0$, we see that $\{a_n\}$ are the Fourier coefficients of f.

Note that when f is Riemann integrable, the coefficients a_n are bounded, and since the factor $e^{-4\pi^2 n^2 t}$ tends to zero extremely fast, the series defining u converges. In fact, in this case, u is twice differentiable and solves equation (7).

The natural question with regard to the boundary condition is the following: do we have $u(x, t) \to f(x)$ as t tends to 0, and in what sense? A simple application of the Parseval identity shows that this limit holds in the mean square sense (Exercise 11). For a better understanding of the properties of our solution (8), we write it as

$$u(x, t) = (f * H_t)(x),$$

where H_t is the **heat kernel for the circle**, given by

$$H_t(x) = \sum_{n=-\infty}^{\infty} e^{-4\pi^2 n^2 t} e^{2\pi i n x}, \tag{9}$$

and where the convolution for functions with period 1 is defined by

$$(f * g)(x) = \int_0^1 f(x-y)g(y)\,dy.$$

An analogy between the heat kernel and the Poisson kernel (of Chapter 2) is given in Exercise 12. However, unlike in the case of the Poisson kernel, there is no elementary formula for the heat kernel. Nevertheless, it turns out that it is a good kernel (in the sense of Chapter 2). The proof is not obvious and requires the use of the celebrated Poisson summation formula, which will be taken up in Chapter 5. As a corollary, we will also find that H_t is everywhere positive, a fact that is also not obvious from its defining expression (9). We can, however, give the following heuristic argument for the positivity of H_t. Suppose that we begin with an initial temperature distribution f which is everywhere ≤ 0. Then it is physically reasonable to expect $u(x,t) \leq 0$ for all t since heat travels from hot to cold. Now

$$u(x,t) = \int_0^1 f(x-y)H_t(y)\,dy.$$

If H_t is negative for some x_0, then we may choose $f \leq 0$ supported near x_0, and this would imply $u(x_0,t) > 0$, which is a contradiction.

5 Exercises

1. Let $\gamma : [a,b] \to \mathbb{R}^2$ be a parametrization for the closed curve Γ.

(a) Prove that γ is a parametrization by arc-length if and only if the length of the curve from $\gamma(a)$ to $\gamma(s)$ is precisely $s-a$, that is,

$$\int_a^s |\gamma'(t)|\,dt = s-a.$$

(b) Prove that any curve Γ admits a parametrization by arc-length. [Hint: If η is any parametrization, let $h(s) = \int_a^s |\eta'(t)|\,dt$ and consider $\gamma = \eta \circ h^{-1}$.]

2. Suppose $\gamma : [a,b] \to \mathbb{R}^2$ is a parametrization for a closed curve Γ, with $\gamma(t) = (x(t), y(t))$.

The **reverse parametrization** of γ is defined by $\gamma^- : [a, b] \to \mathbb{R}^2$ with $\gamma^-(t) = \gamma(b + a - t)$. The image of γ^- is precisely Γ, except that the points $\gamma^-(t)$ and $\gamma(t)$ travel in opposite directions. Thus γ^- "reverses" the orientation of the curve.

(a) Prove that $\int_\gamma (x\,dy - y\,dx) = -\int_{\gamma^-}(x\,dy - y\,dx)$, hence we we may assume (after a possible change in orientation) that

$$\mathcal{A} = \frac{1}{2}\int_a^b (x(s)y'(s) - y(s)x'(s))\,ds = \int_a^b x(s)y'(s)\,ds.$$

(b) Theorem 1.1 can be generalized by considering curves that are merely "piecewise smooth" in the sense that $x(t)$ and $y(t)$ are assumed continuous and, except for finitely many points, they each have derivatives that are piecewise continuous as defined in Section 1 of Chapter 2. Note that the functions $x'(t)$ and $y'(t)$ are then integrable, hence the formula for $\mathcal{A}$ is well-defined. Show that the proof of Theorem 1.1 carries over to piecewise smooth curves. In particular this gives the isoperimetric inequality for polygons.

3. Suppose Γ is a curve in the plane, and that there exists a set of coordinates x and y so that the x-axis divides the curve into the union of the graph of two continuous functions $y = f(x)$ and $y = g(x)$ for $0 \le x \le 1$, and with $f(x) \ge g(x)$ (see Figure 6). Let Ω denote the region between the graphs of these two functions:

$$\Omega = \{(x, y) :\ 0 \le x \le 1 \ \text{ and } \ g(x) \le y \le f(x)\}.$$

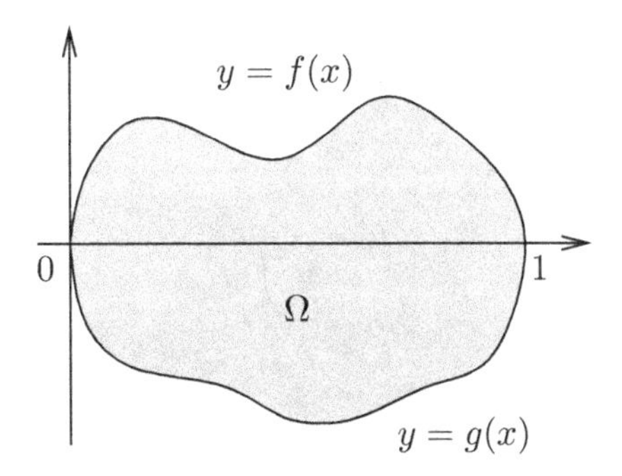

Figure 6. Simple version of the area formula

With the familiar interpretation that the integral $\int h(x)\,dx$ gives the area under the graph of the function h, we see that the area of Ω is $\int_0^1 f(x)\,dx -$

$\int_0^1 g(x)\,dx$. Show that this definition coincides with the area formula $\mathcal{A}$ given in the text, that is,

$$\int_0^1 f(x)\,dx - \int_0^1 g(x)\,dx = \left| - \int_\Gamma y\,dx \right| = \mathcal{A}.$$

Also, note that if the orientation of the curve is chosen so that Ω "lies to the left" of Γ, then the above formula holds without the absolute value signs.

This formula generalizes to any set that can be written as a finite union of domains like Ω above.

4. Observe that with the definition of ℓ and $\mathcal{A}$ given in the text, the isoperimetric inequality continues to hold (with the same proof) even when Γ is not simple.

Show that this stronger version of the isoperimetric inequality is equivalent to Wirtinger's inequality, which says that if f is 2π-periodic, of class C^1, and satisfies $\int_0^{2\pi} f(t)\,dt = 0$, then

$$\int_0^{2\pi} |f(t)|^2\,dt \leq \int_0^{2\pi} |f'(t)|^2\,dt$$

with equality if and only if $f(t) = A\sin t + B\cos t$ (Exercise 11, Chapter 3).
[Hint: In one direction, note that if the length of the curve is 2π and γ is an appropriate arc-length parametrization, then

$$2(\pi - \mathcal{A}) = \int_0^{2\pi} \left[x'(s) + y(s)\right]^2\,ds + \int_0^{2\pi} (y'(s)^2 - y(s)^2)\,ds.$$

A change of coordinates will guarantee $\int_0^{2\pi} y(s)\,ds = 0$. For the other direction, start with a real-valued f satisfying all the hypotheses of Wirtinger's inequality, and construct g, 2π-periodic and so that the term in brackets above vanishes.]

5. Prove that the sequence $\{\gamma_n\}_{n=1}^{\infty}$, where γ_n is the fractional part of

$$\left(\frac{1+\sqrt{5}}{2}\right)^n,$$

is not equidistributed in $[0, 1]$.
[Hint: Show that $U_n = \left(\frac{1+\sqrt{5}}{2}\right)^n + \left(\frac{1-\sqrt{5}}{2}\right)^n$ is the solution of the difference equation $U_{r+1} = U_r + U_{r-1}$ with $U_0 = 2$ and $U_1 = 1$. The U_n satisfy the same difference equation as the Fibonacci numbers.]

6. Let $\theta = p/q$ be a rational number where p and q are relatively prime integers (that is, θ is in lowest form). We assume without loss of generality that $q > 0$. Define a sequence of numbers in $[0, 1)$ by $\xi_n = \langle n\theta \rangle$ where $\langle \cdot \rangle$ denotes the fractional part. Show that the sequence $\{\xi_1, \xi_2, \ldots\}$ is equidistributed on the points of the form

$$0,\ 1/q,\ 2/q,\ \ldots,\ (q-1)/q.$$

In fact, prove that for any $0 \leq a < q$, one has

$$\frac{\#\{n:\ 1 \leq n \leq N,\ \langle n\theta \rangle = a/q\}}{N} = \frac{1}{q} + O\left(\frac{1}{N}\right).$$

[Hint: For each integer $k \geq 0$, there exists a unique integer n with $kq \leq n < (k+1)q$ and so that $\langle n\theta \rangle = a/q$. Why can one assume $k = 0$? Prove the existence of n by using the fact[1] that if p and q are relatively prime, there exist integers x, y such that $xp + yq = 1$. Next, divide N by q with remainder, that is, write $N = \ell q + r$ where $0 \leq \ell$ and $0 \leq r < q$. Establish the inequalities

$$\ell \leq \#\{n:\ 1 \leq n \leq N,\ \langle n\theta \rangle = a/q\} \leq \ell + 1.]$$

7. Prove the second part of Weyl's criterion: if a sequence of numbers $\xi_1, \xi_2, \ldots$ in $[0,1)$ is equidistributed, then for all $k \in \mathbb{Z} - \{0\}$

$$\frac{1}{N}\sum_{n=1}^{N} e^{2\pi i k \xi_n} \to 0 \qquad \text{as } N \to \infty.$$

[Hint: It suffices to show that $\frac{1}{N}\sum_{n=1}^{N} f(\xi_n) \to \int_0^1 f(x)\,dx$ for all continuous f. Prove this first when f is the characteristic function of an interval.]

8. Show that for any $a \neq 0$, and σ with $0 < \sigma < 1$, the sequence $\langle an^\sigma \rangle$ is equidistributed in $[0,1)$.

[Hint: Prove that $\sum_{n=1}^{N} e^{2\pi i b n^\sigma} = O(N^\sigma) + O(N^{1-\sigma})$ if $b \neq 0$.] In fact, note the following

$$\sum_{n=1}^{N} e^{2\pi i b n^\sigma} - \int_1^N e^{2\pi i b x^\sigma}\,dx = O\left(\sum_{n=1}^{N} n^{-1+\sigma}\right).$$

9. In contrast with the result in Exercise 8, prove that $\langle a \log n \rangle$ is *not* equidistributed for any a.

[Hint: Compare the sum $\sum_{n=1}^{N} e^{2\pi i b \log n}$ with the corresponding integral.]

10. Suppose that f is a periodic function on $\mathbb{R}$ of period 1, and $\{\xi_n\}$ is a sequence which is equidistributed in $[0,1)$. Prove that:

---

[1]The elementary results in arithmetic used in this exercise can be found at the beginning of Chapter 8.

(a) If f is continuous and satisfies $\int_0^1 f(x)\,dx = 0$, then

$$\lim_{N\to\infty} \frac{1}{N} \sum_{n=1}^{N} f(x+\xi_n) = 0 \quad \text{uniformly in } x.$$

[Hint: Establish this result first for trigonometric polynomials.]

(b) If f is merely integrable on $[0,1]$ and satisfies $\int_0^1 f(x)\,dx = 0$, then

$$\lim_{N\to\infty} \int_0^1 \left| \frac{1}{N} \sum_{n=1}^{N} f(x+\xi_n) \right|^2 dx = 0.$$

11. Show that if $u(x,t) = (f * H_t)(x)$ where H_t is the heat kernel, and f is Riemann integrable, then

$$\int_0^1 |u(x,t) - f(x)|^2\,dx \to 0 \quad \text{as } t \to 0.$$

12. A change of variables in (8) leads to the solution

$$u(\theta,\tau) = \sum a_n e^{-n^2\tau} e^{in\theta} = (f * h_\tau)(\theta)$$

of the equation

$$\frac{\partial u}{\partial \tau} = \frac{\partial^2 u}{\partial \theta^2} \quad \text{with } 0 \le \theta \le 2\pi \text{ and } \tau > 0,$$

with boundary condition $u(\theta,0) = f(\theta) \sim \sum a_n e^{in\theta}$. Here $h_\tau(\theta) = \sum_{n=-\infty}^{\infty} e^{-n^2\tau} e^{in\theta}$. This version of the heat kernel on $[0,2\pi]$ is the analogue of the Poisson kernel, which can be written as $P_r(\theta) = \sum_{n=-\infty}^{\infty} e^{-|n|\tau} e^{in\theta}$ with $r = e^{-\tau}$ (and so $0 < r < 1$ corresponds to $\tau > 0$).

13. The fact that the kernel $H_t(x)$ is a good kernel, hence $u(x,t) \to f(x)$ at each point of continuity of f, is not easy to prove. This will be shown in the next chapter. However, one can prove directly that $H_t(x)$ is "peaked" at $x = 0$ as $t \to 0$ in the following sense:

(a) Show that $\int_{-1/2}^{1/2} |H_t(x)|^2\,dx$ is of the order of magnitude of $t^{-1/2}$ as $t \to 0$. More precisely, prove that $t^{1/2} \int_{-1/2}^{1/2} |H_t(x)|^2\,dx$ converges to a non-zero limit as $t \to 0$.

(b) Prove that $\int_{-1/2}^{1/2} x^2 |H_t(x)|^2\,dx = O(t^{1/2})$ as $t \to 0$.

[Hint: For (a) compare the sum $\sum_{-\infty}^{\infty} e^{-cn^2t}$ with the integral $\int_{-\infty}^{\infty} e^{-cx^2t}\,dx$ where $c > 0$. For (b) use $x^2 \leq C(\sin \pi x)^2$ for $-1/2 \leq x \leq 1/2$, and apply the mean value theorem to e^{-cx^2t}.]

6 Problems

1.* This problem explores another relationship between the geometry of a curve and Fourier series. The diameter of a closed curve Γ parametrized by $\gamma(t) = (x(t), y(t))$ on $[-\pi, \pi]$ is defined by

$$d = \sup_{P,\,Q\in\Gamma} |P - Q| = \sup_{t_1,\,t_2\in[-\pi,\pi]} |\gamma(t_1) - \gamma(t_2)|.$$

If a_n is the nth Fourier coefficient of $\gamma(t) = x(t) + iy(t)$ and ℓ denotes the length of Γ, then

(a) $2|a_n| \leq d$ for all $n \neq 0$.

(b) $\ell \leq \pi d$, whenever Γ is convex.

Property (a) follows from the fact that $2a_n = \frac{1}{2\pi}\int_{-\pi}^{\pi}[\gamma(t) - \gamma(t + \pi/n)]e^{-int}\,dt$.

The equality $\ell = \pi d$ is satisfied when Γ is a circle, but surprisingly, this is not the only case. In fact, one finds that $\ell = \pi d$ is equivalent to $2|a_1| = d$. We re-parametrize γ so that for each t in $[-\pi, \pi]$ the tangent to the curve makes an angle t with the y-axis. Then, if $a_1 = 1$ we have

$$\gamma'(t) = ie^{it}(1 + r(t)),$$

where r is a real-valued function which satisfies $r(t) + r(t + \pi) = 0$, and $|r(t)| \leq 1$. Figure 7 (a) shows the curve obtained by setting $r(t) = \cos 5t$. Also, Figure 7 (b) consists of the curve where $r(t) = h(3t)$, with $h(s) = -1$ if $-\pi \leq s \leq 0$ and $h(s) = 1$ if $0 < s < \pi$. This curve (which is only piecewise of class C^1) is known as the Reuleaux triangle and is the classical example of a convex curve of constant width which is not a circle.

2.* Here we present an estimate of Weyl which leads to some interesting results.

(a) Let $S_N = \sum_{n=1}^{N} e^{2\pi i f(n)}$. Show that for $H \leq N$, one has

$$|S_N|^2 \leq c\frac{N}{H}\sum_{h=0}^{H}\left|\sum_{n=1}^{N-h} e^{2\pi i(f(n+h)-f(n))}\right|,$$

for some constant $c > 0$ independent of N, H, and f.

(b) Use this estimate to show that the sequence $\langle n^2\gamma\rangle$ is equidistributed in $[0, 1)$ whenever γ is irrational.

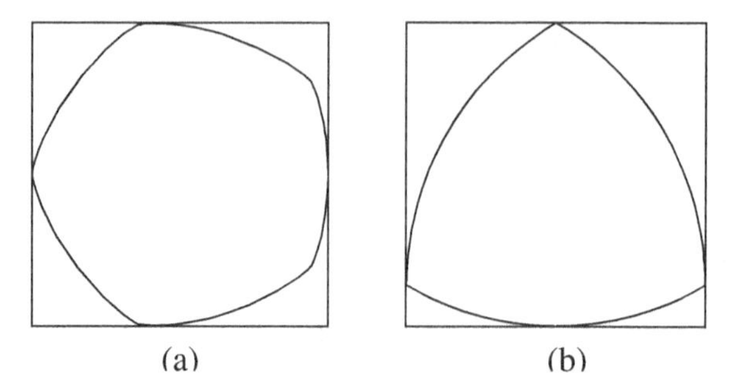

Figure 7. Some curves with maximal length for a given diameter

(c) More generally, show that if $\{\xi_n\}$ is a sequence of real numbers so that for all positive integers h the difference $\langle \xi_{n+h} - \xi_n \rangle$ is equidistributed in $[0, 1)$, then $\langle \xi_n \rangle$ is also equidistributed in $[0, 1)$.

(d) Suppose that $P(x) = c_n x^n + \cdots + c_0$ is a polynomial with real coefficients, where at least one of $c_1, \ldots, c_n$ is irrational. Then the sequence $\langle P(n) \rangle$ is equidistributed in $[0, 1)$.

[Hint: For (a), let $a_n = e^{2\pi i f(n)}$ when $1 \leq n \leq N$ and 0 otherwise. Then write $H \sum_n a_n = \sum_{k=1}^{H} \sum_n a_{n+k}$ and apply the Cauchy-Schwarz inequality. For (b), note that $(n+h)^2\gamma - n^2\gamma = 2nh\gamma + h^2\gamma$, and use the fact that for each integer h, the sequence $\langle 2nh\gamma \rangle$ is equidistributed. Finally, to prove (d), assume first that $P(x) = Q(x) + c_1 x + c_0$ where c_1 is irrational, and estimate the exponential sum $\sum_{n=1}^{N} e^{2\pi i k P(n)}$. Then, argue by induction on the highest degree term which has an irrational coefficient, and use part (c).]

3.* If $\sigma > 0$ is not an integer and $a \neq 0$, then $\langle a n^\sigma \rangle$ is equidistributed in $[0, 1)$. See also Exercise 8.

4. An elementary construction of a continuous but nowhere differentiable function is obtained by "piling up singularities," as follows.

On $[-1, 1]$ consider the function

$$\varphi(x) = |x|$$

and extend φ to $\mathbb{R}$ by requiring it to be periodic of period 2. Clearly, φ is continuous on $\mathbb{R}$ and $|\varphi(x)| \leq 1$ for all x so the function f defined by

$$f(x) = \sum_{n=0}^{\infty} \left(\frac{3}{4}\right)^n \varphi(4^n x)$$

is continuous on $\mathbb{R}$.

(a) Fix $x_0 \in \mathbb{R}$. For every positive integer m, let $\delta_m = \pm\frac{1}{2}4^{-m}$ where the sign is chosen so that no integer lies in between $4^m x_0$ and $4^m(x_0 + \delta_m)$. Consider the quotient

$$\gamma_n = \frac{\varphi(4^n(x_0 + \delta_m)) - \varphi(4^n x_0)}{\delta_m}.$$

Prove that if $n > m$, then $\gamma_n = 0$, and for $0 \leq n \leq m$ one has $|\gamma_n| \leq 4^n$ with $|\gamma_m| = 4^m$.

(b) From the above observations prove the estimate

$$\left|\frac{f(x_0 + \delta_m) - f(x_0)}{\delta_m}\right| \geq \frac{1}{2}(3^m + 1),$$

and conclude that f is not differentiable at x_0.

5. Let f be a Riemann integrable function on the interval $[-\pi, \pi]$. We define the generalized delayed means of the Fourier series of f by

$$\sigma_{N,K} = \frac{S_N + \cdots + S_{N+K-1}}{K}.$$

Note that in particular

$$\sigma_{0,N} = \sigma_N, \qquad \sigma_{N,1} = S_N \qquad \text{and} \qquad \sigma_{N,N} = \Delta_N,$$

where Δ_N are the specific delayed means used in Section 3.

(a) Show that

$$\sigma_{N,K} = \frac{1}{K}\left((N+K)\sigma_{N+K} - N\sigma_N\right),$$

and

$$\sigma_{N,K} = S_N + \sum_{N+1 \leq |\nu| \leq N+K-1} \left(1 - \frac{|\nu| - N}{K}\right) \hat{f}(\nu) e^{i\nu\theta}.$$

From this last expression for $\sigma_{N,K}$ conclude that

$$|\sigma_{N,K} - S_M| \leq \sum_{N+1 \leq |\nu| \leq N+K-1} |\hat{f}(\nu)|$$

for all $N \leq M < N + K$.

(b) Use one of the above formulas and Fejér's theorem to show that with $N = kn$ and $K = n$, then

$$\sigma_{kn,n}(f)(\theta) \to f(\theta) \quad \text{as } n \to \infty$$

whenever f is continuous at θ, and also

$$\sigma_{kn,n}(f)(\theta) \to \frac{f(\theta^+) + f(\theta^-)}{2} \quad \text{as } n \to \infty$$

at a jump discontinuity (refer to the preceding chapters and their exercises for the appropriate definitions and results). In the case when f is continuous on $[-\pi, \pi]$, show that $\sigma_{kn,n}(f) \to f$ uniformly as $n \to \infty$.

(c) Using part (a), show that if $\hat{f}(\nu) = O(1/|\nu|)$ and $kn \leq m < (k+1)n$, we get

$$|\sigma_{kn,n} - S_m| \leq \frac{C}{k} \quad \text{for some constant } C > 0.$$

(d) Suppose that $\hat{f}(\nu) = O(1/|\nu|)$. Prove that if f is continuous at θ then

$$S_N(f)(\theta) \to f(\theta) \quad \text{as } N \to \infty,$$

and if f has a jump discontinuity at θ then

$$S_N(f)(\theta) \to \frac{f(\theta^+) + f(\theta^-)}{2} \quad \text{as } N \to \infty.$$

Also, show that if f is continuous on $[-\pi, \pi]$, then $S_N(f) \to f$ uniformly.

(e) The above arguments show that if $\sum c_n$ is Cesàro summable to s and $c_n = O(1/n)$, then $\sum c_n$ converges to s. This is a weak version of Littlewood's theorem (Problem 3, Chapter 2).

6. Dirichlet's theorem states that the Fourier series of a real continuous periodic function f which has only a finite number of relative maxima and minima converges everywhere to f (and uniformly).

Prove this theorem by showing that such a function satisfies $\hat{f}(n) = O(1/|n|)$. [Hint: Argue as in Exercise 17, Chapter 3; then use conclusion (d) in Problem 5 above.]

5 The Fourier Transform on $\mathbb{R}$

> The theory of Fourier series and integrals has always had major difficulties and necessitated a large mathematical apparatus in dealing with questions of convergence. It engendered the development of methods of summation, although these did not lead to a completely satisfactory solution of the problem.... For the Fourier transform, the introduction of distributions (hence the space $\mathcal{S}$) is inevitable either in an explicit or hidden form.... As a result one may obtain all that is desired from the point of view of the continuity and inversion of the Fourier transform.
>
> *L. Schwartz,* 1950

The theory of Fourier series applies to functions on the circle, or equivalently, periodic functions on $\mathbb{R}$. In this chapter, we develop an analogous theory for functions on the entire real line which are non-periodic. The functions we consider will be suitably "small" at infinity. There are several ways of defining an appropriate notion of "smallness," but it will nevertheless be vital to assume some sort of vanishing at infinity.

On the one hand, recall that the Fourier series of a periodic function associates a sequence of numbers, namely the Fourier coefficients, to that function; on the other hand, given a suitable function f on $\mathbb{R}$, the analogous object associated to f will in fact be another function $\hat{f}$ on $\mathbb{R}$ which is called the Fourier transform of f. Since the Fourier transform of a function on $\mathbb{R}$ is again a function on $\mathbb{R}$, one can observe a symmetry between a function and its Fourier transform, whose analogue is not as apparent in the setting of Fourier series.

Roughly speaking, the Fourier transform is a continuous version of the Fourier coefficients. Recall that the Fourier coefficients a_n of a function f defined on the circle are given by

$$a_n = \int_0^1 f(x)e^{-2\pi inx}\,dx, \tag{1}$$

and then in the appropriate sense we have

$$f(x) = \sum_{n=-\infty}^{\infty} a_n e^{2\pi i n x}. \tag{2}$$

Here we have replaced θ by $2\pi x$, as we have frequently done previously.

Now, consider the following analogy where we replace all of the discrete symbols (such as integers and sums) by their continuous counterparts (such as real numbers and integrals). In other words, given a function f on all of $\mathbb{R}$, we define its Fourier transform by changing the domain of integration from the circle to all of $\mathbb{R}$, and by replacing $n \in \mathbb{Z}$ by $\xi \in \mathbb{R}$ in (1), that is, by setting

$$\hat{f}(\xi) = \int_{-\infty}^{\infty} f(x) e^{-2\pi i x \xi}\, dx. \tag{3}$$

We push our analogy further, and consider the following continuous version of (2): replacing the sum by an integral, and a_n by $\hat{f}(\xi)$, leads to the Fourier inversion formula,

$$f(x) = \int_{-\infty}^{\infty} \hat{f}(\xi) e^{2\pi i x \xi}\, d\xi. \tag{4}$$

Under a suitable hypotheses on f, the identity (4) actually holds, and much of the theory in this chapter aims at proving and exploiting this relation. The validity of the Fourier inversion formula is also suggested by the following simple observation. Suppose f is supported in a finite interval contained in $I = [-L/2, L/2]$, and we expand f in a Fourier series on I. Then, letting L tend to infinity, we are led to (4) (see Exercise 1).

The special properties of the Fourier transform make it an important tool in the study of partial differential equations. For instance, we shall see how the Fourier inversion formula allows us to analyze some equations that are modeled on the real line. In particular, following the ideas developed on the circle, we solve the time-dependent heat equation for an infinite rod and the steady-state heat equation in the upper half-plane.

In the last part of the chapter we discuss further topics related to the Poisson summation formula,

$$\sum_{n \in \mathbb{Z}} f(n) = \sum_{n \in \mathbb{Z}} \hat{f}(n),$$

which gives another remarkable connection between periodic functions (and their Fourier series) and non-periodic functions on the line (and their Fourier transforms). This identity allows us to prove an assertion made in the previous chapter, namely, that the heat kernel $H_t(x)$ satisfies the properties of a good kernel. In addition, the Poisson summation formula arises in many other settings, in particular in parts of number theory, as we shall see in Book II.

We make a final comment about the approach we have chosen. In our study of Fourier series, we found it useful to consider Riemann integrable functions on the circle. In particular, this generality assured us that even functions that had certain discontinuities could be treated by the theory. In contrast, our exposition of the elementary properties of the Fourier transform is stated in terms of the Schwartz space $\mathcal{S}$ of testing functions. These are functions that are indefinitely differentiable and that, together with their derivatives, are rapidly decreasing at infinity. The reliance on this space of functions is a device that allows us to come quickly to the main conclusions, formulated in a direct and transparent fashion. Once this is carried out, we point out some easy extensions to a somewhat wider setting. The more general theory of Fourier transforms (which must necessarily be based on Lebesgue integration) will be treated in Book III.

1 Elementary theory of the Fourier transform

We begin by extending the notion of integration to functions that are defined on the whole real line.

1.1 Integration of functions on the real line

Given the notion of the integral of a function on a closed and bounded interval, the most natural extension of this definition to continuous functions over $\mathbb{R}$ is

$$\int_{-\infty}^{\infty} f(x)\,dx = \lim_{N\to\infty}\int_{-N}^{N} f(x)\,dx.$$

Of course, this limit may not exist. For example, it is clear that if $f(x) = 1$, or even if $f(x) = 1/(1+|x|)$, then the above limit is infinite. A moment's reflection suggests that the limit will exist if we impose on f enough decay as $|x|$ tends to infinity. A useful condition is as follows.

A function f defined on $\mathbb{R}$ is said to be of **moderate decrease** if f is continuous and there exists a constant $A > 0$ so that

$$|f(x)| \le \frac{A}{1+x^2} \qquad \text{for all } x \in \mathbb{R}.$$

This inequality says that f is bounded (by A for instance), and also that it decays at infinity at least as fast as $1/x^2$, since $A/(1+x^2) \leq A/x^2$.

For example, the function $f(x) = 1/(1+|x|^n)$ is of moderate decrease as long as $n \geq 2$. Another example is given by the function $e^{-a|x|}$ for $a > 0$.

We shall denote by $\mathcal{M}(\mathbb{R})$ the set of functions of moderate decrease on $\mathbb{R}$. As an exercise, the reader can check that under the usual addition of functions and multiplication by scalars, $\mathcal{M}(\mathbb{R})$ forms a vector space over $\mathbb{C}$.

We next see that whenever f belongs to $\mathcal{M}(\mathbb{R})$, then we may define

$$\int_{-\infty}^{\infty} f(x)\, dx = \lim_{N\to\infty} \int_{-N}^{N} f(x)\, dx, \tag{5}$$

where the limit now exists. Indeed, for each N the integral $I_N = \int_{-N}^{N} f(x)\, dx$ is well defined because f is continuous. It now suffices to show that $\{I_N\}$ is a Cauchy sequence, and this follows because if $M > N$, then

$$\begin{aligned} |I_M - I_N| &\leq \left| \int_{N\leq|x|\leq M} f(x)\, dx \right| \\ &\leq A \int_{N\leq|x|\leq M} \frac{dx}{x^2} \\ &\leq \frac{2A}{N} \to 0 \quad \text{as } N \to \infty. \end{aligned}$$

Notice we have also proved that $\int_{|x|\geq N} f(x)\, dx \to 0$ as $N \to \infty$. At this point, we remark that we may replace the exponent 2 in the definition of moderate decrease by $1 + \epsilon$ where $\epsilon > 0$; that is,

$$|f(x)| \leq \frac{A}{1+|x|^{1+\epsilon}} \quad \text{for all } x \in \mathbb{R}.$$

This definition would work just as well for the purpose of the theory developed in this chapter. We chose $\epsilon = 1$ merely as a matter of convenience.

We summarize some elementary properties of integration over $\mathbb{R}$ in a proposition.

Proposition 1.1 *The integral of a function of moderate decrease defined by (5) satisfies the following properties:*

(i) *Linearity: if $f, g \in \mathcal{M}(\mathbb{R})$ and $a, b \in \mathbb{C}$, then*

$$\int_{-\infty}^{\infty} (af(x) + bg(x))\, dx = a \int_{-\infty}^{\infty} f(x)\, dx + b \int_{-\infty}^{\infty} g(x)\, dx.$$

(ii) *Translation invariance: for every $h \in \mathbb{R}$ we have*

$$\int_{-\infty}^{\infty} f(x-h)\, dx = \int_{-\infty}^{\infty} f(x)\, dx.$$

(iii) *Scaling under dilations: if $\delta > 0$, then*

$$\delta \int_{-\infty}^{\infty} f(\delta x)\, dx = \int_{-\infty}^{\infty} f(x)\, dx.$$

(iv) *Continuity: if $f \in \mathcal{M}(\mathbb{R})$, then*

$$\int_{-\infty}^{\infty} |f(x-h) - f(x)|\, dx \to 0 \quad \textit{as } h \to 0.$$

We say a few words about the proof. Property (i) is immediate. To verify property (ii), it suffices to see that

$$\int_{-N}^{N} f(x-h)\, dx - \int_{-N}^{N} f(x)\, dx \to 0 \quad \text{as } N \to \infty.$$

Since $\int_{-N}^{N} f(x-h)\, dx = \int_{-N-h}^{N-h} f(x)\, dx$, the above difference is majorized by

$$\left| \int_{-N-h}^{-N} f(x)\, dx \right| + \left| \int_{N-h}^{N} f(x)\, dx \right| \le \frac{A'}{1+N^2}$$

for large N, which tends to 0 as N tends to infinity.

The proof of property (iii) is similar once we observe that $\delta \int_{-N}^{N} f(\delta x)\, dx = \int_{-\delta N}^{\delta N} f(x)\, dx$.

To prove property (iv) it suffices to take $|h| \le 1$. For a preassigned $\epsilon > 0$, we first choose N so large that

$$\int_{|x| \ge N} |f(x)|\, dx \le \epsilon/4 \quad \text{and} \quad \int_{|x| \ge N} |f(x-h)|\, dx \le \epsilon/4.$$

Now with N fixed, we use the fact that since f is continuous, it is uniformly continuous in the interval $[-N-1, N+1]$. Hence

$\sup_{|x|\leq N} |f(x-h) - f(x)| \to 0$ as h tends to 0. So we can take h so small that this supremum is less than $\epsilon/4N$. Altogether, then,

$$\begin{aligned}\int_{-\infty}^{\infty} |f(x-h) - f(x)|\, dx \leq \int_{-N}^{N} |f(x-h) - f(x)|\, dx \\ + \int_{|x|\geq N} |f(x-h)|\, dx \\ + \int_{|x|\geq N} |f(x)|\, dx \\ \leq \epsilon/2 + \epsilon/4 + \epsilon/4 = \epsilon,\end{aligned}$$

and thus conclusion (iv) follows.

1.2 Definition of the Fourier transform

If $f \in \mathcal{M}(\mathbb{R})$, we define its **Fourier transform** for $\xi \in \mathbb{R}$ by

$$\hat{f}(\xi) = \int_{-\infty}^{\infty} f(x) e^{-2\pi i x\xi}\, dx.$$

Of course, $|e^{-2\pi i x\xi}| = 1$, so the integrand is of moderate decrease, and the integral makes sense.

In fact, this last observation implies that $\hat{f}$ is bounded, and moreover, a simple argument shows that $\hat{f}$ is continuous and tends to 0 as $|\xi| \to \infty$ (Exercise 5). However, nothing in the definition above guarantees that $\hat{f}$ is of moderate decrease, or has a specific decay. In particular, it is not clear in this context how to make sense of the integral $\int_{-\infty}^{\infty} \hat{f}(\xi) e^{2\pi i x\xi}\, d\xi$ and the resulting Fourier inversion formula. To remedy this, we introduce a more refined space of functions considered by Schwartz which is very useful in establishing the initial properties of the Fourier transform.

The choice of the Schwartz space is motivated by an important principle which ties the decay of $\hat{f}$ to the continuity and differentiability properties of f (and vice versa): the faster $\hat{f}(\xi)$ decreases as $|\xi| \to \infty$, the "smoother" f must be. An example that reflects this principle is given in Exercise 3. We also note that this relationship between f and $\hat{f}$ is reminiscent of a similar one between the smoothness of a function on the circle and the decay of its Fourier coefficients; see the discussion of Corollary 2.4 in Chapter 2.

1.3 The Schwartz space

The **Schwartz space** on $\mathbb{R}$ consists of the set of all indefinitely differentiable functions f so that f and all its derivatives $f', f'', \ldots, f^{(\ell)}, \ldots,$

are **rapidly decreasing**, in the sense that

$$\sup_{x \in \mathbb{R}} |x|^k |f^{(\ell)}(x)| < \infty \qquad \text{for every } k, \ell \geq 0.$$

We denote this space by $\mathcal{S} = \mathcal{S}(\mathbb{R})$, and again, the reader should verify that $\mathcal{S}(\mathbb{R})$ is a vector space over $\mathbb{C}$. Moreover, if $f \in \mathcal{S}(\mathbb{R})$, we have

$$f'(x) = \frac{df}{dx} \in \mathcal{S}(\mathbb{R}) \quad \text{and} \quad xf(x) \in \mathcal{S}(\mathbb{R}).$$

This expresses the important fact that the Schwartz space is closed under differentiation and multiplication by polynomials.

A simple example of a function in $\mathcal{S}(\mathbb{R})$ is the **Gaussian** defined by

$$f(x) = e^{-x^2},$$

which plays a central role in the theory of the Fourier transform, as well as other fields (for example, probability theory and physics). The reader can check that the derivatives of f are of the form $P(x)e^{-x^2}$ where P is a polynomial, and this immediately shows that $f \in \mathcal{S}(\mathbb{R})$. In fact, e^{-ax^2} belongs to $\mathcal{S}(\mathbb{R})$ whenever $a > 0$. Later, we will normalize the Gaussian by choosing $a = \pi$.

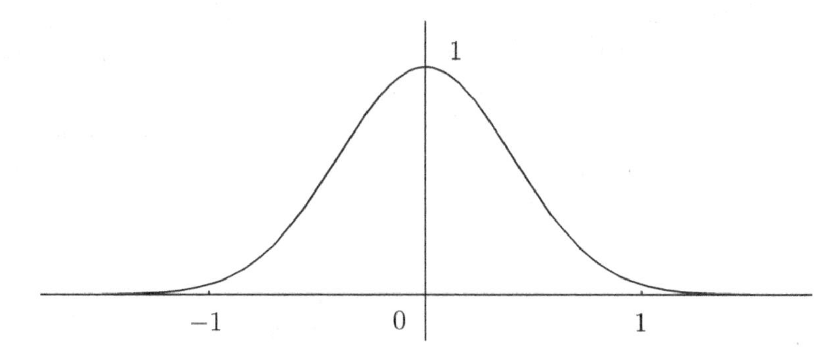

Figure 1. The Gaussian $e^{-\pi x^2}$

An important class of other examples in $\mathcal{S}(\mathbb{R})$ are the "bump functions" which vanish outside bounded intervals (Exercise 4).

As a final remark, note that although $e^{-|x|}$ decreases rapidly at infinity, it is not differentiable at 0 and therefore does not belong to $\mathcal{S}(\mathbb{R})$.

1.4 The Fourier transform on $\mathcal{S}$

The **Fourier transform** of a function $f \in \mathcal{S}(\mathbb{R})$ is defined by

$$\hat{f}(\xi) = \int_{-\infty}^{\infty} f(x) e^{-2\pi i x \xi}\, dx.$$

Some simple properties of the Fourier transform are gathered in the following proposition. We use the notation

$$f(x) \longrightarrow \hat{f}(\xi)$$

to mean that $\hat{f}$ denotes the Fourier transform of f.

Proposition 1.2 *If $f \in \mathcal{S}(\mathbb{R})$ then:*

(i) $f(x+h) \longrightarrow \hat{f}(\xi) e^{2\pi i h \xi}$ *whenever* $h \in \mathbb{R}$.

(ii) $f(x) e^{-2\pi i x h} \longrightarrow \hat{f}(\xi + h)$ *whenever* $h \in \mathbb{R}$.

(iii) $f(\delta x) \longrightarrow \delta^{-1} \hat{f}(\delta^{-1} \xi)$ *whenever* $\delta > 0$.

(iv) $f'(x) \longrightarrow 2\pi i \xi \hat{f}(\xi)$.

(v) $-2\pi i x f(x) \longrightarrow \dfrac{d}{d\xi} \hat{f}(\xi)$.

In particular, except for factors of $2\pi i$, the Fourier transform interchanges differentiation and multiplication by x. This is the key property that makes the Fourier transform a central object in the theory of differential equations. We shall return to this point later.

Proof. Property (i) is an immediate consequence of the translation invariance of the integral, and property (ii) follows from the definition. Also, the third property of Proposition 1.1 establishes (iii).

Integrating by parts gives

$$\int_{-N}^{N} f'(x) e^{-2\pi i x \xi}\, dx = \left[f(x) e^{-2\pi i x \xi} \right]_{-N}^{N} + 2\pi i \xi \int_{-N}^{N} f(x) e^{-2\pi i x \xi}\, dx,$$

so letting N tend to infinity gives (iv).

Finally, to prove property (v), we must show that $\hat{f}$ is differentiable and find its derivative. Let $\epsilon > 0$ and consider

$$\frac{\hat{f}(\xi + h) - \hat{f}(\xi)}{h} - (\widehat{-2\pi i x f})(\xi) = \int_{-\infty}^{\infty} f(x) e^{-2\pi i x \xi} \left[\frac{e^{-2\pi i x h} - 1}{h} + 2\pi i x \right] dx.$$

Since $f(x)$ and $xf(x)$ are of rapid decrease, there exists an integer N so that $\int_{|x|\geq N}|f(x)|\,dx \leq \epsilon$ and $\int_{|x|\geq N}|x|\,|f(x)|\,dx \leq \epsilon$. Moreover, for $|x| \leq N$, there exists h_0 so that $|h| < h_0$ implies

$$\left|\frac{e^{-2\pi ixh}-1}{h}+2\pi ix\right| \leq \frac{\epsilon}{N}.$$

Hence for $|h| < h_0$ we have

$$\begin{aligned}
&\left|\frac{\hat{f}(\xi+h)-\hat{f}(\xi)}{h} - (\widehat{-2\pi ixf})(\xi)\right| \\
&\qquad \leq \int_{-N}^{N}\left|f(x)e^{-2\pi ix\xi}\left[\frac{e^{-2\pi ixh}-1}{h}+2\pi ix\right]\right|\,dx + C\epsilon \\
&\qquad \leq C'\epsilon.
\end{aligned}$$

Theorem 1.3 *If $f \in \mathcal{S}(\mathbb{R})$, then $\hat{f} \in \mathcal{S}(\mathbb{R})$.*

The proof is an easy application of the fact that the Fourier transform interchanges differentiation and multiplication. In fact, note that if $f \in \mathcal{S}(\mathbb{R})$, its Fourier transform $\hat{f}$ is bounded; then also, for each pair of non-negative integers ℓ and k, the expression

$$\xi^k\left(\frac{d}{d\xi}\right)^{\ell}\hat{f}(\xi)$$

is bounded, since by the last proposition, it is the Fourier transform of

$$\frac{1}{(2\pi i)^k}\left(\frac{d}{dx}\right)^{k}[(-2\pi ix)^{\ell}f(x)].$$

The proof of the inversion formula

$$f(x) = \int_{-\infty}^{\infty}\hat{f}(\xi)e^{2\pi ix\xi}\,d\xi \qquad \text{for } f \in \mathcal{S}(\mathbb{R}),$$

which we give in the next section, is based on a careful study of the function e^{-ax^2}, which, as we have already observed, is in $\mathcal{S}(\mathbb{R})$ if $a > 0$.

The Gaussians as good kernels

We begin by considering the case $a = \pi$ because of the normalization:

$$\int_{-\infty}^{\infty} e^{-\pi x^2}\, dx = 1. \tag{6}$$

To see why (6) is true, we use the multiplicative property of the exponential to reduce the calculation to a two-dimensional integral. More precisely, we can argue as follows:

$$\begin{aligned}
\left(\int_{-\infty}^{\infty} e^{-\pi x^2}\, dx\right)^2 &= \int_{-\infty}^{\infty}\int_{-\infty}^{\infty} e^{-\pi(x^2+y^2)}\, dx\, dy \\
&= \int_0^{2\pi}\int_0^{\infty} e^{-\pi r^2} r\, dr\, d\theta \\
&= \int_0^{\infty} 2\pi r e^{-\pi r^2}\, dr \\
&= \left[-e^{-\pi r^2}\right]_0^{\infty} \\
&= 1,
\end{aligned}$$

where we have evaluated the two-dimensional integral using polar coordinates.

The fundamental property of the Gaussian which is of interest to us, and which actually follows from (6), is that $e^{-\pi x^2}$ equals its Fourier transform! We isolate this important result in a theorem.

Theorem 1.4 *If $f(x) = e^{-\pi x^2}$, then $\hat{f}(\xi) = f(\xi)$.*

Proof. Define

$$F(\xi) = \hat{f}(\xi) = \int_{-\infty}^{\infty} e^{-\pi x^2} e^{-2\pi i x\xi}\, dx,$$

and observe that $F(0) = 1$, by our previous calculation. By property (v) in Proposition 1.2, and the fact that $f'(x) = -2\pi x f(x)$, we obtain

$$F'(\xi) = \int_{-\infty}^{\infty} f(x)(-2\pi i x) e^{-2\pi i x\xi}\, dx = i\int_{-\infty}^{\infty} f'(x) e^{-2\pi i x\xi}\, dx.$$

By (iv) of the same proposition, we find that

$$F'(\xi) = i(2\pi i\xi)\hat{f}(\xi) = -2\pi\xi F(\xi).$$

If we define $G(\xi) = F(\xi)e^{\pi\xi^2}$, then from what we have seen above, it follows that $G'(\xi) = 0$, hence G is constant. Since $F(0) = 1$, we conclude that G is identically equal to 1, therefore $F(\xi) = e^{-\pi\xi^2}$, as was to be shown.

The scaling properties of the Fourier transform under dilations yield the following important transformation law, which follows from (iii) in Proposition 1.2 (with δ replaced by $\delta^{-1/2}$).

Corollary 1.5 *If* $\delta > 0$ *and* $K_\delta(x) = \delta^{-1/2}e^{-\pi x^2/\delta}$, *then* $\widehat{K_\delta}(\xi) = e^{-\pi\delta\xi^2}$.

We pause to make an important observation. As δ tends to 0, the function K_δ peaks at the origin, while its Fourier transform $\widehat{K_\delta}$ gets flatter. So in this particular example, we see that K_δ and $\widehat{K_\delta}$ cannot both be localized (that is, concentrated) at the origin. This is an example of a general phenomenon called the Heisenberg uncertainty principle, which we will discuss at the end of this chapter.

We have now constructed a family of good kernels on the real line, analogous to those on the circle considered in Chapter 2. Indeed, with

$$K_\delta(x) = \delta^{-1/2}e^{-\pi x^2/\delta},$$

we have:

(i) $\int_{-\infty}^{\infty} K_\delta(x)\,dx = 1$.

(ii) $\int_{-\infty}^{\infty} |K_\delta(x)|\,dx \leq M$.

(iii) For every $\eta > 0$, we have $\int_{|x|>\eta} |K_\delta(x)|\,dx \to 0$ as $\delta \to 0$.

To prove (i), we may change variables and use (6), or note that the integral equals $\widehat{K_\delta}(0)$, which is 1 by Corollary 1.5. Since $K_\delta \geq 0$, it is clear that property (ii) is also true. Finally we can again change variables to get

$$\int_{|x|>\eta} |K_\delta(x)|\,dx = \int_{|y|>\eta/\delta^{1/2}} e^{-\pi y^2}\,dy \to 0$$

as δ tends to 0. We have thus proved the following result.

Theorem 1.6 *The collection* $\{K_\delta\}_{\delta>0}$ *is a family of good kernels as* $\delta \to 0$.

We next apply these good kernels via the operation of convolution, which is given as follows. If $f, g \in \mathcal{S}(\mathbb{R})$, their **convolution** is defined by

$$(f * g)(x) = \int_{-\infty}^{\infty} f(x-t)g(t)\,dt. \tag{7}$$

For a fixed value of x, the function $f(x-t)g(t)$ is of rapid decrease in t, hence the integral converges.

By the argument in Section 4 of Chapter 2 (with a slight modification), we get the following corollary.

Corollary 1.7 *If $f \in \mathcal{S}(\mathbb{R})$, then*

$$(f * K_\delta)(x) \to f(x) \quad \textit{uniformly in } x \textit{ as } \delta \to 0.$$

Proof. First, we claim that f is uniformly continuous on $\mathbb{R}$. Indeed, given $\epsilon > 0$ there exists $R > 0$ so that $|f(x)| < \epsilon/4$ whenever $|x| \geq R$. Moreover, f is continuous, hence uniformly continuous on the compact interval $[-R, R]$, and together with the previous observation, we can find $\eta > 0$ so that $|f(x) - f(y)| < \epsilon$ whenever $|x - y| < \eta$. Now we argue as usual. Using the first property of good kernels, we can write

$$(f * K_\delta)(x) - f(x) = \int_{-\infty}^{\infty} K_\delta(t)\,[f(x-t) - f(x)]\,dt,$$

and since $K_\delta \geq 0$, we find

$$|(f * K_\delta)(x) - f(x)| \leq \int_{|t|>\eta} + \int_{|t|\leq\eta} K_\delta(t)\,|f(x-t) - f(x)|\,dt.$$

The first integral is small by the third property of good kernels, and the fact that f is bounded, while the second integral is also small since f is uniformly continuous and $\int K_\delta = 1$. This concludes the proof of the corollary.

1.5 The Fourier inversion

The next result is an identity sometimes called the multiplication formula.

Proposition 1.8 *If $f, g \in \mathcal{S}(\mathbb{R})$, then*

$$\int_{-\infty}^{\infty} f(x)\hat{g}(x)\,dx = \int_{-\infty}^{\infty} \hat{f}(y)g(y)\,dy.$$

To prove the proposition, we need to digress briefly to discuss the interchange of the order of integration for double integrals. Suppose $F(x, y)$ is a continuous function in the plane $(x, y) \in \mathbb{R}^2$. We will assume the following decay condition on F:

$$|F(x, y)| \leq A/(1 + x^2)(1 + y^2).$$

Then, we can state that for each x the function $F(x, y)$ is of moderate decrease in y, and similarly for each fixed y the function $F(x, y)$ is of moderate decrease in x. Moreover, the function $F_1(x) = \int_{-\infty}^{\infty} F(x, y)\,dy$ is continuous and of moderate decrease; similarly for the function $F_2(y) = \int_{-\infty}^{\infty} F(x, y)\,dx$. Finally

$$\int_{-\infty}^{\infty} F_1(x)\,dx = \int_{-\infty}^{\infty} F_2(y)\,dy.$$

The proof of these facts may be found in the appendix.

We now apply this to $F(x, y) = f(x)g(y)e^{-2\pi ixy}$. Then $F_1(x) = f(x)\hat{g}(x)$, and $F_2(y) = \hat{f}(y)g(y)$ so

$$\int_{-\infty}^{\infty} f(x)\hat{g}(x)\,dx = \int_{-\infty}^{\infty} \hat{f}(y)g(y)\,dy,$$

which is the assertion of the proposition.

The multiplication formula and the fact that the Gaussian is its own Fourier transform lead to a proof of the first major theorem.

Theorem 1.9 (Fourier inversion) *If $f \in \mathcal{S}(\mathbb{R})$, then*

$$f(x) = \int_{-\infty}^{\infty} \hat{f}(\xi)e^{2\pi ix\xi}\,d\xi.$$

Proof. We first claim that

$$f(0) = \int_{-\infty}^{\infty} \hat{f}(\xi)\,d\xi.$$

Let $G_\delta(x) = e^{-\pi\delta x^2}$ so that $\widehat{G_\delta}(\xi) = K_\delta(\xi)$. By the multiplication formula we get

$$\int_{-\infty}^{\infty} f(x)K_\delta(x)\,dx = \int_{-\infty}^{\infty} \hat{f}(\xi)G_\delta(\xi)\,d\xi.$$

Since K_δ is a good kernel, the first integral goes to $f(0)$ as δ tends to 0. Since the second integral clearly converges to $\int_{-\infty}^{\infty} \hat{f}(\xi)\,d\xi$ as δ tends to 0, our claim is proved. In general, let $F(y) = f(y + x)$ so that

$$f(x) = F(0) = \int_{-\infty}^{\infty} \hat{F}(\xi)\,d\xi = \int_{-\infty}^{\infty} \hat{f}(\xi)e^{2\pi ix\xi}\,d\xi.$$

As the name of Theorem 1.9 suggests, it provides a formula that inverts the Fourier transform; in fact we see that the Fourier transform is its own inverse except for the change of x to $-x$. More precisely, we may define two mappings $\mathcal{F} : \mathcal{S}(\mathbb{R}) \to \mathcal{S}(\mathbb{R})$ and $\mathcal{F}^* : \mathcal{S}(\mathbb{R}) \to \mathcal{S}(\mathbb{R})$ by

$$\mathcal{F}(f)(\xi) = \int_{-\infty}^{\infty} f(x)e^{-2\pi i x\xi}\,dx \quad \text{and} \quad \mathcal{F}^*(g)(x) = \int_{-\infty}^{\infty} g(\xi)e^{2\pi i x\xi}\,d\xi.$$

Thus $\mathcal{F}$ is the Fourier transform, and Theorem 1.9 guarantees that $\mathcal{F}^* \circ \mathcal{F} = I$ on $\mathcal{S}(\mathbb{R})$, where I is the identity mapping. Moreover, since the definitions of $\mathcal{F}$ and $\mathcal{F}^*$ differ only by a sign in the exponential, we see that $\mathcal{F}(f)(y) = \mathcal{F}^*(f)(-y)$, so we also have $\mathcal{F} \circ \mathcal{F}^* = I$. As a consequence, we conclude that $\mathcal{F}^*$ is the inverse of the Fourier transform on $\mathcal{S}(\mathbb{R})$, and we get the following result.

Corollary 1.10 *The Fourier transform is a bijective mapping on the Schwartz space.*

1.6 The Plancherel formula

We need a few further results about convolutions of Schwartz functions. The key fact is that the Fourier transform interchanges convolutions with pointwise products, a result analogous to the situation for Fourier series.

Proposition 1.11 *If $f, g \in \mathcal{S}(\mathbb{R})$ then:*

(i) $f * g \in \mathcal{S}(\mathbb{R})$.

(ii) $f * g = g * f$.

(iii) $\widehat{(f * g)}(\xi) = \hat{f}(\xi)\hat{g}(\xi)$.

Proof. To prove that $f * g$ is rapidly decreasing, observe first that for any $\ell \geq 0$ we have $\sup_x |x|^\ell |g(x-y)| \leq A_\ell (1+|y|)^\ell$, because g is rapidly decreasing (to check this assertion, consider separately the two cases $|x| \leq 2|y|$ and $|x| \geq 2|y|$). From this, we see that

$$\sup_x |x^\ell (f * g)(x)| \leq A_\ell \int_{-\infty}^{\infty} |f(y)|(1+|y|)^\ell\,dy,$$

so that $x^\ell (f * g)(x)$ is a bounded function for every $\ell \geq 0$. These estimates carry over to the derivatives of $f * g$, thereby proving that $f * g \in \mathcal{S}(\mathbb{R})$ because

$$\left(\frac{d}{dx}\right)^k (f * g)(x) = \left(f * \left(\frac{d}{dx}\right)^k g\right)(x) \quad \text{for } k = 1, 2, \ldots.$$

This identity is proved first for $k = 1$ by differentiating under the integral defining $f * g$. The interchange of differentiation and integration is justified in this case by the rapid decrease of dg/dx. The identity then follows for every k by iteration.

For fixed x, the change of variables $x - y = u$ shows that

$$(f * g)(x) = \int_{-\infty}^{\infty} f(x-u)g(u)\,du = (g * f)(x).$$

This change of variables is a composition of two changes, $y \mapsto -y$ and $y \mapsto y - h$ (with $h = x$). For the first one we use the observation that $\int_{-\infty}^{\infty} F(x)\,dx = \int_{-\infty}^{\infty} F(-x)\,dx$ for any Schwartz function F, and for the second, we apply (ii) of Proposition 1.1

Finally, consider $F(x, y) = f(y)g(x - y)e^{-2\pi i x\xi}$. Since f and g are rapidly decreasing, considering separately the two cases $|x| \leq 2|y|$ and $|x| \geq 2|y|$, we see that the discussion of the change of order of integration after Proposition 1.8 applies to F. In this case $F_1(x) = (f * g)(x)e^{-2\pi i x\xi}$, and $F_2(y) = f(y)e^{-2\pi i y\xi}\hat{g}(\xi)$. Thus $\int_{-\infty}^{\infty} F_1(x)\,dx = \int_{-\infty}^{\infty} F_2(y)\,dy$, which implies (iii). The proposition is therefore proved.

We now use the properties of convolutions of Schwartz functions to prove the main result of this section. The result we have in mind is the analogue for functions on $\mathbb{R}$ of Parseval's identity for Fourier series.

The Schwartz space can be equipped with a Hermitian inner product

$$(f, g) = \int_{-\infty}^{\infty} f(x)\overline{g(x)}\,dx$$

whose associated norm is

$$\|f\| = \left(\int_{-\infty}^{\infty} |f(x)|^2\,dx\right)^{1/2}.$$

The second major theorem in the theory states that the Fourier transform is a unitary transformation on $\mathcal{S}(\mathbb{R})$.

Theorem 1.12 (Plancherel) *If $f \in \mathcal{S}(\mathbb{R})$ then $\|\hat{f}\| = \|f\|$.*

Proof. If $f \in \mathcal{S}(\mathbb{R})$ define $f^\flat(x) = \overline{f(-x)}$. Then $\widehat{f^\flat}(\xi) = \overline{\hat{f}(\xi)}$. Now let $h = f * f^\flat$. Clearly, we have

$$\hat{h}(\xi) = |\hat{f}(\xi)|^2 \quad \text{and} \quad h(0) = \int_{-\infty}^{\infty} |f(x)|^2\,dx.$$

The theorem now follows from the inversion formula applied with $x = 0$, that is,

$$\int_{-\infty}^{\infty} \hat{h}(\xi)\, d\xi = h(0).$$

1.7 Extension to functions of moderate decrease

In the previous sections, we have limited our assertion of the Fourier inversion and Plancherel formulas to the case when the function involved belonged to the Schwartz space. It does not really involve further ideas to extend these results to functions of moderate decrease, once we make the additional assumption that the Fourier transform of the function under consideration is also of moderate decrease. Indeed, the key observation, which is easy to prove, is that the convolution $f * g$ of two functions f and g of moderate decrease is again a function of moderate decrease (Exercise 7); also $\widehat{f * g} = \hat{f}\hat{g}$. Moreover, the multiplication formula continues to hold, and we deduce the Fourier inversion and Plancherel formulas when f and $\hat{f}$ are both of moderate decrease.

This generalization, although modest in scope, is nevertheless useful in some circumstances.

1.8 The Weierstrass approximation theorem

We now digress briefly by further exploiting our good kernels to prove the Weierstrass approximation theorem. This result was already alluded to in Chapter 2.

Theorem 1.13 *Let f be a continuous function on the closed and bounded interval $[a, b] \subset \mathbb{R}$. Then, for any $\epsilon > 0$, there exists a polynomial P such that*

$$\sup_{x \in [a,b]} |f(x) - P(x)| < \epsilon.$$

In other words, f can be uniformly approximated by polynomials.

Proof. Let $[-M, M]$ denote any interval that contains $[a, b]$ in its interior, and let g be a continuous function on $\mathbb{R}$ that equals 0 outside $[-M, M]$ and equals f in $[a, b]$. For example, extend f as follows: from b to M define g by a straight line segment going from $f(b)$ to 0, and from a to $-M$ by a straight line segment from $f(a)$ also to 0. Let B be a bound for g, that is, $|g(x)| \leq B$ for all x. Then, since $\{K_\delta\}$ is a family of good kernels, and g is continuous with compact support, we may argue as in the proof of Corollary 1.7 to see that $g * K_\delta$ converges uniformly to g as δ tends to 0. In fact, we choose δ_0 so that

$$|g(x) - (g * K_{\delta_0})(x)| < \epsilon/2 \quad \text{for all } x \in \mathbb{R}.$$

Now, we recall that e^x is given by the power series expansion $e^x = \sum_{n=0}^{\infty} x^n/n!$ which converges uniformly in every compact interval of $\mathbb{R}$. Therefore, there exists an integer N so that

$$|K_{\delta_0}(x) - R(x)| \leq \frac{\epsilon}{4MB} \quad \text{for all } x \in [-2M, 2M]$$

where $R(x) = \delta_0^{-1/2} \sum_{n=0}^{N} \frac{(-\pi x^2/\delta_0)^n}{n!}$. Then, recalling that g vanishes outside the interval $[-M, M]$, we have that for all $x \in [-M, M]$

$$\begin{aligned}
|(g * K_{\delta_0})(x) - (g * R)(x)| &= \left| \int_{-M}^{M} g(t) \left[K_{\delta_0}(x-t) - R(x-t)\right] dt \right| \\
&\leq \int_{-M}^{M} |g(t)|\, |K_{\delta_0}(x-t) - R(x-t)|\, dt \\
&\leq 2MB \sup_{z \in [-2M, 2M]} |K_{\delta_0}(z) - R(z)| \\
&< \epsilon/2.
\end{aligned}$$

Therefore, the triangle inequality implies that $|g(x) - (g * R)(x)| < \epsilon$ whenever $x \in [-M, M]$, hence $|f(x) - (g * R)(x)| < \epsilon$ when $x \in [a, b]$.

Finally, note that $g * R$ is a polynomial in the x variable. Indeed, by definition we have $(g * R)(x) = \int_{-M}^{M} g(t)R(x-t)\, dt$, and $R(x-t)$ is a polynomial in x since it can be expressed, after several expansions, as $R(x-t) = \sum_n a_n(t)x^n$ where the sum is finite. This concludes the proof of the theorem.

2 Applications to some partial differential equations

We mentioned earlier that a crucial property of the Fourier transform is that it interchanges differentiation and multiplication by polynomials. We now use this crucial fact together with the Fourier inversion theorem to solve some specific partial differential equations.

2.1 The time-dependent heat equation on the real line

In Chapter 4 we considered the heat equation on the circle. Here we study the analogous problem on the real line.

Consider an infinite rod, which we model by the real line, and suppose that we are given an initial temperature distribution $f(x)$ on the rod at time $t = 0$. We wish now to determine the temperature $u(x,t)$ at a point x at time $t > 0$. Considerations similar to the ones given in Chapter 1 show that when u is appropriately normalized, it solves the following partial differential equation:

$$\frac{\partial u}{\partial t} = \frac{\partial^2 u}{\partial x^2}, \tag{8}$$

called the **heat equation.** The initial condition we impose is $u(x, 0) = f(x)$.

Just as in the case of the circle, the solution is given in terms of a convolution. Indeed, define the **heat kernel** of the line by

$$\mathcal{H}_t(x) = K_\delta(x), \quad \text{with } \delta = 4\pi t,$$

so that

$$\mathcal{H}_t(x) = \frac{1}{(4\pi t)^{1/2}} e^{-x^2/4t} \quad \text{and} \quad \hat{\mathcal{H}}_t(\xi) = e^{-4\pi^2 t\xi^2}.$$

Taking the Fourier transform of equation (8) in the x variable (formally) leads to

$$\frac{\partial \hat{u}}{\partial t}(\xi, t) = -4\pi^2\xi^2\hat{u}(\xi, t).$$

Fixing ξ, this is an ordinary differential equation in the variable t (with unknown $\hat{u}(\xi, \cdot)$), so there exists a constant $A(\xi)$ so that

$$\hat{u}(\xi, t) = A(\xi)e^{-4\pi^2\xi^2 t}.$$

We may also take the Fourier transform of the initial condition and obtain $\hat{u}(\xi, 0) = \hat{f}(\xi)$, hence $A(\xi) = \hat{f}(\xi)$. This leads to the following theorem.

Theorem 2.1 *Given $f \in \mathcal{S}(\mathbb{R})$, let*

$$u(x,t) = (f * \mathcal{H}_t)(x) \quad \textit{for } t > 0$$

where $\mathcal{H}_t$ is the heat kernel. Then:

(i) *The function u is C^2 when $x \in \mathbb{R}$ and $t > 0$, and u solves the heat equation.*

(ii) *$u(x,t) \to f(x)$ uniformly in x as $t \to 0$. Hence if we set $u(x,0) = f(x)$, then u is continuous on the closure of the upper half-plane $\overline{\mathbb{R}^2_+} = \{(x,t) : x \in \mathbb{R},\ t \geq 0\}$.*

(iii) *$\int_{-\infty}^{\infty} |u(x,t) - f(x)|^2\, dx \to 0$ as $t \to 0$.*

Proof. Because $u = f * \mathcal{H}_t$, taking the Fourier transform in the x-variable gives $\hat{u} = \hat{f}\hat{\mathcal{H}}_t$, and so $\hat{u}(\xi,t) = \hat{f}(\xi)e^{-4\pi^2\xi^2 t}$. The Fourier inversion formula gives

$$u(x,t) = \int_{-\infty}^{\infty} \hat{f}(\xi)e^{-4\pi^2 t\xi^2} e^{2\pi i \xi x}\, d\xi.$$

By differentiating under the integral sign, one verifies (i). In fact, one observes that u is indefinitely differentiable. Note that (ii) is an immediate consequence of Corollary 1.7. Finally, by Plancherel's formula, we have

$$\begin{aligned}\int_{-\infty}^{\infty} |u(x,t) - f(x)|^2\, dx &= \int_{-\infty}^{\infty} |\hat{u}(\xi,t) - \hat{f}(\xi)|^2\, d\xi \\ &= \int_{-\infty}^{\infty} |\hat{f}(\xi)|^2\, |e^{-4\pi^2 t\xi^2} - 1|\, d\xi.\end{aligned}$$

To see that this last integral goes to 0 as $t \to 0$, we argue as follows: since $|e^{-4\pi^2 t\xi^2} - 1| \leq 2$ and $f \in \mathcal{S}(\mathbb{R})$, we can find N so that

$$\int_{|\xi| \geq N} |\hat{f}(\xi)|^2 |e^{-4\pi^2 t\xi^2} - 1|\, d\xi < \epsilon,$$

and for all small t we have $\sup_{|\xi| \leq N} |\hat{f}(\xi)|^2 |e^{-4\pi^2 t\xi^2} - 1| < \epsilon/2N$ since $\hat{f}$ is bounded. Thus

$$\int_{|\xi| \leq N} |\hat{f}(\xi)|^2\, |e^{-4\pi^2 t\xi^2} - 1|\, d\xi < \epsilon \quad \text{for all small } t.$$

This completes the proof of the theorem.

The above theorem guarantees the existence of a solution to the heat equation with initial data f. This solution is also unique, if uniqueness is formulated appropriately. In this regard, we note that $u = f * \mathcal{H}_t$, $f \in \mathcal{S}(\mathbb{R})$, satisfies the following additional property.

Corollary 2.2 *$u(\cdot, t)$ belongs to $\mathcal{S}(\mathbb{R})$ uniformly in t, in the sense that for any $T > 0$*

$$\sup_{\substack{x \in \mathbb{R} \\ 0 < t < T}} |x|^k \left| \frac{\partial^\ell}{\partial x^\ell} u(x,t) \right| < \infty \quad \textit{for each } k, \ell \geq 0. \tag{9}$$

Proof. This result is a consequence of the following estimate:

$$\begin{aligned} |u(x,t)| &\leq \int_{|y|\leq|x|/2} |f(x-y)|\mathcal{H}_t(y)\,dy + \int_{|y|\geq|x|/2} |f(x-y)|\mathcal{H}_t(y)\,dy \\ &\leq \frac{C_N}{(1+|x|)^N} + \frac{C}{\sqrt{t}}e^{-cx^2/t}. \end{aligned}$$

Indeed, since f is rapidly decreasing, we have $|f(x-y)| \leq C_N/(1+|x|)^N$ when $|y| \leq |x|/2$. Also, if $|y| \geq |x|/2$ then $\mathcal{H}_t(y) \leq Ct^{-1/2}e^{-cx^2/t}$, and we obtain the above inequality. Consequently, we see that $u(x,t)$ is rapidly decreasing uniformly for $0 < t < T$.

The same argument can be applied to the derivatives of u in the x variable since we may differentiate under the integral sign and apply the above estimate with f replaced by f', and so on.

This leads to the following uniqueness theorem.

Theorem 2.3 *Suppose $u(x,t)$ satisfies the following conditions:*

(i) *u is continuous on the closure of the upper half-plane.*

(ii) *u satisfies the heat equation for $t > 0$.*

(iii) *u satisfies the boundary condition $u(x,0) = 0$.*

(iv) *$u(\cdot,t) \in \mathcal{S}(\mathbb{R})$ uniformly in t, as in (9).*

Then, we conclude that $u = 0$.

Below we use the abbreviations $\partial_x^\ell u$ and $\partial_t u$ to denote $\partial^\ell u/\partial x^\ell$ and $\partial u/\partial t$, respectively.

Proof. We define the energy at time t of the solution $u(x,t)$ by

$$E(t) = \int_{\mathbb{R}} |u(x,t)|^2\,dx.$$

Clearly $E(t) \geq 0$. Since $E(0) = 0$ it suffices to show that E is a decreasing function, and this is achieved by proving that $dE/dt \leq 0$. The assumptions on u allow us to differentiate $E(t)$ under the integral sign

$$\frac{dE}{dt} = \int_{\mathbb{R}} [\partial_t u(x,t)\overline{u}(x,t) + u(x,t)\partial_t\overline{u}(x,t)]\,dx.$$

But u satisfies the heat equation, therefore $\partial_t u = \partial_x^2 u$ and $\partial_t \overline{u} = \partial_x^2\overline{u}$, so that after an integration by parts, where we use the fact that u and its

x derivatives decrease rapidly as $|x| \to \infty$, we find

$$\begin{aligned}
\frac{dE}{dt} &= \int_{\mathbb{R}} \left[\partial_x^2 u(x,t)\overline{u}(x,t) + u(x,t)\partial_x^2 \overline{u}(x,t)\right] dx \\
&= -\int_{\mathbb{R}} \left[\partial_x u(x,t)\partial_x \overline{u}(x,t) + \partial_x u(x,t)\partial_x \overline{u}(x,t)\right] dx \\
&= -2\int_{\mathbb{R}} |\partial_x u(x,t)|^2 \, dx \\
&\leq 0,
\end{aligned}$$

as claimed. Thus $E(t) = 0$ for all t, hence $u = 0$.

Another uniqueness theorem for the heat equation, with a less restrictive assumption than (9), can be found in Problem 6. Examples when uniqueness fails are given in Exercise 12 and Problem 4.

2.2 The steady-state heat equation in the upper half-plane

The equation we are now concerned with is

$$\triangle u = \frac{\partial^2 u}{\partial x^2} + \frac{\partial^2 u}{\partial y^2} = 0 \tag{10}$$

in the upper half-plane $\mathbb{R}^2_+ = \{(x,y) : x \in \mathbb{R},\ y > 0\}$. The boundary condition we require is $u(x,0) = f(x)$. The operator $\triangle$ is the Laplacian and the above partial differential equation describes the steady-state heat distribution in $\mathbb{R}^2_+$ subject to $u = f$ on the boundary. The kernel that solves this problem is called the **Poisson kernel** for the upper half-plane, and is given by

$$\mathcal{P}_y(x) = \frac{1}{\pi}\frac{y}{x^2 + y^2} \qquad \text{where } x \in \mathbb{R} \text{ and } y > 0.$$

This is the analogue of the Poisson kernel for the disc discussed in Section 5.4 of Chapter 2.

Note that for each fixed y the kernel $\mathcal{P}_y$ is only of moderate decrease as a function of x, so we will use the theory of the Fourier transform appropriate for these types of functions (see Section 1.7).

We proceed as in the case of the time-dependent heat equation, by taking the Fourier transform of equation (10) (formally) in the x variable, thereby obtaining

$$-4\pi^2\xi^2 \hat{u}(\xi, y) + \frac{\partial^2 \hat{u}}{\partial y^2}(\xi, y) = 0$$

with the boundary condition $\hat{u}(\xi, 0) = \hat{f}(\xi)$. The general solution of this ordinary differential equation in y (with ξ fixed) takes the form

$$\hat{u}(\xi, y) = A(\xi)e^{-2\pi|\xi|y} + B(\xi)e^{2\pi|\xi|y}.$$

If we disregard the second term because of its rapid exponential increase we find, after setting $y = 0$, that

$$\hat{u}(\xi, y) = \hat{f}(\xi)e^{-2\pi|\xi|y}.$$

Therefore u is given in terms of the convolution of f with a kernel whose Fourier transform is $e^{-2\pi|\xi|y}$. This is precisely the Poisson kernel given above, as we prove next.

Lemma 2.4 *The following two identities hold:*

$$\int_{-\infty}^{\infty} e^{-2\pi|\xi|y} e^{2\pi i\xi x}\, d\xi = \mathcal{P}_y(x),$$

$$\int_{-\infty}^{\infty} \mathcal{P}_y(x) e^{-2\pi i x\xi}\, dx = e^{-2\pi|\xi|y}.$$

Proof. The first formula is fairly straightforward since we can split the integral from $-\infty$ to 0 and 0 to ∞. Then, since $y > 0$ we have

$$\int_0^{\infty} e^{-2\pi\xi y} e^{2\pi i\xi x}\, d\xi = \int_0^{\infty} e^{2\pi i(x+iy)\xi}\, d\xi = \left[\frac{e^{2\pi i(x+iy)\xi}}{2\pi i(x+iy)}\right]_0^{\infty} = -\frac{1}{2\pi i(x+iy)},$$

and similarly,

$$\int_{-\infty}^{0} e^{2\pi\xi y} e^{2\pi i\xi x}\, d\xi = \frac{1}{2\pi i(x-iy)}.$$

Therefore

$$\int_{-\infty}^{\infty} e^{-2\pi|\xi|y} e^{2\pi i\xi x}\, d\xi = \frac{1}{2\pi i(x-iy)} - \frac{1}{2\pi i(x+iy)} = \frac{y}{\pi(x^2+y^2)}.$$

The second formula is now a consequence of the Fourier inversion theorem applied in the case when f and $\hat{f}$ are of moderate decrease.

Lemma 2.5 *The Poisson kernel is a good kernel on* $\mathbb{R}$ *as* $y \to 0$.

Proof. Setting $\xi = 0$ in the second formula of the lemma shows that $\int_{-\infty}^{\infty} \mathcal{P}_y(x)\, dx = 1$, and clearly $\mathcal{P}_y(x) \geq 0$, so it remains to check the last property of good kernels. Given a fixed $\delta > 0$, we may change variables $u = x/y$ so that

$$\int_{\delta}^{\infty} \frac{y}{x^2 + y^2}\, dx = \int_{\delta/y}^{\infty} \frac{du}{1+u^2} = [\arctan u]_{\delta/y}^{\infty} = \pi/2 - \arctan(\delta/y),$$

and this quantity goes to 0 as $y \to 0$. Since $\mathcal{P}_y(x)$ is an even function, the proof is complete.

The following theorem establishes the existence of a solution to our problem.

Theorem 2.6 *Given $f \in \mathcal{S}(\mathbb{R})$, let $u(x, y) = (f * \mathcal{P}_y)(x)$. Then:*

(i) *$u(x, y)$ is C^2 in $\mathbb{R}^2_+$ and $\Delta u = 0$.*

(ii) *$u(x, y) \to f(x)$ uniformly as $y \to 0$.*

(iii) *$\int_{-\infty}^{\infty} |u(x, y) - f(x)|^2\, dx \to 0$ as $y \to 0$.*

(iv) *If $u(x, 0) = f(x)$, then u is continuous on the closure $\overline{\mathbb{R}^2_+}$ of the upper half-plane, and vanishes at infinity in the sense that*

$$u(x, y) \to 0 \qquad \text{as } |x| + y \to \infty.$$

Proof. The proofs of parts (i), (ii), and (iii) are similar to the case of the heat equation, and so are left to the reader. Part (iv) is a consequence of two easy estimates whenever f is of moderate decrease. First, we have

$$|(f * \mathcal{P}_y)(x)| \leq C\left(\frac{1}{(1+x^2)} + \frac{y}{x^2+y^2}\right)$$

which is proved (as in the case of the heat equation) by splitting the integral $\int_{-\infty}^{\infty} f(x-t)\mathcal{P}_y(t)\, dt$ into the part where $|t| \leq |x|/2$ and the part where $|t| \geq |x|/2$. Also, we have $|(f * \mathcal{P}_y)(x)| \leq C/y$, since $\sup_x \mathcal{P}_y(x) \leq c/y$.

Using the first estimate when $|x| \geq |y|$ and the second when $|x| \leq |y|$ gives the desired decrease at infinity.

We next show that the solution is essentially unique.

Theorem 2.7 *Suppose u is continuous on the closure of the upper half-plane $\overline{\mathbb{R}^2_+}$, satisfies $\Delta u = 0$ for $(x, y) \in \mathbb{R}^2_+$, $u(x, 0) = 0$, and $u(x, y)$ vanishes at infinity. Then $u = 0$.*

A simple example shows that a condition concerning the decay of u at infinity is needed: take $u(x, y) = y$. Clearly u satisfies the steady-state heat equation and vanishes on the real line, yet u is not identically zero.

The proof of the theorem relies on a basic fact about harmonic functions, which are functions satisfying $\triangle u = 0$. The fact is that the value of a harmonic function at a point equals its average value around any circle centered at that point.

Lemma 2.8 (Mean-value property) *Suppose Ω is an open set in $\mathbb{R}^2$ and let u be a function of class C^2 with $\triangle u = 0$ in Ω. If the closure of the disc centered at (x, y) and of radius R is contained in Ω, then*

$$u(x, y) = \frac{1}{2\pi}\int_0^{2\pi} u(x + r\cos\theta, y + r\sin\theta)\, d\theta$$

for all $0 \leq r \leq R$.

Proof. Let $U(r, \theta) = u(x + r\cos\theta, y + r\sin\theta)$. Expressing the Laplacian in polar coordinates, the equation $\triangle u = 0$ then implies

$$0 = \frac{\partial^2 U}{\partial\theta^2} + r\frac{\partial}{\partial r}\left(r\frac{\partial U}{\partial r}\right).$$

If we define $F(r) = \frac{1}{2\pi}\int_0^{2\pi} U(r, \theta)\, d\theta$, the above gives

$$r\frac{\partial}{\partial r}\left(r\frac{\partial F}{\partial r}\right) = \frac{1}{2\pi}\int_0^{2\pi} -\frac{\partial^2 U}{\partial\theta^2}(r, \theta)\, d\theta.$$

The integral of $\partial^2 U/\partial\theta^2$ over the circle vanishes since $\partial U/\partial\theta$ is periodic, hence $r\frac{\partial}{\partial r}\left(r\frac{\partial F}{\partial r}\right) = 0$, and consequently $r\partial F/\partial r$ must be constant. Evaluating this expression at $r = 0$ we find that $\partial F/\partial r = 0$. Thus F is constant, but since $F(0) = u(x, y)$, we finally find that $F(r) = u(x, y)$ for all $0 \leq r \leq R$, which is the mean-value property.

Finally, note that the argument above is implicit in the proof of Theorem 5.7, Chapter 2.

To prove Theorem 2.7 we argue by contradiction. Considering separately the real and imaginary parts of u, we may suppose that u itself is real-valued, and is somewhere strictly positive, say $u(x_0, y_0) > 0$ for some $x_0 \in \mathbb{R}$ and $y_0 > 0$. We shall see that this leads to a contradiction. First, since u vanishes at infinity, we can find a large semi-disc of radius R, $D_R^+ = \{(x, y) : x^2 + y^2 \leq R,\ \ y \geq 0\}$ outside of which $u(x, y) \leq \frac{1}{2}u(x_0, y_0)$. Next, since u is continuous in D_R^+, it attains its maximum M there, so there exists a point $(x_1, y_1) \in D_R^+$ with $u(x_1, y_1) = M$, while

$u(x, y) \leq M$ in the semi-disc; also, since $u(x, y) \leq \frac{1}{2}u(x_0, y_0) \leq M/2$ outside of the semi-disc, we have $u(x, y) \leq M$ throughout the entire upper half-plane. Now the mean-value property for harmonic functions implies

$$u(x_1, y_1) = \frac{1}{2\pi} \int_0^{2\pi} u(x_1 + \rho \cos\theta, y_1 + \rho \sin\theta)\, d\theta$$

whenever the circle of integration lies in the upper half-plane. In particular, this equation holds if $0 < \rho < y_1$. Since $u(x_1, y_1)$ equals the maximum value M, and $u(x_1 + \rho\cos\theta, y_1 + \rho\sin\theta) \leq M$, it follows by continuity that $u(x_1 + \rho\cos\theta, y_1 + \rho\sin\theta) = M$ on the whole circle. For otherwise $u(x, y) \leq M - \epsilon$, on an arc of length $\delta > 0$ on the circle, and this would give

$$\frac{1}{2\pi} \int_0^{2\pi} u(x_1 + \rho\cos\theta, y_1 + \rho\sin\theta)\, d\theta \leq M - \frac{\epsilon\delta}{2\pi} < M,$$

contradicting the fact that $u(x_1, y_1) = M$. Now letting $\rho \to y_1$, and using the continuity of u again, we see that this implies $u(x_1, 0) = M > 0$, which contradicts the fact that $u(x, 0) = 0$ for all x.

3 The Poisson summation formula

The definition of the Fourier transform was motivated by the desire for a continuous version of Fourier series, applicable to functions defined on the real line. We now show that there exists a further remarkable connection between the analysis of functions on the circle and related functions on $\mathbb{R}$.

Given a function $f \in \mathcal{S}(\mathbb{R})$ on the real line, we can construct a new function on the circle by the recipe

$$F_1(x) = \sum_{n=-\infty}^{\infty} f(x + n).$$

Since f is rapidly decreasing, the series converges absolutely and uniformly on every compact subset of $\mathbb{R}$, so F_1 is continuous. Note that $F_1(x + 1) = F_1(x)$ because passage from n to $n + 1$ in the above sum merely shifts the terms on the series defining $F_1(x)$. Hence F_1 is periodic with period 1. The function F_1 is called the **periodization** of f.

There is another way to arrive at a "periodic version" of f, this time by Fourier analysis. Start with the identity

$$f(x) = \int_{-\infty}^{\infty} \hat{f}(\xi) e^{2\pi i \xi x}\, d\xi,$$

and consider its discrete analogue, where the integral is replaced by a sum

$$F_2(x) = \sum_{n=-\infty}^{\infty} \hat{f}(n)e^{2\pi inx}.$$

Once again, the sum converges absolutely and uniformly since $\hat{f}$ belongs to the Schwartz space, hence F_2 is continuous. Moreover, F_2 is also periodic of period 1 since this is the case for each one of the exponentials $e^{2\pi inx}$.

The fundamental fact is that these two approaches, which produce F_1 and F_2, actually lead to the same function.

Theorem 3.1 (Poisson summation formula) *If $f \in \mathcal{S}(\mathbb{R})$, then*

$$\sum_{n=-\infty}^{\infty} f(x+n) = \sum_{n=-\infty}^{\infty} \hat{f}(n)e^{2\pi inx}.$$

In particular, setting $x = 0$ we have

$$\sum_{n=-\infty}^{\infty} f(n) = \sum_{n=-\infty}^{\infty} \hat{f}(n).$$

In other words, the Fourier coefficients of the periodization of f are given precisely by the values of the Fourier transform of f on the integers.

Proof. To check the first formula it suffices, by Theorem 2.1 in Chapter 2, to show that both sides (which are continuous) have the same Fourier coefficients (viewed as functions on the circle). Clearly, the mth Fourier coefficient of the right-hand side is $\hat{f}(m)$. For the left-hand side we have

$$\begin{aligned}
\int_0^1 \left(\sum_{n=-\infty}^{\infty} f(x+n) \right) e^{-2\pi imx}\, dx &= \sum_{n=-\infty}^{\infty} \int_0^1 f(x+n)e^{-2\pi imx}\, dx \\
&= \sum_{n=-\infty}^{\infty} \int_n^{n+1} f(y)e^{-2\pi imy}\, dy \\
&= \int_{-\infty}^{\infty} f(y)e^{-2\pi imy}\, dy \\
&= \hat{f}(m),
\end{aligned}$$

where the interchange of the sum and integral is permissible since f is rapidly decreasing. This completes the proof of the theorem.

We observe that the theorem extends to the case when we merely assume that both f and $\hat{f}$ are of moderate decrease; the proof is in fact unchanged.

It turns out that the operation of periodization is important in a number of questions, even when the Poisson summation formula does not apply. We give an example by considering the elementary function $f(x) = 1/x$, $x \neq 0$. The result is that $\sum_{n=-\infty}^{\infty} 1/(x+n)$, when summed symmetrically, gives the partial fraction decomposition of the cotangent function. In fact this sum equals $\pi \cot \pi x$, when x is not an integer. Similarly with $f(x) = 1/x^2$, we get $\sum_{n=-\infty}^{\infty} 1/(x+n)^2 = \pi^2/(\sin \pi x)^2$, whenever $x \notin \mathbb{Z}$ (see Exercise 15).

3.1 Theta and zeta functions

We define the **theta function** $\vartheta(s)$ for $s > 0$ by

$$\vartheta(s) = \sum_{n=-\infty}^{\infty} e^{-\pi n^2 s}.$$

The condition on s ensures the absolute convergence of the series. A crucial fact about this special function is that it satisfies the following functional equation.

Theorem 3.2 $s^{-1/2}\vartheta(1/s) = \vartheta(s)$ *whenever* $s > 0$.

The proof of this identity consists of a simple application of the Poisson summation formula to the pair

$$f(x) = e^{-\pi s x^2} \quad \text{and} \quad \hat{f}(\xi) = s^{-1/2} e^{-\pi \xi^2/s}.$$

The theta function $\vartheta(s)$ also extends to complex values of s when $\mathrm{Re}(s) > 0$, and the functional equation is still valid then. The theta function is intimately connected with an important function in number theory, the **zeta function** $\zeta(s)$ defined for $\mathrm{Re}(s) > 1$ by

$$\zeta(s) = \sum_{n=1}^{\infty} \frac{1}{n^s}.$$

Later we will see that this function carries essential information about the prime numbers (see Chapter 8).

It also turns out that ζ, ϑ, and another important function Γ are related by the following identity:

$$\pi^{-s/2}\Gamma(s/2)\zeta(s) = \frac{1}{2}\int_0^{\infty} t^{s/2-1}(\vartheta(t) - 1)\, dt,$$

which is valid for $s > 1$ (Exercises 17 and 18).

Returning to the function ϑ, define the generalization $\Theta(z|\tau)$ given by

$$\Theta(z|\tau) = \sum_{n=-\infty}^{\infty} e^{i\pi n^2 \tau} e^{2\pi i n z}$$

whenever $\text{Im}(\tau) > 0$ and $z \in \mathbb{C}$. Taking $z = 0$ and $\tau = is$ we get $\Theta(z|\tau) = \vartheta(s)$.

3.2 Heat kernels

Another application related to the Poisson summation formula and the theta function is the time-dependent heat equation on the circle. A solution to the equation

$$\frac{\partial u}{\partial t} = \frac{\partial^2 u}{\partial x^2}$$

subject to $u(x,0) = f(x)$, where f is periodic of period 1, was given in the previous chapter by

$$u(x,t) = (f * H_t)(x)$$

where $H_t(x)$ is the heat kernel on the circle, that is,

$$H_t(x) = \sum_{n=-\infty}^{\infty} e^{-4\pi^2 n^2 t} e^{2\pi i n x}.$$

Note in particular that with our definition of the generalized theta function in the previous section, we have $\Theta(x|4\pi i t) = H_t(x)$. Also, recall that the heat equation on $\mathbb{R}$ gave rise to the heat kernel

$$\mathcal{H}_t(x) = \frac{1}{(4\pi t)^{1/2}} e^{-x^2/4t}$$

where $\hat{\mathcal{H}}_t(\xi) = e^{-4\pi^2 \xi^2 t}$. The fundamental relation between these two objects is an immediate consequence of the Poisson summation formula:

Theorem 3.3 *The heat kernel on the circle is the periodization of the heat kernel on the real line:*

$$H_t(x) = \sum_{n=-\infty}^{\infty} \mathcal{H}_t(x+n).$$

Although the proof that $\mathcal{H}_t$ is a good kernel on $\mathbb{R}$ was fairly straightforward, we left open the harder problem that H_t is a good kernel on the circle. The above results allow us to resolve this matter.

Corollary 3.4 *The kernel $H_t(x)$ is a good kernel for $t \to 0$.*

Proof. We already observed that $\int_{|x|\leq 1/2} H_t(x)\,dx = 1$. Now note that $H_t \geq 0$, which is immediate from the above formula since $\mathcal{H}_t \geq 0$. Finally, we claim that when $|x| \leq 1/2$,

$$H_t(x) = \mathcal{H}_t(x) + \mathcal{E}_t(x),$$

where the error satisfies $|\mathcal{E}_t(x)| \leq c_1 e^{-c_2/t}$ with $c_1, c_2 > 0$ and $0 < t \leq 1$. To see this, note again that the formula in the theorem gives

$$H_t(x) = \mathcal{H}_t(x) + \sum_{|n|\geq 1} \mathcal{H}_t(x+n);$$

therefore, since $|x| \leq 1/2$,

$$\mathcal{E}_t(x) = \frac{1}{\sqrt{4\pi t}} \sum_{|n|\geq 1} e^{-(x+n)^2/4t} \leq Ct^{-1/2} \sum_{n\geq 1} e^{-cn^2/t}.$$

Note that $n^2/t \geq n^2$ and $n^2/t \geq 1/t$ whenever $0 < t \leq 1$, so $e^{-cn^2/t} \leq e^{-\frac{c}{2}n^2} e^{-\frac{c}{2}\frac{1}{t}}$. Hence

$$|\mathcal{E}_t(x)| \leq Ct^{-1/2} e^{-\frac{c}{2}\frac{1}{t}} \sum_{n\geq 1} e^{-\frac{c}{2}n^2} \leq c_1 e^{-c_2/t}.$$

The proof of the claim is complete, and as a result $\int_{|x|\leq 1/2} |\mathcal{E}_t(x)|\,dx \to 0$ as $t \to 0$. It is now clear that H_t satisfies

$$\int_{\eta<|x|\leq 1/2} |H_t(x)|\,dx \to 0 \quad \text{as } t \to 0,$$

because $\mathcal{H}_t$ does.

3.3 Poisson kernels

In a similar manner to the discussion above about the heat kernels, we state the relation between the Poisson kernels for the disc and the upper half-plane where

$$P_r(\theta) = \frac{1-r^2}{1-2r\cos\theta + r^2} \quad \text{and} \quad \mathcal{P}_y(x) = \frac{1}{\pi}\frac{y}{y^2+x^2}.$$

Theorem 3.5 *$P_r(2\pi x) = \sum_{n\in\mathbb{Z}} \mathcal{P}_y(x+n)$ where $r = e^{-2\pi y}$.*

This is again an immediate corollary of the Poisson summation formula applied to $f(x) = \mathcal{P}_y(x)$ and $\hat{f}(\xi) = e^{-2\pi|\xi|y}$. Of course, here we use the Poisson summation formula under the assumptions that f and $\hat{f}$ are of moderate decrease.

4 The Heisenberg uncertainty principle

The mathematical thrust of the principle can be formulated in terms of a relation between a function and its Fourier transform. The basic underlying law, formulated in its vaguest and most general form, states that a function and its Fourier transform cannot both be essentially localized. Somewhat more precisely, if the "preponderance" of the mass of a function is concentrated in an interval of length L, then the preponderance of the mass of its Fourier transform cannot lie in an interval of length essentially smaller than L^{-1}. The exact statement is as follows.

Theorem 4.1 *Suppose ψ is a function in $\mathcal{S}(\mathbb{R})$ which satisfies the normalizing condition $\int_{-\infty}^{\infty} |\psi(x)|^2\,dx = 1$. Then*

$$\left(\int_{-\infty}^{\infty} x^2|\psi(x)|^2\,dx\right)\left(\int_{-\infty}^{\infty} \xi^2|\hat{\psi}(\xi)|^2\,d\xi\right) \geq \frac{1}{16\pi^2},$$

and equality holds if and only if $\psi(x) = Ae^{-Bx^2}$ where $B > 0$ and $|A|^2 = \sqrt{2B/\pi}$.

In fact, we have

$$\left(\int_{-\infty}^{\infty} (x-x_0)^2|\psi(x)|^2\,dx\right)\left(\int_{-\infty}^{\infty} (\xi-\xi_0)^2|\hat{\psi}(\xi)|^2\,d\xi\right) \geq \frac{1}{16\pi^2}$$

for every $x_0, \xi_0 \in \mathbb{R}$.

Proof. The second inequality actually follows from the first by replacing $\psi(x)$ by $e^{-2\pi i x\xi_0}\psi(x+x_0)$ and changing variables. To prove the first inequality, we argue as follows. Beginning with our normalizing assumption $\int |\psi|^2 = 1$, and recalling that ψ and ψ' are rapidly decreasing, an integration by parts gives

$$\begin{aligned}
1 &= \int_{-\infty}^{\infty} |\psi(x)|^2\,dx \\
&= -\int_{-\infty}^{\infty} x\frac{d}{dx}|\psi(x)|^2\,dx \\
&= -\int_{-\infty}^{\infty} \left(x\psi'(x)\overline{\psi(x)} + x\overline{\psi'(x)}\psi(x)\right)\,dx.
\end{aligned}$$

The last identity follows because $|\psi|^2 = \psi\overline{\psi}$. Therefore

$$\begin{aligned} 1 &\leq 2 \int_{-\infty}^{\infty} |x|\, |\psi(x)|\, |\psi'(x)|\, dx \\ &\leq 2 \left(\int_{-\infty}^{\infty} x^2 |\psi(x)|^2\, dx \right)^{1/2} \left(\int_{-\infty}^{\infty} |\psi'(x)|^2\, dx \right)^{1/2}, \end{aligned}$$

where we have used the Cauchy-Schwarz inequality. The identity

$$\int_{-\infty}^{\infty} |\psi'(x)|^2\, dx = 4\pi^2 \int_{-\infty}^{\infty} \xi^2 |\hat{\psi}(\xi)|^2\, d\xi,$$

which holds because of the properties of the Fourier transform and the Plancherel formula, concludes the proof of the inequality in the theorem.

If equality holds, then we must also have equality where we applied the Cauchy-Schwarz inequality, and as a result we find that $\psi'(x) = \beta x \psi(x)$ for some constant β. The solutions to this equation are $\psi(x) = Ae^{\beta x^2/2}$, where A is constant. Since we want ψ to be a Schwartz function, we must take $\beta = -2B < 0$, and since we impose the condition $\int_{-\infty}^{\infty} |\psi(x)|^2\, dx = 1$ we find that $|A|^2 = \sqrt{2B/\pi}$, as was to be shown.

The precise assertion contained in Theorem 4.1 first came to light in the study of quantum mechanics. It arose when one considered the extent to which one could simultaneously locate the position and momentum of a particle. Assuming we are dealing with (say) an electron that travels along the real line, then according to the laws of physics, matters are governed by a "state function" ψ, which we can assume to be in $\mathcal{S}(\mathbb{R})$, and which is normalized according to the requirement that

$$\int_{-\infty}^{\infty} |\psi(x)|^2\, dx = 1. \tag{11}$$

The position of the particle is then determined not as a definite point x; instead its probable location is given by the rules of quantum mechanics as follows:

- *The probability that the particle is located in the interval* (a, b) *is* $\int_a^b |\psi(x)|^2\, dx$.

According to this law we can calculate the probable location of the particle with the aid of ψ: in fact, there may be only a small probability that the particle is located in a given interval (a', b'), but nevertheless it is somewhere on the real line since $\int_{-\infty}^{\infty} |\psi(x)|^2\, dx = 1$.

In addition to the **probability density** $|\psi(x)|^2 dx$, there is the **expectation** of where the particle might be. This expectation is the best guess of the position of the particle, given its probability distribution determined by $|\psi(x)|^2 dx$, and is the quantity defined by

$$\overline{x} = \int_{-\infty}^{\infty} x|\psi(x)|^2\, dx. \tag{12}$$

Why is this our best guess? Consider the simpler (idealized) situation where we are given that the particle can be found at only finitely many different points, $x_1, x_2, \ldots, x_N$ on the real axis, with p_i the probability that the particle is at x_i, and $p_1 + p_2 + \cdots + p_N = 1$. Then, if we knew nothing else, and were forced to make one choice as to the position of the particle, we would naturally take $\overline{x} = \sum_{i=1}^N x_i p_i$, which is the appropriate weighted average of the possible positions. The quantity (12) is clearly the general (integral) version of this.

We next come to the notion of **variance**, which in our terminology is the **uncertainty** attached to our expectation. Having determined that the expected position of the particle is $\overline{x}$ (given by (12)), the resulting uncertainty is the quantity

$$\int_{-\infty}^{\infty} (x - \overline{x})^2|\psi(x)|^2\, dx. \tag{13}$$

Notice that if ψ is highly concentrated near $\overline{x}$, it means that there is a high probability that x is near $\overline{x}$, and so (13) is small, because most of the contribution to the integral takes place for values of x near $\overline{x}$. Here we have a small uncertainty. On the other hand, if $\psi(x)$ is rather flat (that is, the probability distribution $|\psi(x)|^2 dx$ is not very concentrated), then the integral (13) is rather big, because large values of $(x - \overline{x})^2$ will come into play, and as a result the uncertainty is relatively large.

It is also worthwhile to observe that the expectation $\overline{x}$ is that choice for which the uncertainty $\int_{-\infty}^{\infty}(x - \overline{x})^2|\psi(x)|^2\, dx$ is the smallest. Indeed, if we try to minimize this quantity by equating to 0 its derivative with respect to $\overline{x}$, we find that $2\int_{-\infty}^{\infty}(x - \overline{x})|\psi(x)|^2\, dx = 0$, which gives (12).

So far, we have discussed the "expectation" and "uncertainty" related to the position of the particle. Of equal relevance are the corresponding notions regarding its momentum. The corresponding rule of quantum mechanics is:

- *The probability that the momentum ξ of the particle belongs to the interval (a, b) is $\int_a^b |\hat{\psi}(\xi)|^2\, d\xi$ where $\hat{\psi}$ is the Fourier transform of ψ.*

Combining these two laws with Theorem 4.1 gives $1/16\pi^2$ as the lower bound for the product of the uncertainty of the position and the uncertainty of the momentum of a particle. So the more certain we are about the location of the particle, the less certain we can be about its momentum, and vice versa. However, we have simplified the statement of the two laws by rescaling to change the units of measurement. Actually, there enters a fundamental but small physical number $\hbar$ called Planck's constant. When properly taken into account, the physical conclusion is

$$\text{(uncertainty of position)}\times\text{(uncertainty of momentum)} \geq \hbar/16\pi^2.$$

5 Exercises

1. Corollary 2.3 in Chapter 2 leads to the following simplified version of the Fourier inversion formula. Suppose f is a continuous function supported on an interval $[-M, M]$, whose Fourier transform $\hat{f}$ is of moderate decrease.

(a) Fix L with $L/2 > M$, and show that $f(x) = \sum a_n(L)e^{2\pi inx/L}$ where

$$a_n(L) = \frac{1}{L}\int_{-L/2}^{L/2} f(x)e^{-2\pi inx/L}\,dx = \frac{1}{L}\hat{f}(n/L).$$

Alternatively, we may write $f(x) = \delta\sum_{n=-\infty}^{\infty}\hat{f}(n\delta)e^{2\pi in\delta x}$ with $\delta = 1/L$.

(b) Prove that if F is continuous and of moderate decrease, then

$$\int_{-\infty}^{\infty} F(\xi)\,d\xi = \lim_{\substack{\delta\to 0\\ \delta>0}} \delta\sum_{n=-\infty}^{\infty} F(\delta n).$$

(c) Conclude that $f(x) = \displaystyle\int_{-\infty}^{\infty}\hat{f}(\xi)e^{2\pi ix\xi}\,d\xi$.

[Hint: For (a), note that the Fourier series of f on $[-L/2, L/2]$ converges absolutely. For (b), first approximate the integral by $\int_{-N}^{N} F$ and the sum by $\delta\sum_{|n|\leq N/\delta} F(n\delta)$. Then approximate the second integral by Riemann sums.]

2. Let f and g be the functions defined by

$$f(x) = \chi_{[-1,1]}(x) = \begin{cases} 1 & \text{if } |x|\leq 1,\\ 0 & \text{otherwise,}\end{cases} \quad\text{and}\quad g(x) = \begin{cases} 1-|x| & \text{if } |x|\leq 1,\\ 0 & \text{otherwise.}\end{cases}$$

Although f is not continuous, the integral defining its Fourier transform still makes sense. Show that

$$\hat{f}(\xi) = \frac{\sin 2\pi\xi}{\pi\xi} \quad\text{and}\quad \hat{g}(\xi) = \left(\frac{\sin\pi\xi}{\pi\xi}\right)^2,$$

with the understanding that $\hat{f}(0) = 2$ and $\hat{g}(0) = 1$.

3. The following exercise illustrates the principle that the decay of $\hat{f}$ is related to the continuity properties of f.

(a) Suppose that f is a function of moderate decrease on $\mathbb{R}$ whose Fourier transform $\hat{f}$ is continuous and satisfies

$$\hat{f}(\xi) = O\left(\frac{1}{|\xi|^{1+\alpha}}\right) \quad \text{as } |\xi| \to \infty$$

for some $0 < \alpha < 1$. Prove that f satisfies a Hölder condition of order α, that is, that

$$|f(x+h) - f(x)| \leq M|h|^\alpha \quad \text{for some } M > 0 \text{ and all } x, h \in \mathbb{R}.$$

(b) Let f be a continuous function on $\mathbb{R}$ which vanishes for $|x| \geq 1$, with $f(0) = 0$, and which is equal to $1/\log(1/|x|)$ for all x in a neighborhood of the origin. Prove that $\hat{f}$ is not of moderate decrease. In fact, there is no $\epsilon > 0$ so that $\hat{f}(\xi) = O(1/|\xi|^{1+\epsilon})$ as $|\xi| \to \infty$.

[Hint: For part (a), use the Fourier inversion formula to express $f(x+h) - f(x)$ as an integral involving $\hat{f}$, and estimate this integral separately for ξ in the two ranges $|\xi| \leq 1/|h|$ and $|\xi| \geq 1/|h|$.]

4. Bump functions. Examples of compactly supported functions in $\mathcal{S}(\mathbb{R})$ are very handy in many applications in analysis. Some examples are:

(a) Suppose $a < b$, and f is the function such that $f(x) = 0$ if $x \leq a$ or $x \geq b$ and

$$f(x) = e^{-1/(x-a)} e^{-1/(b-x)} \quad \text{if } a < x < b.$$

Show that f is indefinitely differentiable on $\mathbb{R}$.

(b) Prove that there exists an indefinitely differentiable function F on $\mathbb{R}$ such that $F(x) = 0$ if $x \leq a$, $F(x) = 1$ if $x \geq b$, and F is strictly increasing on $[a, b]$.

(c) Let $\delta > 0$ be so small that $a + \delta < b - \delta$. Show that there exists an indefinitely differentiable function g such that g is 0 if $x \leq a$ or $x \geq b$, g is 1 on $[a + \delta, b - \delta]$, and g is strictly monotonic on $[a, a + \delta]$ and $[b - \delta, b]$.

[Hint: For (b) consider $F(x) = c\int_{-\infty}^{x} f(t)\,dt$ where c is an appropriate constant.]

5. Suppose f is continuous and of moderate decrease.

(a) Prove that $\hat{f}$ is continuous and $\hat{f}(\xi) \to 0$ as $|\xi| \to \infty$.

(b) Show that if $\hat{f}(\xi) = 0$ for all ξ, then f is identically 0.

[Hint: For part (a), show that $\hat{f}(\xi) = \frac{1}{2}\int_{-\infty}^{\infty}[f(x) - f(x - 1/(2\xi))]e^{-2\pi i x\xi}\,dx$. For part (b), verify that the multiplication formula $\int f(x)\hat{g}(x)\,dx = \int \hat{f}(y)g(y)\,dy$ still holds whenever $g \in \mathcal{S}(\mathbb{R})$.]

6. The function $e^{-\pi x^2}$ is its own Fourier transform. Generate other functions that (up to a constant multiple) are their own Fourier transforms. What must the constant multiples be? To decide this, prove that $\mathcal{F}^4 = I$. Here $\mathcal{F}(f) = \hat{f}$ is the Fourier transform, $\mathcal{F}^4 = \mathcal{F} \circ \mathcal{F} \circ \mathcal{F} \circ \mathcal{F}$, and I is the identity operator $(If)(x) = f(x)$ (see also Problem 7).

7. Prove that the convolution of two functions of moderate decrease is a function of moderate decrease.

[Hint: Write

$$\int f(x-y)g(y)\,dy = \int_{|y|\leq|x|/2} + \int_{|y|\geq|x|/2}.$$

In the first integral $f(x-y) = O(1/(1+x^2))$ while in the second integral $g(y) = O(1/(1+x^2))$.]

8. Prove that f is continuous, of moderate decrease, and $\int_{-\infty}^{\infty} f(y)e^{-y^2}e^{2xy}dy = 0$ for all $x \in \mathbb{R}$, then $f = 0$.

[Hint: Consider $f * e^{-x^2}$.]

9. If f is of moderate decrease, then

$$\int_{-R}^{R}\left(1 - \frac{|\xi|}{R}\right)\hat{f}(\xi)e^{2\pi i x\xi}\,d\xi = (f * \mathcal{F}_R)(x), \tag{14}$$

where the Fejér kernel on the real line is defined by

$$\mathcal{F}_R(t) = \begin{cases} R\left(\dfrac{\sin \pi t R}{\pi t R}\right)^2 & \text{if } t \neq 0, \\ R & \text{if } t = 0. \end{cases}$$

Show that $\{\mathcal{F}_R\}$ is a family of good kernels as $R \to \infty$, and therefore (14) tends uniformly to $f(x)$ as $R \to \infty$. This is the analogue of Fejér's theorem for Fourier series in the context of the Fourier transform.

10. Below is an outline of a different proof of the Weierstrass approximation theorem.

Define the **Landau** kernels by

$$L_n(x) = \begin{cases} \dfrac{(1-x^2)^n}{c_n} & \text{if } -1 \leq x \leq 1, \\ 0 & \text{if } |x| \geq 1, \end{cases}$$

where c_n is chosen so that $\int_{-\infty}^{\infty} L_n(x)\,dx = 1$. Prove that $\{L_n\}_{n\geq 0}$ is a family of good kernels as $n \to \infty$. As a result, show that if f is a continuous function supported in $[-1/2, 1/2]$, then $(f * L_n)(x)$ is a sequence of polynomials on $[-1/2, 1/2]$ which converges uniformly to f.

[Hint: First show that $c_n \geq 2/(n+1)$.]

11. Suppose that u is the solution to the heat equation given by $u = f * \mathcal{H}_t$ where $f \in \mathcal{S}(\mathbb{R})$. If we also set $u(x,0) = f(x)$, prove that u is continuous on the closure of the upper half-plane, and vanishes at infinity, that is,

$$u(x,t) \to 0 \quad \text{as } |x| + t \to \infty.$$

[Hint: To prove that u vanishes at infinity, show that (i) $|u(x,t)| \leq C/\sqrt{t}$ and (ii) $|u(x,t)| \leq C/(1+|x|^2) + Ct^{-1/2}e^{-cx^2/t}$. Use (i) when $|x| \leq t$, and (ii) otherwise.]

12. Show that the function defined by

$$u(x,t) = \frac{x}{t}\mathcal{H}_t(x)$$

satisfies the heat equation for $t > 0$ and $\lim_{t\to 0} u(x,t) = 0$ for every x, but u is *not* continuous at the origin.

[Hint: Approach the origin with (x,t) on the parabola $x^2/4t = c$ where c is a constant.]

13. Prove the following uniqueness theorem for harmonic functions in the strip $\{(x,y) : 0 < y < 1,\ -\infty < x < \infty\}$: if u is harmonic in the strip, continuous on its closure with $u(x,0) = u(x,1) = 0$ for all $x \in \mathbb{R}$, and u vanishes at infinity, then $u = 0$.

14. Prove that the periodization of the Fejér kernel $\mathcal{F}_N$ on the real line (Exercise 9) is equal to the Fejér kernel for periodic functions of period 1. In other words,

$$\sum_{n=-\infty}^{\infty} \mathcal{F}_N(x+n) = F_N(x),$$

when $N \geq 1$ is an integer, and where

$$F_N(x) = \sum_{n=-N}^{N} \left(1 - \frac{|n|}{N}\right) e^{2\pi i n x} = \frac{1}{N}\frac{\sin^2(N\pi x)}{\sin^2(\pi x)}.$$

15. This exercise provides another example of periodization.

(a) Apply the Poisson summation formula to the function g in Exercise 2 to obtain

$$\sum_{n=-\infty}^{\infty} \frac{1}{(n+\alpha)^2} = \frac{\pi^2}{(\sin \pi\alpha)^2}$$

whenever α is real, but not equal to an integer.

(b) Prove as a consequence that

$$\sum_{n=-\infty}^{\infty} \frac{1}{(n+\alpha)} = \frac{\pi}{\tan \pi\alpha} \tag{15}$$

whenever α is real but not equal to an integer. [Hint: First prove it when $0 < \alpha < 1$. To do so, integrate the formula in (a). What is the precise meaning of the series on the left-hand side of (15)? Evaluate at $\alpha = 1/2$.]

16. The Dirichlet kernel on the real line is defined by

$$\int_{-R}^{R} \hat{f}(\xi) e^{2\pi i x\xi}\, d\xi = (f * \mathcal{D}_R)(x) \quad \text{so that} \quad \mathcal{D}_R(x) = \widehat{\chi_{[-R,R]}}(x) = \frac{\sin(2\pi R x)}{\pi x}.$$

Also, the modified Dirichlet kernel for periodic functions of period 1 is defined by

$$D_N^*(x) = \sum_{|n| \leq N-1} e^{2\pi i n x} + \frac{1}{2}(e^{-2\pi i N x} + e^{2\pi i N x}).$$

Show that the result in Exercise 15 gives

$$\sum_{n=-\infty}^{\infty} \mathcal{D}_N(x+n) = D_N^*(x),$$

where $N \geq 1$ is an integer, and the infinite series must be summed symmetrically. In other words, the periodization of $\mathcal{D}_N$ is the modified Dirichlet kernel D_N^*.

17. The **gamma function** is defined for $s > 0$ by

$$\Gamma(s) = \int_0^\infty e^{-x} x^{s-1}\, dx.$$

(a) Show that for $s > 0$ the above integral makes sense, that is, that the following two limits exist:

$$\lim_{\substack{\delta \to 0 \\ \delta > 0}} \int_\delta^1 e^{-x} x^{s-1}\, dx \quad \text{and} \quad \lim_{A \to \infty} \int_1^A e^{-x} x^{s-1}\, dx.$$

(b) Prove that $\Gamma(s+1) = s\Gamma(s)$ whenever $s > 0$, and conclude that for every integer $n \geq 1$ we have $\Gamma(n+1) = n!$.

(c) Show that

$$\Gamma\left(\frac{1}{2}\right) = \sqrt{\pi} \quad \text{and} \quad \Gamma\left(\frac{3}{2}\right) = \frac{\sqrt{\pi}}{2}.$$

[Hint: For (c), use $\int_{-\infty}^{\infty} e^{-\pi x^2}\, dx = 1$.]

18. The **zeta function** is defined for $s > 1$ by $\zeta(s) = \sum_{n=1}^{\infty} 1/n^s$. Verify the identity

$$\pi^{-s/2}\Gamma(s/2)\zeta(s) = \frac{1}{2}\int_0^{\infty} t^{\frac{s}{2}-1}(\vartheta(t) - 1)\, dt \qquad \text{whenever } s > 1$$

where Γ and ϑ are the gamma and theta functions, respectively:

$$\Gamma(s) = \int_0^{\infty} e^{-t} t^{s-1}\, dt \quad \text{and} \quad \vartheta(s) = \sum_{n=-\infty}^{\infty} e^{-\pi n^2 s}.$$

More about the zeta function and its relation to the prime number theorem can be found in Book II.

19. The following is a variant of the calculation of $\zeta(2m) = \sum_{n=1}^{\infty} 1/n^{2m}$ found in Problem 4, Chapter 3.

(a) Apply the Poisson summation formula to $f(x) = t/(\pi(x^2 + t^2))$ and $\hat{f}(\xi) = e^{-2\pi t|\xi|}$ where $t > 0$ in order to get

$$\frac{1}{\pi}\sum_{n=-\infty}^{\infty} \frac{t}{t^2 + n^2} = \sum_{n=-\infty}^{\infty} e^{-2\pi t|n|}.$$

(b) Prove the following identity valid for $0 < t < 1$:

$$\frac{1}{\pi}\sum_{n=-\infty}^{\infty} \frac{t}{t^2 + n^2} = \frac{1}{\pi t} + \frac{2}{\pi}\sum_{m=1}^{\infty}(-1)^{m+1}\zeta(2m)t^{2m-1}$$

as well as

$$\sum_{n=-\infty}^{\infty} e^{-2\pi t|n|} = \frac{2}{1 - e^{-2\pi t}} - 1.$$

(c) Use the fact that

$$\frac{z}{e^z - 1} = 1 - \frac{z}{2} + \sum_{m=1}^{\infty} \frac{B_{2m}}{(2m)!} z^{2m},$$

where B_k are the Bernoulli numbers to deduce from the above formula,

$$2\zeta(2m) = (-1)^{m+1} \frac{(2\pi)^{2m}}{(2m)!} B_{2m}.$$

20. The following results are relevant in information theory when one tries to recover a signal from its samples.

Suppose f is of moderate decrease and that its Fourier transform $\hat{f}$ is supported in $I = [-1/2, 1/2]$. Then, f is entirely determined by its restriction to $\mathbb{Z}$. This means that if g is another function of moderate decrease whose Fourier transform is supported in I and $f(n) = g(n)$ for all $n \in \mathbb{Z}$, then $f = g$. More precisely:

(a) Prove that the following reconstruction formula holds:

$$f(x) = \sum_{n=-\infty}^{\infty} f(n) K(x - n) \quad \text{where } K(y) = \frac{\sin \pi y}{\pi y}.$$

Note that $K(y) = O(1/|y|)$ as $|y| \to \infty$.

(b) If $\lambda > 1$, then

$$f(x) = \sum_{n=-\infty}^{\infty} \frac{1}{\lambda} f\left(\frac{n}{\lambda}\right) K_\lambda\left(x - \frac{n}{\lambda}\right) \quad \text{where } K_\lambda(y) = \frac{\cos \pi y - \cos \pi \lambda y}{\pi^2(\lambda - 1) y^2}.$$

Thus, if one samples f "more often," the series in the reconstruction formula converges faster since $K_\lambda(y) = O(1/|y|^2)$ as $|y| \to \infty$. Note that $K_\lambda(y) \to K(y)$ as $\lambda \to 1$.

(c) Prove that $\displaystyle\int_{-\infty}^{\infty} |f(x)|^2 \, dx = \sum_{n=-\infty}^{\infty} |f(n)|^2$.

[Hint: For part (a) show that if χ is the characteristic function of I, then $\hat{f}(\xi) = \chi(\xi) \sum_{n=-\infty}^{\infty} f(n) e^{-2\pi i n \xi}$. For (b) use the function in Figure 2 instead of $\chi(\xi)$.]

21. Suppose that f is continuous on $\mathbb{R}$. Show that f and $\hat{f}$ cannot both be compactly supported unless $f = 0$. This can be viewed in the same spirit as the uncertainty principle.

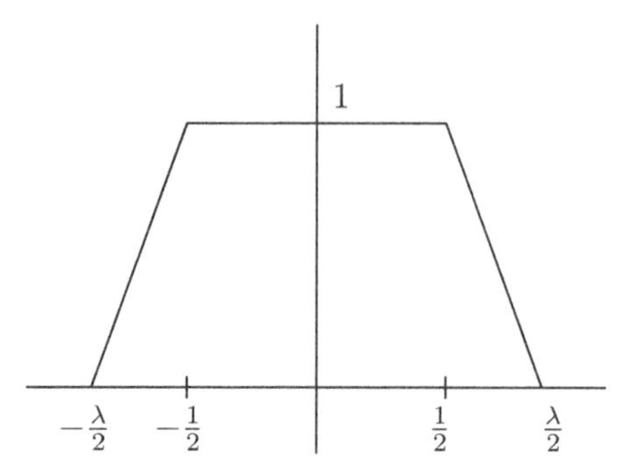

Figure 2. The function in Exercise 20

[Hint: Assume f is supported in $[0, 1/2]$. Expand f in a Fourier series in the interval $[0, 1]$, and note that as a result, f is a trigonometric polynomial.]

22. The heuristic assertion stated before Theorem 4.1 can be made precise as follows. If F is a function on $\mathbb{R}$, then we say that the preponderance of its mass is contained in an interval I (centered at the origin) if

$$\int_I x^2|F(x)|^2\,dx \geq \frac{1}{2}\int_{\mathbb{R}} x^2|F(x)|^2\,dx. \tag{16}$$

Now suppose $f \in \mathcal{S}$, and (16) holds with $F = f$ and $I = I_1$; also with $F = \hat{f}$ and $I = I_2$. Then if L_j denotes the length of I_j, we have

$$L_1 L_2 \geq \frac{1}{2\pi}.$$

A similar conclusion holds if the intervals are not necessarily centered at the origin.

23. The Heisenberg uncertainty principle can be formulated in terms of the operator $L = -\frac{d^2}{dx^2} + x^2$, which acts on Schwartz functions by the formula

$$L(f) = -\frac{d^2 f}{dx^2} + x^2 f.$$

This operator, sometimes called the **Hermite operator**, is the quantum analogue of the harmonic oscillator. Consider the usual inner product on $\mathcal{S}$ given by

$$(f, g) = \int_{-\infty}^{\infty} f(x)\overline{g(x)}\,dx \quad \text{whenever } f, g \in \mathcal{S}.$$

(a) Prove that the Heisenberg uncertainty principle implies

$$(Lf, f) \geq (f, f) \qquad \text{for all } f \in \mathcal{S}.$$

This is usually denoted by $L \geq I$. [Hint: Integrate by parts.]

(b) Consider the operators A and A^* defined on $\mathcal{S}$ by

$$A(f) = \frac{df}{dx} + xf \qquad \text{and} \qquad A^*(f) = -\frac{df}{dx} + xf.$$

The operators A and A^* are sometimes called the **annihilation** and **creation** operators, respectively. Prove that for all $f, g \in \mathcal{S}$ we have

(i) $(Af, g) = (f, A^*g)$,

(ii) $(Af, Af) = (A^*Af, f) \geq 0$,

(iii) $A^*A = L - I$.

In particular, this again shows that $L \geq I$.

(c) Now for $t \in \mathbb{R}$, let

$$A_t(f) = \frac{df}{dx} + txf \qquad \text{and} \qquad A_t^*(f) = -\frac{df}{dx} + txf.$$

Use the fact that $(A_t^* A_t f, f) \geq 0$ to give another proof of the Heisenberg uncertainty principle which says that whenever $\int_{-\infty}^{\infty} |f(x)|^2\, dx = 1$ then

$$\left(\int_{-\infty}^{\infty} x^2 |f(x)|^2\, dx\right)\left(\int_{-\infty}^{\infty} \left|\frac{df}{dx}\right|^2 dx\right) \geq 1/4.$$

[Hint: Think of $(A_t^* A_t f, f)$ as a quadratic polynomial in t.]

6 Problems

1. The equation

$$x^2 \frac{\partial^2 u}{\partial x^2} + ax\frac{\partial u}{\partial x} = \frac{\partial u}{\partial t} \tag{17}$$

with $u(x, 0) = f(x)$ for $0 < x < \infty$ and $t > 0$ is a variant of the heat equation which occurs in a number of applications. To solve (17), make the change of variables $x = e^{-y}$ so that $-\infty < y < \infty$. Set $U(y, t) = u(e^{-y}, t)$ and $F(y) = f(e^{-y})$. Then the problem reduces to the equation

$$\frac{\partial^2 U}{\partial y^2} + (1 - a)\frac{\partial U}{\partial y} = \frac{\partial U}{\partial t},$$

with $U(y,0) = F(y)$. This can be solved like the usual heat equation (the case $a = 1$) by taking the Fourier transform in the y variable. One must then compute the integral $\int_{-\infty}^{\infty} e^{(-4\pi^2\xi^2+(1-a)2\pi i\xi)t} e^{2\pi i\xi v}\, d\xi$. Show that the solution of the original problem is then given by

$$u(x,t) = \frac{1}{(4\pi t)^{1/2}} \int_0^\infty e^{-(\log(v/x)+(1-a)t)^2/(4t)} f(v)\, \frac{dv}{v}.$$

2. The **Black-Scholes** equation from finance theory is

$$\frac{\partial V}{\partial t} + rs\frac{\partial V}{\partial s} + \frac{\sigma^2 s^2}{2}\frac{\partial^2 V}{\partial s^2} - rV = 0, \qquad 0 < t < T, \tag{18}$$

subject to the "final" boundary condition $V(s,T) = F(s)$. An appropriate change of variables reduces this to the equation in Problem 1. Alternatively, the substitution $V(s,t) = e^{ax+b\tau}U(x,\tau)$ where $x = \log s$, $\tau = \frac{\sigma^2}{2}(T-t)$, $a = \frac{1}{2} - \frac{r}{\sigma^2}$, and $b = -\left(\frac{1}{2} + \frac{r}{\sigma^2}\right)^2$ reduces (18) to the one-dimensional heat equation with the initial condition $U(x,0) = e^{-ax}F(e^x)$. Thus a solution to the Black-Scholes equation is

$$V(s,t) = \frac{e^{-r(T-t)}}{\sqrt{2\pi\sigma^2(T-t)}} \int_0^\infty e^{-\frac{(\log(s/s^*)+(r-\sigma^2/2)(T-t))^2}{2\sigma^2(T-t)}} F(s^*)\, ds^*.$$

3. * **The Dirichlet problem in a strip**. Consider the equation $\Delta u = 0$ in the horizontal strip

$$\{(x,y) : 0 < y < 1,\ -\infty < x < \infty\}$$

with boundary conditions $u(x,0) = f_0(x)$ and $u(x,1) = f_1(x)$, where f_0 and f_1 are both in the Schwartz space.

(a) Show (formally) that if u is a solution to this problem, then

$$\hat{u}(\xi,y) = A(\xi)e^{2\pi\xi y} + B(\xi)e^{-2\pi\xi y}.$$

Express A and B in terms of $\widehat{f_0}$ and $\widehat{f_1}$, and show that

$$\hat{u}(\xi,y) = \frac{\sinh(2\pi(1-y)\xi)}{\sinh(2\pi\xi)}\widehat{f_0}(\xi) + \frac{\sinh(2\pi y\xi)}{\sinh(2\pi\xi)}\widehat{f_0}(\xi).$$

(b) Prove as a result that

$$\int_{-\infty}^{\infty} |u(x,y) - f_0(x)|^2\, dx \to 0 \qquad \text{as } y \to 0$$

and

$$\int_{-\infty}^{\infty} |u(x,y) - f_1(x)|^2\, dx \to 0 \qquad \text{as } y \to 1.$$

(c) If $\Phi(\xi) = (\sinh 2\pi a\xi)/(\sinh 2\pi\xi)$, with $0 \leq a < 1$, then Φ is the Fourier transform of φ where

$$\varphi(x) = \frac{\sin \pi a}{2} \cdot \frac{1}{\cosh \pi x + \cos \pi a}.$$

This can be shown, for instance, by using contour integration and the residue formula from complex analysis (see Book II, Chapter 3).

(d) Use this result to express u in terms of Poisson-like integrals involving f_0 and f_1 as follows:

$$u(x, y) = \frac{\sin \pi y}{2}\left(\int_{-\infty}^{\infty} \frac{f_0(x-t)}{\cosh \pi t - \cos \pi y}\, dt + \int_{-\infty}^{\infty} \frac{f_1(x-t)}{\cosh \pi t + \cos \pi y}\, dt\right).$$

(e) Finally, one can check that the function $u(x, y)$ defined by the above expression is harmonic in the strip, and converges uniformly to $f_0(x)$ as $y \to 0$, and to $f_1(x)$ as $y \to 1$. Moreover, one sees that $u(x, y)$ vanishes at infinity, that is, $\lim_{|x|\to\infty} u(x, y) = 0$, uniformly in y.

In Exercise 12, we gave an example of a function that satisfies the heat equation in the upper half-plane, with boundary value 0, but which was not identically 0. We observed in this case that u was in fact not continuous up to the boundary.

In Problem 4 we exhibit examples illustrating non-uniqueness, but this time with continuity up to the boundary $t = 0$. These examples satisfy a growth condition at infinity, namely $|u(x, t)| \leq Ce^{cx^{2+\epsilon}}$, for any $\epsilon > 0$. Problems 5 and 6 show that under the more restrictive growth condition $|u(x, t)| \leq Ce^{cx^2}$, uniqueness does hold.

4.* If g is a smooth function on $\mathbb{R}$, define the formal power series

$$u(x, t) = \sum_{n=0}^{\infty} g^{(n)}(t) \frac{x^{2n}}{(2n!)}. \tag{19}$$

(a) Check formally that u solves the heat equation.

(b) For $a > 0$, consider the function defined by

$$g(t) = \begin{cases} e^{-t^{-a}} & \text{if } t > 0 \\ 0 & \text{if } t \leq 0. \end{cases}$$

One can show that there exists $0 < \theta < 1$ depending on a so that

$$|g^{(k)}(t)| \leq \frac{k!}{(\theta t)^k} e^{-\frac{1}{2}t^{-a}} \qquad \text{for } t > 0.$$

(c) As a result, for each x and t the series (19) converges; u solves the heat equation; u vanishes for $t = 0$; and u satisfies the estimate $|u(x, t)| \leq Ce^{c|x|^{2a/(a-1)}}$ for some constants $C, c > 0$.

(d) Conclude that for every $\epsilon > 0$ there exists a non-zero solution to the heat equation which is continuous for $x \in \mathbb{R}$ and $t \geq 0$, which satisfies $u(x,0) = 0$ and $|u(x,t)| \leq Ce^{c|x|^{2+\epsilon}}$.

5.* The following "maximum principle" for solutions of the heat equation will be used in the next problem.

Theorem. *Suppose that $u(x,t)$ is a real-valued solution of the heat equation in the upper half-plane, which is continuous on its closure. Let R denote the rectangle*

$$R = \{(x,y) \in \mathbb{R}^2 : \; a \leq x \leq b, \; 0 \leq t \leq c\}$$

and $\partial' R$ be the part of the boundary of R which consists of the two vertical sides and its base on the line $t = 0$ (see Figure 3). *Then*

$$\min_{(x,t)\in\partial' R} u(x,t) = \min_{(x,t)\in R} u(x,t) \quad \textit{and} \quad \max_{(x,t)\in\partial' R} u(x,t) = \max_{(x,t)\in R} u(x,t).$$

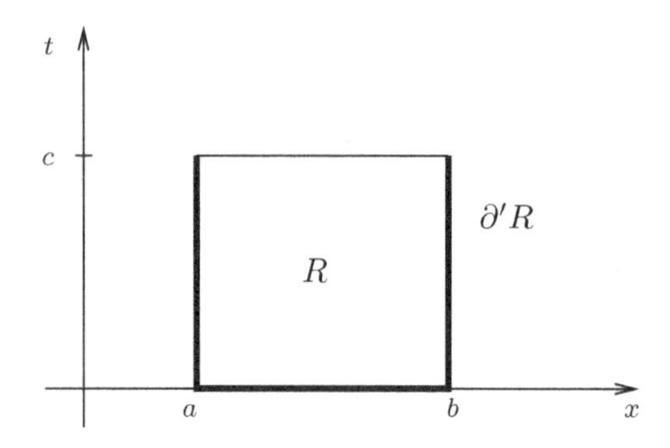

Figure 3. The rectangle R and part of its boundary $\partial' R$

The steps leading to a proof of this result are outlined below.

(a) Show that it suffices to prove that if $u \geq 0$ on $\partial' R$, then $u \geq 0$ in R.

(b) For $\epsilon > 0$, let

$$v(x,t) = u(x,t) + \epsilon t.$$

Then, v has a minimum on R, say at (x_1, t_1). Show that $x_1 = a$ or b, or else $t_1 = 0$. To do so, suppose on the contrary that $a < x_1 < b$ and $0 < t_1 \leq c$, and prove that $v_{xx}(x_1, t_1) - v_t(x_1, t_1) \leq -\epsilon$. However, show also that the left-hand side must be non-negative.

(c) Deduce from (b) that $u(x,t) \geq \epsilon(t_1 - t)$ for any $(x,t) \in R$ and let $\epsilon \to 0$.

6.* The examples in Problem 4 are optimal in the sense of the following uniqueness theorem due to Tychonoff.

Theorem. *Suppose $u(x,t)$ satisfies the following conditions:*

(i) *$u(x,t)$ solves the heat equation for all $x \in \mathbb{R}$ and and all $t > 0$.*

(ii) *$u(x,t)$ is continuous for all $x \in \mathbb{R}$ and $0 \leq t \leq c$.*

(iii) *$u(x,0) = 0$.*

(iv) *$|u(x,t)| \leq Me^{ax^2}$ for some M, a, and all $x \in \mathbb{R}$, $0 \leq t < c$.*

Then u is identically equal to 0.

7.* The **Hermite functions** $h_k(x)$ are defined by the generating identity

$$\sum_{k=0}^{\infty} h_k(x)\frac{t^k}{k!} = e^{-(x^2/2-2tx+t^2)}.$$

(a) Show that an alternate definition of the Hermite functions is given by the formula

$$h_k(x) = (-1)^k e^{x^2/2}\left(\frac{d}{dx}\right)^k e^{-x^2}.$$

[Hint: Write $e^{-(x^2/2-2tx+t^2)} = e^{x^2/2}e^{-(x-t)^2}$ and use Taylor's formula.] Conclude from the above expression that each $h_k(x)$ is of the form $P_k(x)e^{-x^2/2}$, where P_k is a polynomial of degree k. In particular, the Hermite functions belong to the Schwartz space and $h_0(x) = e^{-x^2/2}$, $h_1(x) = 2xe^{-x^2/2}$.

(b) Prove that the family $\{h_k\}_{k=0}^{\infty}$ is complete in the sense that if f is a Schwartz function, and

$$(f, h_k) = \int_{-\infty}^{\infty} f(x)h_k(x)\,dx = 0 \qquad \text{for all } k \geq 0,$$

then $f = 0$. [Hint: Use Exercise 8.]

(c) Define $h_k^*(x) = h_k((2\pi)^{1/2}x)$. Then

$$\widehat{h_k^*}(\xi) = (-i)^k h_k^*(\xi).$$

Therefore, each h_k^* is an eigenfunction for the Fourier transform.

(d) Show that h_k is an eigenfunction for the operator defined in Exercise 23, and in fact, prove that

$$Lh_k = (2k+1)h_k.$$

In particular, we conclude that the functions h_k are mutually orthogonal for the L^2 inner product on the Schwartz space.

(e) Finally, show that $\int_{-\infty}^{\infty}[h_k(x)]^2\,dx = \pi^{1/2}2^k k!$. [Hint: Square the generating relation.]

8.* To refine the results in Chapter 4, and to prove that

$$f_\alpha(x) = \sum_{n=0}^{\infty} 2^{-n\alpha} e^{2\pi i 2^n x}$$

is nowhere differentiable even in the case $\alpha = 1$, we need to consider a variant of the delayed means $\triangle_N$, which in turn will be analyzed by the Poisson summation formula.

(a) Fix an indefinitely differentiable function Φ satisfying

$$\Phi(\xi) = \begin{cases} 1 & \text{when } |\xi| \le 1, \\ 0 & \text{when } |\xi| \ge 2. \end{cases}$$

By the Fourier inversion formula, there exists $\varphi \in \mathcal{S}$ so that $\hat{\varphi}(\xi) = \Phi(\xi)$. Let $\varphi_N(x) = N\varphi(Nx)$ so that $\widehat{\varphi_N}(\xi) = \Phi(\xi/N)$. Finally, set

$$\tilde{\triangle}_N(x) = \sum_{n=-\infty}^{\infty} \varphi_N(x+n).$$

Observe by the Poisson summation formula that $\tilde{\triangle}_N(x) = \sum_{n=-\infty}^{\infty} \Phi(n/N)e^{2\pi i n x}$, thus $\tilde{\triangle}_N$ is a trigonometric polynomial of degree $\le 2N$, with terms whose coefficients are 1 when $|n| \le N$. Let

$$\tilde{\triangle}_N(f) = f * \tilde{\triangle}_N.$$

Note that

$$S_N(f_\alpha) = \tilde{\triangle}_{N'}(f_\alpha)$$

where N' is the largest integer of the form 2^k with $N' \le N$.

(b) If we set $\tilde{\triangle}_N(x) = \varphi_N(x) + E_N(x)$ where

$$E_N(x) = \sum_{|n|\ge 1} \varphi_N(x+n),$$

then one sees that:

(i) $\sup_{|x|\le 1/2} |E_N'(x)| \to 0$ as $N \to \infty$.

(ii) $|\tilde{\triangle}_N'(x)| \le cN^2$.

(iii) $|\tilde{\triangle}_N'(x)| \le c/(N|x|^3)$, for $|x| \le 1/2$.

Moreover, $\int_{|x|\le 1/2} \tilde{\triangle}_N'(x)\,dx = 0$, and $-\int_{|x|\le 1/2} x\tilde{\triangle}_N'(x)\,dx \to 1$ as $N \to \infty$.

(c) The above estimates imply that if $f'(x_0)$ exists, then

$$(f * \tilde{\triangle}_N')(x_0 + h_N) \to f'(x_0) \qquad \text{as } N \to \infty,$$

whenever $|h_N| \le C/N$. Then, conclude that both the real and imaginary parts of f_1 are nowhere differentiable, as in the proof given in Section 3, Chapter 4.

6 The Fourier Transform on $\mathbb{R}^d$

> It occurred to me that in order to improve treatment planning one had to know the distribution of the attenuation coefficient of tissues in the body. This information would be useful for diagnostic purposes and would constitute a tomogram or series of tomograms.
>
> It was immediately evident that the problem was a mathematical one. If a fine beam of gamma rays of intensity I_0 is incident on the body and the emerging density is I, then the measurable quantity g equals $\log(I_0/I) = \int_L f\, ds$, where f is the variable absorption coefficient along the line L. Hence if f is a function of two dimensions, and g is known for all lines intersecting the body, the question is, can f be determined if g is known?
>
> Fourteen years would elapse before I learned that Radon had solved this problem in 1917.
>
> *A. M. Cormack*, 1979

The previous chapter introduced the theory of the Fourier transform on $\mathbb{R}$ and illustrated some of its applications to partial differential equations. Here, our aim is to present an analogous theory for functions of several variables.

After a brief review of some relevant notions in $\mathbb{R}^d$, we begin with some general facts about the Fourier transform on the Schwartz space $\mathcal{S}(\mathbb{R}^d)$. Fortunately, the main ideas and techniques have already been considered in the one-dimensional case. In fact, with the appropriate notation, the statements (and proofs) of the key theorems, such as the Fourier inversion and Plancherel formulas, remain unchanged.

Next, we highlight the connection to some higher dimensional problems in mathematical physics, and in particular we investigate the wave equation in d dimensions, with a detailed analysis in the cases $d = 3$ and $d = 2$. At this stage, we discover a rich interplay between the Fourier transform and rotational symmetry, that arises only in $\mathbb{R}^d$ when $d \geq 2$.

Finally, the chapter ends with a discussion of the Radon transform. This topic is of substantial interest in its own right, but in addition has significant relevance in its application to the use of X-ray scans as well as to other parts of mathematics.

1 Preliminaries

The setting in this chapter will be $\mathbb{R}^d$, the vector space[1] of all d-tuples of real numbers $(x_1, \ldots, x_d)$ with $x_i \in \mathbb{R}$. Addition of vectors is componentwise, and so is multiplication by real scalars. Given $x = (x_1, \ldots, x_d) \in \mathbb{R}^d$ we define

$$|x| = (x_1^2 + \cdots + x_d^2)^{1/2},$$

so that $|x|$ is simply the length of the vector x in the usual Euclidean norm. In fact, we equip $\mathbb{R}^d$ with the standard inner product defined by

$$x \cdot y = x_1 y_1 + \cdots + x_d y_d,$$

so that $|x|^2 = x \cdot x$. We use the notation $x \cdot y$ in place of (x, y) of Chapter 3.

Given a d-tuple $\alpha = (\alpha_1, \ldots, \alpha_d)$ of non-negative integers (sometimes called a **multi-index**), the monomial x^α is defined by

$$x^\alpha = x_1^{\alpha_1} x_2^{\alpha_2} \cdots x_d^{\alpha_d}.$$

Similarly, we define the differential operator $(\partial/\partial x)^\alpha$ by

$$\left(\frac{\partial}{\partial x}\right)^\alpha = \left(\frac{\partial}{\partial x_1}\right)^{\alpha_1} \left(\frac{\partial}{\partial x_2}\right)^{\alpha_2} \cdots \left(\frac{\partial}{\partial x_d}\right)^{\alpha_d} = \frac{\partial^{|\alpha|}}{\partial x_1^{\alpha_1} \cdots \partial x_d^{\alpha_d}},$$

where $|\alpha| = \alpha_1 + \cdots + \alpha_d$ is the order of the multi-index α.

1.1 Symmetries

Analysis in $\mathbb{R}^d$, and in particular the theory of the Fourier transform, is shaped by three important groups of symmetries of the underlying space:

(i) Translations

(ii) Dilations

(iii) Rotations

[1]See Chapter 3 for a brief review of vector spaces and inner products. Here we find it convenient to use lower case letters such as x (as opposed to X) to designate points in $\mathbb{R}^d$. Also, we use $|\cdot|$ instead of $\|\cdot\|$ to denote the Euclidean norm.

We have seen that translations $x \mapsto x + h$, with $h \in \mathbb{R}^d$ fixed, and dilations $x \mapsto \delta x$, with $\delta > 0$, play an important role in the one-dimensional theory. In $\mathbb{R}$, the only two rotations are the identity and multiplication by -1. However, in $\mathbb{R}^d$ with $d \geq 2$ there are more rotations, and the understanding of the interaction between the Fourier transform and rotations leads to fruitful insights regarding spherical symmetries.

A **rotation** in $\mathbb{R}^d$ is a linear transformation $R : \mathbb{R}^d \to \mathbb{R}^d$ which preserves the inner product. In other words,

- $R(ax + by) = aR(x) + bR(y)$ for all $x, y \in \mathbb{R}^d$ and $a, b \in \mathbb{R}$.
- $R(x) \cdot R(y) = x \cdot y$ for all $x, y \in \mathbb{R}^d$.

Equivalently, this last condition can be replaced by $|R(x)| = |x|$ for all $x \in \mathbb{R}^d$, or $R^t = R^{-1}$ where R^t and R^{-1} denote the transpose and inverse of R, respectively.[2] In particular, we have $\det(R) = \pm 1$, where $\det(R)$ is the determinant of R. If $\det(R) = 1$ we say that R is a **proper rotation**; otherwise, we say that R is an **improper rotation**.

EXAMPLE 1. On the real line $\mathbb{R}$, there are two rotations: the identity which is proper, and multiplication by -1 which is improper.

EXAMPLE 2. The rotations in the plane $\mathbb{R}^2$ can be described in terms of complex numbers. We identify $\mathbb{R}^2$ with $\mathbb{C}$ by assigning the point (x, y) to the complex number $z = x + iy$. Under this identification, all proper rotations are of the form $z \mapsto ze^{i\varphi}$ for some $\varphi \in \mathbb{R}$, and all improper rotations are of the form $z \mapsto \overline{z}e^{i\varphi}$ for some $\varphi \in \mathbb{R}$ (here, $\overline{z} = x - iy$ denotes the complex conjugate of z). See Exercise 1 for the argument leading to this result.

EXAMPLE 3. Euler gave the following very simple geometric description of rotations in $\mathbb{R}^3$. Given a proper rotation R, there exists a unit vector γ so that:

(i) R fixes γ, that is, $R(\gamma) = \gamma$.

(ii) If $\mathcal{P}$ denotes the plane passing through the origin and perpendicular to γ, then $R : \mathcal{P} \to \mathcal{P}$, and the restriction of R to $\mathcal{P}$ is a rotation in $\mathbb{R}^2$.

---

[2]Recall that the transpose of a linear operator $A : \mathbb{R}^d \to \mathbb{R}^d$ is the linear operator $B : \mathbb{R}^d \to \mathbb{R}^d$ which satisfies $A(x) \cdot y = x \cdot B(y)$ for all $x, y \in \mathbb{R}^d$. We write $B = A^t$. The inverse of A (when it exists) is the linear operator $C : \mathbb{R}^d \to \mathbb{R}^d$ with $A \circ C = C \circ A = I$ (where I is the identity), and we write $C = A^{-1}$.

Geometrically, the vector γ gives the direction of the axis of rotation. A proof of this fact is given in Exercise 2. Finally, if R is improper, then $-R$ is proper (since in $\mathbb{R}^3$ $\det(-R) = -\det(R)$), so R is the composition of a proper rotation and a symmetry with respect to the origin.

EXAMPLE 4. Given two *orthonormal* bases $\{e_1, \ldots, e_d\}$ and $\{e_1', \ldots, e_d'\}$ in $\mathbb{R}^d$, we can define a rotation R by letting $R(e_i) = e_i'$ for $i = 1, \ldots, d$. Conversely, if R is a rotation and $\{e_1, \ldots, e_d\}$ is an orthonormal basis, then $\{e_1', \ldots, e_d'\}$, where $e_j' = R(e_j)$, is another orthonormal basis.

1.2 Integration on $\mathbb{R}^d$

Since we shall be dealing with functions on $\mathbb{R}^d$, we will need to discuss some aspects of integration of such functions. A more detailed review of integration on $\mathbb{R}^d$ is given in the appendix.

A continuous complex-valued function f on $\mathbb{R}^d$ is said to be **rapidly decreasing** if for every multi-index α the function $|x^\alpha f(x)|$ is bounded. Equivalently, a continuous function is of rapid decrease if

$$\sup_{x \in \mathbb{R}^d} |x|^k |f(x)| < \infty \qquad \text{for every } k = 0, 1, 2, \ldots.$$

Given a function of rapid decrease, we define

$$\int_{\mathbb{R}^d} f(x)\, dx = \lim_{N \to \infty} \int_{Q_N} f(x)\, dx,$$

where Q_N denotes the closed cube centered at the origin, with sides of length N parallel to the coordinate axis, that is,

$$Q_N = \{x \in \mathbb{R}^d : \ |x_i| \leq N/2 \quad \text{for } i = 1, \ldots, d\}.$$

The integral over Q_N is a multiple integral in the usual sense of Riemann integration. That the limit exists follows from the fact that the integrals $I_N = \int_{Q_N} f(x)\, dx$ form a Cauchy sequence as N tends to infinity.

Two observations are in order. First, we may replace the square Q_N by the ball $B_N = \{x \in \mathbb{R}^d : |x| \leq N\}$ without changing the definition. Second, we do not need the full force of rapid decrease to show that the limit exists. In fact it suffices to assume that f is continuous and

$$\sup_{x \in \mathbb{R}^d} |x|^{d+\epsilon} |f(x)| < \infty \qquad \text{for some } \epsilon > 0.$$

For example, functions of moderate decrease on $\mathbb{R}$ correspond to $\epsilon = 1$. In keeping with this we define functions of **moderate decrease** on $\mathbb{R}^d$ as those that are continuous and satisfy the above inequality with $\epsilon = 1$.

The interaction of integration with the three important groups of symmetries is as follows: if f is of moderate decrease, then

(i) $\displaystyle\int_{\mathbb{R}^d} f(x+h)\,dx = \int_{\mathbb{R}^d} f(x)\,dx$ for all $h \in \mathbb{R}^d$,

(ii) $\displaystyle\delta^d \int_{\mathbb{R}^d} f(\delta x)\,dx = \int_{\mathbb{R}^d} f(x)\,dx$ for all $\delta > 0$,

(iii) $\displaystyle\int_{\mathbb{R}^d} f(R(x))\,dx = \int_{\mathbb{R}^d} f(x)\,dx$ for every rotation R.

Polar coordinates

It will be convenient to introduce polar coordinates in $\mathbb{R}^d$ and find the corresponding integration formula. We begin with two examples which correspond to the case $d = 2$ and $d = 3$. (A more elaborate discussion applying to all d is contained in the appendix.)

EXAMPLE 1. In $\mathbb{R}^2$, polar coordinates are given by (r, θ) with $r \geq 0$ and $0 \leq \theta < 2\pi$. The Jacobian of the change of variables is equal to r, so that

$$\int_{\mathbb{R}^2} f(x)\,dx = \int_0^{2\pi} \int_0^{\infty} f(r\cos\theta, r\sin\theta)\, r\,dr\,d\theta.$$

Now we may write a point on the unit circle S^1 as $\gamma = (\cos\theta, \sin\theta)$, and given a function g on the circle, we define its integral over S^1 by

$$\int_{S^1} g(\gamma)\,d\sigma(\gamma) = \int_0^{2\pi} g(\cos\theta, \sin\theta)\,d\theta.$$

With this notation we then have

$$\int_{\mathbb{R}^2} f(x)\,dx = \int_{S^1} \int_0^{\infty} f(r\gamma)\, r\,dr\,d\sigma(\gamma).$$

EXAMPLE 2. In $\mathbb{R}^3$ one uses spherical coordinates given by

$$\begin{cases} x_1 &= r\sin\theta\cos\varphi, \\ x_2 &= r\sin\theta\sin\varphi, \\ x_3 &= r\cos\theta, \end{cases}$$

where $0 < r$, $0 \leq \theta \leq \pi$ and $0 \leq \varphi \leq 2\pi$. The Jacobian of the change of variables is $r^2 \sin\theta$ so that

$$\int_{\mathbb{R}^3} f(x)\,dx =$$
$$\int_0^{2\pi}\int_0^{\pi}\int_0^{\infty} f(r\sin\theta\cos\varphi, r\sin\theta\sin\varphi, r\cos\theta)r^2 dr \sin\theta\,d\theta\,d\varphi.$$

If g is a function on the unit sphere $S^2 = \{x \in \mathbb{R}^3 : |x| = 1\}$, and $\gamma = (\sin\theta\cos\varphi, \sin\theta\sin\varphi, \cos\theta)$, we define the surface element $d\sigma(\gamma)$ by

$$\int_{S^2} g(\gamma)\,d\sigma(\gamma) = \int_0^{2\pi}\int_0^{\pi} g(\gamma)\sin\theta\,d\theta\,d\varphi.$$

As a result,

$$\int_{\mathbb{R}^3} f(x)\,dx = \int_{S^2}\int_0^{\infty} f(r\gamma)\,r^2\,dr\,d\sigma(\gamma).$$

In general, it is possible to write any point in $\mathbb{R}^d - \{0\}$ uniquely as

$$x = r\gamma$$

where γ lies on the unit sphere $S^{d-1} \subset \mathbb{R}^d$ and $r > 0$. Indeed, take $r = |x|$ and $\gamma = x/|x|$. Thus one may proceed as in the cases $d = 2$ or $d = 3$ to define spherical coordinates. The formula we shall use is

$$\int_{\mathbb{R}^d} f(x)\,dx = \int_{S^{d-1}}\int_0^{\infty} f(r\gamma)\,r^{d-1}\,dr\,d\sigma(\gamma),$$

whenever f is of moderate decrease. Here $d\sigma(\gamma)$ denotes the surface element on the sphere S^{d-1} obtained from the spherical coordinates.

2 Elementary theory of the Fourier transform

The **Schwartz space** $\mathcal{S}(\mathbb{R}^d)$ (sometimes abbreviated as $\mathcal{S}$) consists of all indefinitely differentiable functions f on $\mathbb{R}^d$ such that

$$\sup_{x\in\mathbb{R}^d} \left| x^\alpha \left(\frac{\partial}{\partial x}\right)^\beta f(x)\right| < \infty,$$

for every multi-index α and β. In other words, f and all its derivatives are required to be rapidly decreasing.

EXAMPLE 1. An example of a function in $\mathcal{S}(\mathbb{R}^d)$ is the d-dimensional Gaussian given by $e^{-\pi|x|^2}$. The theory in Chapter 5 already made clear the central role played by this function in the case $d = 1$.

The **Fourier transform** of a Schwartz function f is defined by

$$\hat{f}(\xi) = \int_{\mathbb{R}^d} f(x)e^{-2\pi i x\cdot\xi}\, dx, \qquad \text{for } \xi \in \mathbb{R}^d.$$

Note the resemblance with the formula in one-dimension, except that we are now integrating on $\mathbb{R}^d$, and the product of x and ξ is replaced by the inner product of the two vectors.

We now list some simple properties of the Fourier transform. In the next proposition the arrow indicates that we have taken the Fourier transform, so $F(x) \longrightarrow G(\xi)$ means that $G(\xi) = \hat{F}(\xi)$.

Proposition 2.1 *Let* $f \in \mathcal{S}(\mathbb{R}^d)$.

(i) $f(x + h) \longrightarrow \hat{f}(\xi)e^{2\pi i\xi\cdot h}$ *whenever* $h \in \mathbb{R}^d$.

(ii) $f(x)e^{-2\pi i x h} \longrightarrow \hat{f}(\xi + h)$ *whenever* $h \in \mathbb{R}^d$.

(iii) $f(\delta x) \longrightarrow \delta^{-d}\hat{f}(\delta^{-1}\xi)$ *whenever* $\delta > 0$.

(iv) $\left(\frac{\partial}{\partial x}\right)^{\alpha} f(x) \longrightarrow (2\pi i\xi)^{\alpha}\hat{f}(\xi)$.

(v) $(-2\pi i x)^{\alpha} f(x) \longrightarrow \left(\frac{\partial}{\partial \xi}\right)^{\alpha} \hat{f}(\xi)$.

(vi) $f(Rx) \longrightarrow \hat{f}(R\xi)$ *whenever R is a rotation.*

The first five properties are proved in the same way as in the one-dimensional case. To verify the last property, simply change variables $y = Rx$ in the integral. Then, recall that $|\det(R)| = 1$, and $R^{-1}y \cdot \xi = y \cdot R\xi$, because R is a rotation.

Properties (iv) and (v) in the proposition show that, up to factors of $2\pi i$, the Fourier transform interchanges differentiation and multiplication by monomials. This motivates the definition of the Schwartz space and leads to the next corollary.

Corollary 2.2 *The Fourier transform maps* $\mathcal{S}(\mathbb{R}^d)$ *to itself.*

At this point we disgress to observe a simple fact concerning the interplay between the Fourier transform and rotations. We say that a function f is **radial** if it depends only on $|x|$; in other words, f is radial if there is a function $f_0(u)$, defined for $u \geq 0$, such that $f(x) = f_0(|x|)$. We note that f is radial if and only if $f(Rx) = f(x)$ for every rotation R. In one direction, this is obvious since $|Rx| = |x|$. Conversely, suppose that $f(Rx) = f(x)$, for all rotations R. Now define f_0 by

$$f_0(u) = \begin{cases} f(0) & \text{if } u = 0, \\ f(x) & \text{if } |x| = u. \end{cases}$$

Note that f_0 is well defined, since if x and x' are points with $|x| = |x'|$ there is always a rotation R so that $x' = Rx$.

Corollary 2.3 *The Fourier transform of a radial function is radial.*

This follows at once from property (vi) in the last proposition. Indeed, the condition $f(Rx) = f(x)$ for all R implies that $\hat{f}(R\xi) = \hat{f}(\xi)$ for all R, thus $\hat{f}$ is radial whenever f is.

An example of a radial function in $\mathbb{R}^d$ is the Gaussian $e^{-\pi|x|^2}$. Also, we observe that when $d = 1$, the radial functions are precisely the even functions, that is, those for which $f(x) = f(-x)$.

After these preliminaries, we retrace the steps taken in the previous chapter to obtain the Fourier inversion formula and Plancherel theorem for $\mathbb{R}^d$.

Theorem 2.4 *Suppose* $f \in \mathcal{S}(\mathbb{R}^d)$. *Then*

$$f(x) = \int_{\mathbb{R}^d} \hat{f}(\xi) e^{2\pi i x \cdot \xi}\, d\xi.$$

Moreover

$$\int_{\mathbb{R}^d} |\hat{f}(\xi)|^2\, d\xi = \int_{\mathbb{R}^d} |f(x)|^2\, dx.$$

The proof proceeds in the following stages.

Step 1. The Fourier transform of $e^{-\pi|x|^2}$ is $e^{-\pi|\xi|^2}$. To prove this, notice that the properties of the exponential functions imply that

$$e^{-\pi|x|^2} = e^{-\pi x_1^2} \cdots e^{-\pi x_d^2} \quad \text{and} \quad e^{-2\pi i x \cdot \xi} = e^{-2\pi i x_1 \cdot \xi_1} \cdots e^{-2\pi i x_d \cdot \xi_d},$$

so that the integrand in the Fourier transform is a product of d functions, each depending on the variable x_j $(1 \leq j \leq d)$ only. Thus the assertion follows by writing the integral over $\mathbb{R}^d$ as a series of repeated integrals, each taken over $\mathbb{R}$. For example, when $d = 2$,

$$\begin{aligned}\int_{\mathbb{R}^2} e^{-\pi|x|^2} e^{-2\pi i x\cdot\xi}\, dx &= \int_{\mathbb{R}} e^{-\pi x_2^2} e^{-2\pi i x_2\cdot\xi_2} \left(\int_{\mathbb{R}} e^{-\pi x_1^2} e^{-2\pi i x_1\cdot\xi_1}\, dx_1 \right) dx_2 \\ &= \int_{\mathbb{R}} e^{-\pi x_2^2} e^{-2\pi i x_2\cdot\xi_2} e^{-\pi\xi_1^2}\, dx_2 \\ &= e^{-\pi\xi_1^2} e^{-\pi\xi_2^2} \\ &= e^{-\pi|\xi|^2}.\end{aligned}$$

As a consequence of Proposition 2.1, applied with $\delta^{1/2}$ instead of δ, we find that $\widehat{(e^{-\pi\delta|x|^2})} = \delta^{-d/2} e^{-\pi|\xi|^2/\delta}$.

Step 2. The family $K_\delta(x) = \delta^{-d/2} e^{-\pi|x|^2/\delta}$ is a family of good kernels in $\mathbb{R}^d$. By this we mean that

(i) $\displaystyle\int_{\mathbb{R}^d} K_\delta(x)\, dx = 1,$

(ii) $\displaystyle\int_{\mathbb{R}^d} |K_\delta(x)|\, dx \le M$ (in fact $K_\delta(x) \ge 0$),

(iii) For every $\eta > 0$, $\displaystyle\int_{|x|\ge\eta} |K_\delta(x)|\, dx \to 0$ as $\delta \to 0$.

The proofs of these assertions are almost identical to the case $d = 1$. As a result

$$\int_{\mathbb{R}^d} K_\delta(x) F(x)\, dx \to F(0) \quad \text{as } \delta \to 0$$

when F is a Schwartz function, or more generally when F is bounded and continuous at the origin.

Step 3. The multiplication formula

$$\int_{\mathbb{R}^d} f(x)\hat{g}(x)\, dx = \int_{\mathbb{R}^d} \hat{f}(y) g(y)\, dy$$

holds whenever f and g are in $\mathcal{S}$. The proof requires the evaluation of the integral of $f(x)g(y)e^{-2\pi i x\cdot y}$ over $(x, y) \in \mathbb{R}^{2d} = \mathbb{R}^d \times \mathbb{R}^d$ as a repeated integral, with each separate integration taken over $\mathbb{R}^d$. The justification is similar to that in the proof of Proposition 1.8 in the previous chapter. (See the appendix.)

The Fourier inversion is then a simple consequence of the multiplication formula and the family of good kernels K_δ, as in Chapter 5. It also follows that the Fourier transform $\mathcal{F}$ is a bijective map of $\mathcal{S}(\mathbb{R}^d)$ to itself, whose inverse is

$$\mathcal{F}^*(g)(x) = \int_{\mathbb{R}^d} g(\xi) e^{2\pi i x \cdot \xi}\, d\xi.$$

Step 4. Next we turn to the convolution, defined by

$$(f * g)(x) = \int_{\mathbb{R}^d} f(y) g(x - y)\, dy, \qquad f, g \in \mathcal{S}.$$

We have that $f * g \in \mathcal{S}(\mathbb{R}^d)$, $f * g = g * f$, and $\widehat{(f * g)}(\xi) = \hat{f}(\xi)\hat{g}(\xi)$. The argument is similar to that in one-dimension. The calculation of the Fourier transform of $f * g$ involves an integration of $f(y)g(x - y)e^{-2\pi i x \cdot \xi}$ (over $\mathbb{R}^{2d} = \mathbb{R}^d \times \mathbb{R}^d$) expressed as a repeated integral.

Then, following the same argument in the previous chapter, we obtain the d-dimensional Plancherel formula, thereby concluding the proof of Theorem 2.4.

3 The wave equation in $\mathbb{R}^d \times \mathbb{R}$

Our next goal is to apply what we have learned about the Fourier transform to the study of the wave equation. Here, we once again simplify matters by restricting ourselves to functions in the Schwartz class $\mathcal{S}$. We note that in any further analysis of the wave equation it is important to allow functions that have much more general behavior, and in particular that may be discontinuous. However, what we lose in generality by only considering Schwartz functions, we gain in transparency. Our study in this restricted context will allow us to explain certain basic ideas in their simplest form.

3.1 Solution in terms of Fourier transforms

The motion of a vibrating string satisfies the equation

$$\frac{\partial^2 u}{\partial x^2} = \frac{1}{c^2}\frac{\partial^2 u}{\partial t^2},$$

which we referred to as the one-dimensional wave equation.

A natural generalization of this equation to d space variables is

$$\frac{\partial^2 u}{\partial x_1^2} + \cdots + \frac{\partial^2 u}{\partial x_d^2} = \frac{1}{c^2}\frac{\partial^2 u}{\partial t^2}. \tag{1}$$

In fact, it is known that in the case $d = 3$, this equation determines the behavior of electromagnetic waves in vacuum (with $c =$ speed of light).

Also, this equation describes the propagation of sound waves. Thus (1) is called the **d-dimensional wave equation**.

Our first observation is that we may assume $c = 1$, since we can rescale the variable t if necessary. Also, if we define the **Laplacian** in d dimensions by

$$\triangle = \frac{\partial^2}{\partial x_1^2} + \cdots + \frac{\partial^2}{\partial x_d^2},$$

then the wave equation can be rewritten as

$$\triangle u = \frac{\partial^2 u}{\partial t^2}. \tag{2}$$

The goal of this section is to find a solution to this equation, subject to the initial conditions

$$u(x,0) = f(x) \quad \text{and} \quad \frac{\partial u}{\partial t}(x,0) = g(x),$$

where $f, g \in \mathcal{S}(\mathbb{R}^d)$. This is called the **Cauchy problem** for the wave equation.

Before solving this problem, we note that while we think of the variable t as time, we do not restrict ourselves to $t > 0$. As we will see, the solution we obtain makes sense for all $t \in \mathbb{R}$. This is a manifestation of the fact that the wave equation can be reversed in time (unlike the heat equation).

A formula for the solution of our problem is given in the next theorem. The heuristic argument which leads to this formula is important since, as we have already seen, it applies to some other boundary value problems as well.

Suppose u solves the Cauchy problem for the wave equation. The technique employed consists of taking the Fourier transform of the equation and of the initial conditions, with respect to the space variables $x_1, \ldots, x_d$. This reduces the problem to an ordinary differential equation in the time variable. Indeed, recalling that differentiation with respect to x_j becomes multiplication by $2\pi i \xi_j$, and the differentiation with respect to t commutes with the Fourier transform in the space variables, we find that (2) becomes

$$-4\pi^2 |\xi|^2 \hat{u}(\xi, t) = \frac{\partial^2 \hat{u}}{\partial t^2}(\xi, t).$$

For each fixed $\xi \in \mathbb{R}^d$, this is an ordinary differential equation in t whose solution is given by

$$\hat{u}(\xi, t) = A(\xi) \cos(2\pi|\xi|t) + B(\xi) \sin(2\pi|\xi|t),$$

where for each ξ, $A(\xi)$ and $B(\xi)$ are unknown constants to be determined by the initial conditions. In fact, taking the Fourier transform (in x) of the initial conditions yields

$$\hat{u}(\xi, 0) = \hat{f}(\xi) \quad \text{and} \quad \frac{\partial \hat{u}}{\partial t}(\xi, 0) = \hat{g}(\xi).$$

We may now solve for $A(\xi)$ and $B(\xi)$ to obtain

$$A(\xi) = \hat{f}(\xi) \quad \text{and} \quad 2\pi|\xi|B(\xi) = \hat{g}(\xi).$$

Therefore, we find that

$$\hat{u}(\xi, t) = \hat{f}(\xi)\cos(2\pi|\xi|t) + \hat{g}(\xi)\frac{\sin(2\pi|\xi|t)}{2\pi|\xi|},$$

and the solution u is given by taking the inverse Fourier transform in the ξ variables. This formal derivation then leads to a precise existence theorem for our problem.

Theorem 3.1 *A solution of the Cauchy problem for the wave equation is*

$$u(x,t) = \int_{\mathbb{R}^d} \left[\hat{f}(\xi)\cos(2\pi|\xi|t) + \hat{g}(\xi)\frac{\sin(2\pi|\xi|t)}{2\pi|\xi|}\right] e^{2\pi i x\cdot\xi}\, d\xi. \tag{3}$$

Proof. We first verify that u solves the wave equation. This is straightforward once we note that we can differentiate in x and t under the integral sign (because f and g are both Schwartz functions) and therefore u is at least C^2. On the one hand we differentiate the exponential with respect to the x variables to get

$$\Delta u(x,t) = \int_{\mathbb{R}^d} \left[\hat{f}(\xi)\cos(2\pi|\xi|t) + \hat{g}(\xi)\frac{\sin(2\pi|\xi|t)}{2\pi|\xi|}\right] (-4\pi^2|\xi|^2) e^{2\pi i x\cdot\xi}\, d\xi,$$

while on the other hand we differentiate the terms in brackets with respect to t twice to get

$$\frac{\partial^2 u}{\partial t^2}(x,t) =$$

$$\int_{\mathbb{R}^d} \left[-4\pi^2|\xi|^2\hat{f}(\xi)\cos(2\pi|\xi|t) - 4\pi^2|\xi|^2\hat{g}(\xi)\frac{\sin(2\pi|\xi|t)}{2\pi|\xi|}\right] e^{2\pi i x\cdot\xi}\, d\xi.$$

This shows that u solves equation (2). Setting $t = 0$ we get

$$u(x,0) = \int_{\mathbb{R}^d} \hat{f}(\xi) e^{2\pi i x\cdot\xi}\, d\xi = f(x)$$

by the Fourier inversion theorem. Finally, differentiating once with respect to t, setting $t = 0$, and using the Fourier inversion shows that

$$\frac{\partial u}{\partial t}(x, 0) = g(x).$$

Thus u also verifies the initial conditions, and the proof of the theorem is complete.

As the reader will note, both $\hat{f}(\xi)\cos(2\pi|\xi|t)$ and $\hat{g}(\xi)\frac{\sin(2\pi|\xi|t)}{2\pi|\xi|}$ are functions in $\mathcal{S}$, assuming as we do that f and g are in $\mathcal{S}$. This is because both $\cos u$ and $(\sin u)/u$ are even functions that are indefinitely differentiable.

Having proved the existence of a solution to the Cauchy problem for the wave equation, we raise the question of uniqueness. Are there solutions to the problem

$$\triangle u = \frac{\partial^2 u}{\partial t^2} \quad \text{subject to} \quad u(x,0) = f(x) \quad \text{and} \quad \frac{\partial u}{\partial t}(x,0) = g(x),$$

other than the one given by the formula in the theorem? In fact the answer is, as expected, no. The proof of this fact, which will not be given here (but see Problem 3), can be based on a conservation of energy argument. This is a local counterpart of a global conservation of energy statement which we will now present.

We observed in Exercise 10, Chapter 3, that in the one-dimensional case, the total energy of the vibrating string is conserved in time. The analogue of this fact holds in higher dimensions as well. Define the **energy** of a solution by

$$E(t) = \int_{\mathbb{R}^d} \left|\frac{\partial u}{\partial t}\right|^2 + \left|\frac{\partial u}{\partial x_1}\right|^2 + \cdots + \left|\frac{\partial u}{\partial x_d}\right|^2 dx.$$

Theorem 3.2 *If u is the solution of the wave equation given by formula (3), then $E(t)$ is conserved, that is,*

$$E(t) = E(0), \qquad \textit{for all } t \in \mathbb{R}.$$

The proof requires the following lemma.

Lemma 3.3 *Suppose a and b are complex numbers and α is real. Then*

$$|a\cos\alpha + b\sin\alpha|^2 + |-a\sin\alpha + b\cos\alpha|^2 = |a|^2 + |b|^2.$$

This follows directly because $e_1 = (\cos\alpha, \sin\alpha)$ and $e_2 = (-\sin\alpha, \cos\alpha)$ are a pair of orthonormal vectors, hence with $Z = (a, b) \in \mathbb{C}^2$, we have

$$|Z|^2 = |Z \cdot e_1|^2 + |Z \cdot e_2|^2,$$

where $\cdot$ represents the inner product in $\mathbb{C}^2$.

Now by Plancherel's theorem,

$$\int_{\mathbb{R}^d} \left|\frac{\partial u}{\partial t}\right|^2 dx = \int_{\mathbb{R}^d} \left|-2\pi|\xi|\hat{f}(\xi)\sin(2\pi|\xi|t) + \hat{g}(\xi)\cos(2\pi|\xi|t)\right|^2 d\xi.$$

Similarly,

$$\int_{\mathbb{R}^d} \sum_{j=1}^{d} \left|\frac{\partial u}{\partial x_j}\right|^2 dx = \int_{\mathbb{R}^d} \left|2\pi|\xi|\hat{f}(\xi)\cos(2\pi|\xi|t) + \hat{g}(\xi)\sin(2\pi|\xi|t)\right|^2 d\xi.$$

We now apply the lemma with

$$a = 2\pi|\xi|\hat{f}(\xi), \qquad b = \hat{g}(\xi) \quad \text{and} \quad \alpha = 2\pi|\xi|t.$$

The result is that

$$\begin{aligned} E(t) &= \int_{\mathbb{R}^d} \left|\frac{\partial u}{\partial t}\right|^2 + \left|\frac{\partial u}{\partial x_1}\right|^2 + \cdots + \left|\frac{\partial u}{\partial x_d}\right|^2 dx \\ &= \int_{\mathbb{R}^d} (4\pi^2|\xi|^2|\hat{f}(\xi)|^2 + |\hat{g}(\xi)|^2)\, d\xi, \end{aligned}$$

which is clearly independent of t. Thus Theorem 3.2 is proved.

The drawback with formula (3), which does give the solution of the wave equation, is that it is quite indirect, involving the calculation of the Fourier transforms of f and g, and then a further inverse Fourier transform. However, for every dimension d there is a more explicit formula. This formula is very simple when $d = 1$ and a little less so when $d = 3$. More generally, the formula is "elementary" whenever d is odd, and more complicated when d is even (see Problems 4 and 5).

In what follows we consider the cases $d = 1$, $d = 3$, and $d = 2$, which together give a picture of the general situation. Recall that in Chapter 1, when discussing the wave equation over the interval $[0, L]$, we found that the solution is given by d'Alembert's formula

$$u(x,t) = \frac{f(x+t) + f(x-t)}{2} + \frac{1}{2}\int_{x-t}^{x+t} g(y)\, dy. \tag{4}$$

with the interpretation that both f and g are extended outside $[0, L]$ by making them *odd* in $[-L, L]$, and periodic on the real line, with period $2L$. The same formula (4) holds for the solution of the wave equation when $d = 1$ and when the initial data are functions in $\mathcal{S}(\mathbb{R})$. In fact, this follows directly from (3) if we note that

$$\cos(2\pi|\xi|t) = \frac{1}{2}(e^{2\pi i|\xi|t} + e^{-2\pi i|\xi|t})$$

and

$$\frac{\sin(2\pi|\xi|t)}{2\pi|\xi|} = \frac{1}{4\pi i|\xi|}(e^{2\pi i|\xi|t} - e^{-2\pi i|\xi|t}).$$

Finally, we note that the two terms that appear in d'Alembert's formula (4) consist of appropriate averages. Indeed, the first term is precisely the average of f over the two points that are the boundary of the interval $[x-t, x+t]$; the second term is, up to a factor of t, the mean value of g over this interval, that is, $(1/2t)\int_{x-t}^{x+t} g(y)\,dy$. This suggests a generalization to higher dimensions, where we might expect to write the solution of our problem as averages of the initial data. This is in fact the case, and we now treat in detail the particular situation $d = 3$.

3.2 The wave equation in $\mathbb{R}^3 \times \mathbb{R}$

If S^2 denotes the unit sphere in $\mathbb{R}^3$, we define the **spherical mean** of the function f over the sphere of radius t centered at x by

$$M_t(f)(x) = \frac{1}{4\pi}\int_{S^2} f(x - t\gamma)\,d\sigma(\gamma), \tag{5}$$

where $d\sigma(\gamma)$ is the element of surface area for S^2. Since 4π is the area of the unit sphere, we can interpret $M_t(f)$ as the average value of f over the sphere centered at x of radius t.

Lemma 3.4 *If $f \in \mathcal{S}(\mathbb{R}^3)$ and t is fixed, then $M_t(f) \in \mathcal{S}(\mathbb{R}^3)$. Moreover, $M_t(f)$ is indefinitely differentiable in t, and each t-derivative also belongs to $\mathcal{S}(\mathbb{R}^3)$.*

Proof. Let $F(x) = M_t(f)(x)$. To show that F is rapidly decreasing, start with the inequality $|f(x)| \leq A_N/(1 + |x|^N)$ which holds for every fixed $N \geq 0$. As a simple consequence, whenever t is fixed, we have

$$|f(x - \gamma t)| \leq A'_N/(1 + |x|^N) \quad \text{for all } \gamma \in S^2.$$

To see this consider separately the cases when $|x| \leq 2|t|$, and $|x| > 2|t|$. Therefore, by integration

$$|F(x)| \leq A'_N/(1 + |x|^N),$$

and since this holds for every N, the function F is rapidly decreasing. One next observes that F is indefinitely differentiable, and

$$\left(\frac{\partial}{\partial x}\right)^\alpha F(x) = M_t(f^{(\alpha)})(x) \tag{6}$$

where $f^{(\alpha)}(x) = (\partial/\partial x)^\alpha f$. It suffices to prove this when $(\partial/\partial x)^\alpha = \partial/\partial x_k$, and then proceed by induction to get the general case. Furthermore, it is enough to take $k = 1$. Now

$$\frac{F(x_1 + h, x_2, x_3) - F(x_1, x_2, x_3)}{h} = \frac{1}{4\pi} \int_{S^2} g_h(\gamma)\, d\sigma(\gamma)$$

where

$$g_h(\gamma) = \frac{f(x + e_1 h - \gamma t) - f(x - \gamma t)}{h},$$

and $e_1 = (1, 0, 0)$. Now, it suffices to observe that $g_h \to \frac{\partial}{\partial x_1} f(x - \gamma t)$ as $h \to 0$ uniformly in γ. As a result, we find that (6) holds, and by the first argument, it follows that $\left(\frac{\partial}{\partial x}\right)^\alpha F(x)$ is also rapidly decreasing, hence $F \in \mathcal{S}$. The same argument applies to each t-derivative of $M_t(f)$.

The basic fact about integration on spheres that we shall need is the following Fourier transform formula.

Lemma 3.5 $\displaystyle \frac{1}{4\pi} \int_{S^2} e^{-2\pi i \xi \cdot \gamma}\, d\sigma(\gamma) = \frac{\sin(2\pi|\xi|)}{2\pi|\xi|}.$

This formula, as we shall see in the following section, is connected to the fact that the Fourier transform of a radial function is radial.

Proof. Note that the integral on the left is radial in ξ. Indeed, if R is a rotation then

$$\int_{S^2} e^{-2\pi i R(\xi)\cdot\gamma}\, d\sigma(\gamma) = \int_{S^2} e^{-2\pi i \xi \cdot R^{-1}(\gamma)}\, d\sigma(\gamma) = \int_{S^2} e^{-2\pi i \xi\cdot\gamma}\, d\sigma(\gamma)$$

because we may change variables $\gamma \to R^{-1}(\gamma)$. (For this, see formula (4) in the appendix.) So if $|\xi| = \rho$, it suffices to prove the lemma with

$\xi = (0, 0, \rho)$. If $\rho = 0$, the lemma is obvious. If $\rho > 0$, we choose spherical coordinates to find that the left-hand side is equal to

$$\frac{1}{4\pi} \int_0^{2\pi} \int_0^{\pi} e^{-2\pi i \rho \cos\theta} \sin\theta \, d\theta \, d\varphi.$$

The change of variables $u = -\cos\theta$ gives

$$\begin{aligned} \frac{1}{4\pi} \int_0^{2\pi} \int_0^{\pi} e^{-2\pi i \rho \cos\theta} \sin\theta \, d\theta \, d\varphi &= \frac{1}{2} \int_0^{\pi} e^{-2\pi i \rho \cos\theta} \sin\theta \, d\theta \\ &= \frac{1}{2} \int_{-1}^{1} e^{2\pi i \rho u} \, du \\ &= \frac{1}{4\pi i \rho} \left[e^{2\pi i \rho u} \right]_{-1}^{1} \\ &= \frac{\sin(2\pi\rho)}{2\pi\rho}, \end{aligned}$$

and the formula is proved.

By the defining formula (5) we may interpret $M_t(f)$ as a convolution of the function f with the element $d\sigma$, and since the Fourier transform interchanges convolutions with products, we are led to believe that $\widehat{M_t(f)}$ is the product of the corresponding Fourier transforms. Indeed, we have the identity

$$\widehat{M_t(f)}(\xi) = \hat{f}(\xi) \frac{\sin(2\pi|\xi|t)}{2\pi|\xi|t}. \tag{7}$$

To see this, write

$$\widehat{M_t(f)}(\xi) = \int_{\mathbb{R}^3} e^{-2\pi i x \cdot \xi} \left(\frac{1}{4\pi} \int_{S^2} f(x - \gamma t) \, d\sigma(\gamma) \right) dx,$$

and note that we may interchange the order of integration and make a simple change of variables to achieve the desired identity.

As a result, we find that the solution of our problem may be expressed by using the spherical means of the initial data.

Theorem 3.6 *The solution when $d = 3$ of the Cauchy problem for the wave equation*

$$\Delta u = \frac{\partial^2 u}{\partial t^2} \quad \textit{subject to} \quad u(x, 0) = f(x) \quad \textit{and} \quad \frac{\partial u}{\partial t}(x, 0) = g(x)$$

is given by

$$u(x, t) = \frac{\partial}{\partial t}(t M_t(f)(x)) + t M_t(g)(x).$$

Proof. Consider first the problem

$$\triangle u = \frac{\partial^2 u}{\partial t^2} \quad \text{subject to} \quad u(x,0) = 0 \quad \text{and} \quad \frac{\partial u}{\partial t}(x,0) = g(x).$$

Then by Theorem 3.1, we know that its solution u_1 is given by

$$\begin{aligned} u_1(x,t) &= \int_{\mathbb{R}^3} \left[\hat{g}(\xi)\frac{\sin(2\pi|\xi|t)}{2\pi|\xi|}\right] e^{2\pi i x\cdot\xi}\, d\xi \\ &= t\int_{\mathbb{R}^3} \left[\hat{g}(\xi)\frac{\sin(2\pi|\xi|t)}{2\pi|\xi|t}\right] e^{2\pi i x\cdot\xi}\, d\xi \\ &= tM_t(g)(x), \end{aligned}$$

where we have used (7) applied to g, and the Fourier inversion formula.

According to Theorem 3.1 again, the solution to the problem

$$\triangle u = \frac{\partial^2 u}{\partial t^2} \quad \text{subject to} \quad u(x,0) = f(x) \quad \text{and} \quad \frac{\partial u}{\partial t}(x,0) = 0$$

is given by

$$\begin{aligned} u_2(x,t) &= \int_{\mathbb{R}^3} \left[\hat{f}(\xi)\cos(2\pi|\xi|t)\right] e^{2\pi i x\cdot\xi}\, d\xi \\ &= \frac{\partial}{\partial t}\left(t\int_{\mathbb{R}^3} \left[\hat{f}(\xi)\frac{\sin(2\pi|\xi|t)}{2\pi|\xi|t}\right] e^{2\pi i x\cdot\xi}\, d\xi\right) \\ &= \frac{\partial}{\partial t}(tM_t(f)(x)). \end{aligned}$$

We may now superpose these two solutions to obtain $u = u_1 + u_2$ as the solution of our original problem.

Huygens principle

The solutions to the wave equation in one and three dimensions are given, respectively, by

$$u(x,t) = \frac{f(x+t)+f(x-t)}{2} + \frac{1}{2}\int_{x-t}^{x+t} g(y)\, dy$$

and

$$u(x,t) = \frac{\partial}{\partial t}(tM_t(f)(x)) + tM_t(g)(x).$$

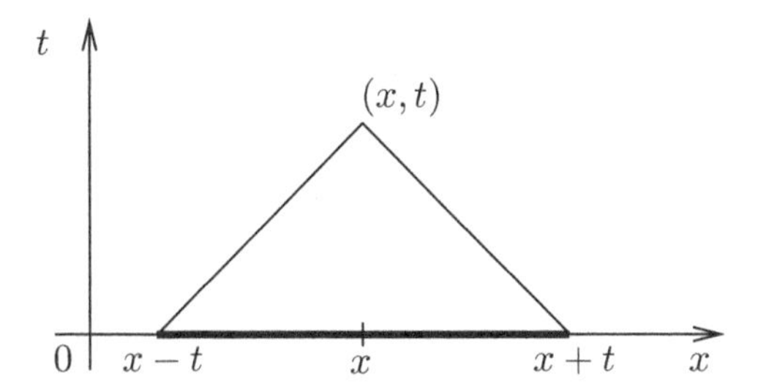

Figure 1. Huygens principle, $d = 1$

We observe that in the one-dimensional problem, the value of the solution at (x, t) depends only on the values of f and g in the interval centered at x of length $2t$, as shown in Figure 1.

If in addition $g = 0$, then the solution depends only on the data at the two boundary points of this interval. In three dimensions, this boundary dependence always holds. More precisely, the solution $u(x, t)$ depends only on the values of f and g in an immediate neighborhood of the sphere centered at x and of radius t. This situation is depicted in Figure 2, where we have drawn the cone originating at (x, t) and with its base the ball centered at x of radius t. This cone is called the **backward light cone** originating at (x, t).

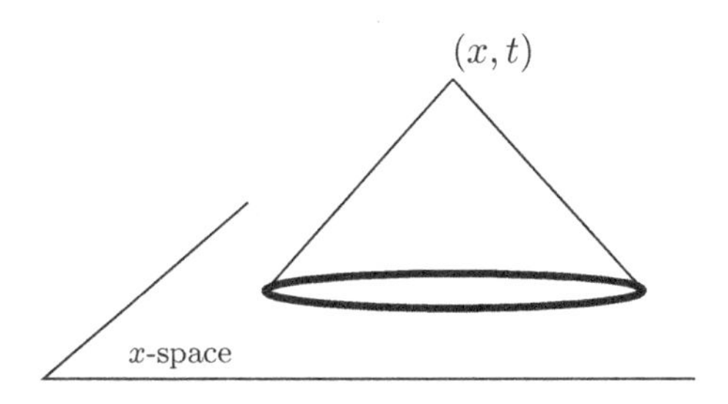

Figure 2. Backward light cone originating at (x, t)

Alternatively, the data at a point x_0 in the plane $t = 0$ influences the solution only on the boundary of a cone originating at x_0, called the **forward light cone** and depicted in Figure 3.

This phenomenon, known as the **Huygens principle**, is immediate from the formulas for u given above.

Another important aspect of the wave equation connected with these

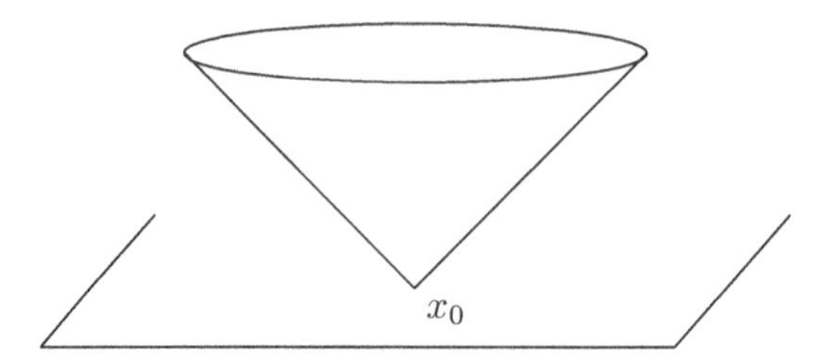

Figure 3. The forward light cone originating at x_0

considerations is that of the **finite speed of propagation**. (In the case where $c = 1$, the speed is 1.) This means that if we have an initial disturbance localized at $x = x_0$, then after a finite time t, its effects will have propagated only inside the ball centered at x_0 of radius $|t|$. To state this precisely, suppose the initial conditions f and g are supported in the ball of radius δ, centered at x_0 (think of δ as small). Then $u(x, t)$ is supported in the ball of radius $|t| + \delta$ centered at x_0. This assertion is clear from the above discussion.

3.3 The wave equation in $\mathbb{R}^2 \times \mathbb{R}$: descent

It is a remarkable fact that the solution of the wave equation in three dimensions leads to a solution of the wave equation in two dimensions. Define the corresponding means by

$$\widetilde{M}_t(F)(x) = \frac{1}{2\pi} \int_{|y| \leq 1} F(x - ty)(1 - |y|^2)^{-1/2}\, dy.$$

Theorem 3.7 *A solution of the Cauchy problem for the wave equation in two dimensions with initial data $f, g \in \mathcal{S}(\mathbb{R}^2)$ is given by*

$$u(x, t) = \frac{\partial}{\partial t}(t\widetilde{M}_t(f)(x)) + t\widetilde{M}_t(g)(x). \tag{8}$$

Notice the difference between this case and the case $d = 3$. Here, u at (x, t) depends on f and g in the whole disc (of radius $|t|$ centered at x), and not just on the values of the initial data near the boundary of that disc.

Formally, the identity in the theorem arises as follows. If we start with an initial pair of functions f and g in $\mathcal{S}(\mathbb{R}^2)$, we may consider the corresponding functions $\tilde{f}$ and $\tilde{g}$ on $\mathbb{R}^3$ that are merely extensions of f and g that are constant in the x_3 variable, that is,

$$\tilde{f}(x_1, x_2, x_3) = f(x_1, x_2) \quad \text{and} \quad \tilde{g}(x_1, x_2, x_3) = g(x_1, x_2).$$

Now, if $\tilde{u}$ is the solution (given in the previous section) of the 3-dimensional wave equation with initial data $\tilde{f}$ and $\tilde{g}$, then one can expect that $\tilde{u}$ is also constant in x_3 so that $\tilde{u}$ satisfies the 2-dimensional wave equation. A difficulty with this argument is that $\tilde{f}$ and $\tilde{g}$ are not rapidly decreasing since they are constant in x_3, so that our previous methods do not apply. However, it is easy to modify the argument so as to obtain a proof of Theorem 3.7.

We fix $T > 0$ and consider a function $\eta(x_3)$ that is in $\mathcal{S}(\mathbb{R})$, such that $\eta(x_3) = 1$ if $|x_3| \leq 3T$. The trick is to truncate $\tilde{f}$ and $\tilde{g}$ in the x_3-variable, and consider instead

$$\tilde{f}^\flat(x_1, x_2, x_3) = f(x_1, x_2)\eta(x_3) \quad \text{and} \quad \tilde{g}^\flat(x_1, x_2, x_3) = g(x_1, x_2)\eta(x_3).$$

Now both $\tilde{f}^\flat$ and $\tilde{g}^\flat$ are in $\mathcal{S}(\mathbb{R}^3)$, so Theorem 3.6 provides a solution $\tilde{u}^\flat$ of the wave equation with initial data $\tilde{f}^\flat$ and $\tilde{g}^\flat$. It is easy to see from the formula that $\tilde{u}^\flat(x, t)$ is independent of x_3, whenever $|x_3| \leq T$ and $|t| \leq T$. In particular, if we define $u(x_1, x_2, t) = \tilde{u}^\flat(x_1, x_2, 0, t)$, then u satisfies the 2-dimensional wave equation when $|t| \leq T$. Since T is arbitrary, u is a solution to our problem, and it remains to see why u has the desired form.

By definition of the spherical coordinates, we recall that the integral of a function H over the sphere S^2 is given by

$$\frac{1}{4\pi}\int_{S^2} H(\gamma)\, d\sigma(\gamma) = \frac{1}{4\pi}\int_0^{2\pi}\int_0^{\pi} H(\sin\theta\cos\varphi, \sin\theta\sin\varphi, \cos\theta)\sin\theta\, d\theta\, d\varphi.$$

If H does not depend on the last variable, that is, $H(x_1, x_2, x_3) = h(x_1, x_2)$ for some function h of two variables, then

$$M_t(H)(x_1, x_2, 0) = \frac{1}{4\pi}\int_0^{2\pi}\int_0^{\pi} h(x_1 - t\sin\theta\cos\varphi, x_2 - t\sin\theta\sin\varphi)\sin\theta\, d\theta\, d\varphi.$$

To calculate this last integral, we split the θ-integral from 0 to $\pi/2$ and then $\pi/2$ to π. By making the change of variables $r = \sin\theta$, we find, after a final change to polar coordinates, that

$$\begin{aligned} M_t(H)(x_1, x_2, 0) &= \frac{1}{2\pi}\int_{|y|\leq 1} h(x - ty)(1 - |y|^2)^{-1/2}\, dy \\ &= \widetilde{M}_t(h)(x_1, x_2). \end{aligned}$$

Applying this to $H = \tilde{f}^\flat$, $h = f$, and $H = \tilde{g}^\flat$, $h = g$, we find that u is given by the formula (8), and the proof of Theorem 3.7 is complete.

Remark. In the case of general d, the solution of the wave equation shares many of the properties we have discussed in the special cases $d = 1, 2$, and 3.

- At a given time t, the initial data at a point x only affects the solution u in a specific region. When $d > 1$ is odd, the data influences only the points on the boundary of the forward light cone originating at x, while when $d = 1$ or d is even, it affects all points of the forward light cone. Alternatively, the solution at a point (x, t) depends only on the data at the base of the backward light cone originating at (x, t). In fact, when $d > 1$ is odd, only the data in an immediate neighborhood of the boundary of the base will influence $u(x, t)$.

- Waves propagate with finite speed: if the initial data is supported in a bounded set, then the support of the solution u spreads with velocity 1 (or more generally c, if the wave equation is not normalized).

We can illustrate some of these facts by the following observation about the different behavior of the propagation of waves in three and two dimensions. Since the propagation of light is governed by the three-dimensional wave equation, if at $t = 0$ a light flashes at the origin, the following happens: any observer will see the flash (after a finite amount of time) only for an instant. In contrast, consider what happens in two dimensions. If we drop a stone in a lake, any point on the surface will begin (after some time) to undulate; although the amplitude of the oscillations will decrease over time, the undulations will continue (in principle) indefinitely.

The difference in character of the formulas for the solutions of the wave equation when $d = 1$ and $d = 3$ on the one hand, and $d = 2$ on the other hand, illustrates a general principle in d-dimensional Fourier analysis: a significant number of formulas that arise are simpler in the case of odd dimensions, compared to the corresponding situations in even dimensions. We will see several further examples of this below.

4 Radial symmetry and Bessel functions

We observed earlier that the Fourier transform of a radial function in $\mathbb{R}^d$ is also radial. In other words, if $f(x) = f_0(|x|)$ for some f_0, then

$\hat{f}(\xi) = F_0(|\xi|)$ for some F_0. A natural problem is to determine a relation between f_0 and F_0.

This problem has a simple answer in dimensions one and three. If $d = 1$ the relation we seek is

$$F_0(\rho) = 2\int_0^\infty \cos(2\pi\rho r) f_0(r)\, dr. \tag{9}$$

If we recall that $\mathbb{R}$ has only two rotations, the identity and multiplication by -1, we find that a function is radial precisely when it is even. Having made this observation it is easy to see that if f is radial, and $|\xi| = \rho$, then

$$\begin{aligned} F_0(\rho) = \hat{f}(|\xi|) &= \int_{-\infty}^{\infty} f(x) e^{-2\pi i x|\xi|}\, dx \\ &= \int_0^\infty f_0(r)(e^{-2\pi i r|\xi|} + e^{2\pi i r|\xi|})\, dr \\ &= 2\int_0^\infty \cos(2\pi\rho r) f_0(r)\, dr. \end{aligned}$$

In the case $d = 3$, the relation between f_0 and F_0 is also quite simple and given by the formula

$$F_0(\rho) = 2\rho^{-1}\int_0^\infty \sin(2\pi\rho r) f_0(r) r\, dr. \tag{10}$$

The proof of this identity is based on the formula for the Fourier transform of the surface element $d\sigma$ given in Lemma 3.5:

$$\begin{aligned} F_0(\rho) = \hat{f}(\xi) &= \int_{\mathbb{R}^3} f(x) e^{-2\pi i x\cdot\xi}\, dx \\ &= \int_0^\infty f_0(r) \int_{S^2} e^{-2\pi i r\gamma\cdot\xi} d\sigma(\gamma) r^2\, dr \\ &= \int_0^\infty f_0(r) \frac{2\sin(2\pi\rho r)}{\rho r} r^2\, dr \\ &= 2\rho^{-1}\int_0^\infty \sin(2\pi\rho r) f_0(r) r\, dr. \end{aligned}$$

More generally, the relation between f_0 and F_0 has a nice description in terms of a family of special functions that arise naturally in problems that exhibit radial symmetry.

The **Bessel function** of order $n \in \mathbb{Z}$, denoted $J_n(\rho)$, is defined as the nth Fourier coefficient of the function $e^{i\rho\sin\theta}$. So

$$J_n(\rho) = \frac{1}{2\pi}\int_0^{2\pi} e^{i\rho\sin\theta} e^{-in\theta}\, d\theta,$$

therefore

$$e^{i\rho\sin\theta} = \sum_{n=-\infty}^{\infty} J_n(\rho)e^{in\theta}.$$

As a result of this definition, we find that when $d = 2$, the relation between f_0 and F_0 is

$$F_0(\rho) = 2\pi \int_0^\infty J_0(2\pi r\rho)f_0(r)r\,dr. \tag{11}$$

Indeed, since $\hat{f}(\xi)$ is radial we take $\xi = (0, -\rho)$ so that

$$\begin{aligned}
\hat{f}(\xi) &= \int_{\mathbb{R}^2} f(x)e^{2\pi i x\cdot(0,\rho)}\,dx \\
&= \int_0^{2\pi}\int_0^\infty f_0(r)e^{2\pi i r\rho\sin\theta} r\,dr\,d\theta \\
&= 2\pi \int_0^\infty J_0(2\pi r\rho)f_0(r)r\,dr,
\end{aligned}$$

as desired.

In general, there are corresponding formulas relating f_0 and F_0 in $\mathbb{R}^d$ in terms of Bessel functions of order $d/2 - 1$ (see Problem 2). In even dimensions, these are the Bessel functions we have defined above. For odd dimensions, we need a more general definition of Bessel functions to encompass half-integral orders. Note that the formulas for the Fourier transform of radial functions give another illustration of the differences between odd and even dimensions. When $d = 1$ or $d = 3$ (as well as $d > 3$, d odd) the formulas are in terms of elementary functions, but this is not the case when d is even.

5 The Radon transform and some of its applications

Invented by Johann Radon in 1917, the integral transform we discuss next has many applications in mathematics and other sciences, including a significant achievement in medicine. To motivate the definitions and the central problem of reconstruction, we first present the close connection between the Radon transform and the development of X-ray scans (or CAT scans) in the theory of medical imaging. The solution of the reconstruction problem, and the introduction of new algorithms and faster computers, all contributed to a rapid development of computerized tomography. In practice, X-ray scans provide a "picture" of an internal organ, one that helps to detect and locate many types of abnormalities.

After a brief description of X-ray scans in two dimensions, we define the X-ray transform and formulate the basic problem of inverting this mapping. Although this problem has an explicit solution in $\mathbb{R}^2$, it is more complicated than the analogous problem in three dimensions, hence we give a complete solution of the reconstruction problem only in $\mathbb{R}^3$. Here we have another example where results are simpler in the odd-dimensional case than in the even-dimensional situation.

5.1 The X-ray transform in $\mathbb{R}^2$

Consider a two-dimensional object $\mathcal{O}$ lying in the plane $\mathbb{R}^2$, which we may think of as a planar cross section of a human organ.

First, we assume that $\mathcal{O}$ is homogeneous, and suppose that a very narrow beam of X-ray photons traverses this object.

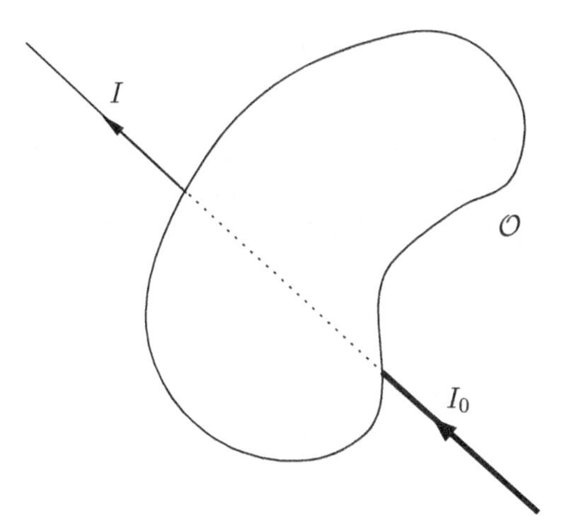

Figure 4. Attenuation of an X-ray beam

If I_0 and I denote the intensity of the beam before and after passing through $\mathcal{O}$, respectively, the following relation holds:

$$I = I_0 e^{-d\rho}.$$

Here d is the distance traveled by the beam in the object, and ρ denotes the **attenuation** coefficient (or absorption coefficient), which depends on the density and other physical characteristics of $\mathcal{O}$. If the object is not homogeneous, but consists of two materials with attenuation coefficients ρ_1 and ρ_2, then the observed decrease in the intensity of the beam is given by

$$I = I_0 e^{-d_1\rho_1 - d_2\rho_2}$$

where d_1 and d_2 denote the distances traveled by the beam in each material. In the case of an arbitrary object whose density and physical characteristics vary from point to point, the attenuation factor is a function ρ in $\mathbb{R}^2$, and the above relations become

$$I = I_0 e^{\int_L \rho}.$$

Here L is the line in $\mathbb{R}^2$ traced by the beam, and $\int_L \rho$ denotes the line integral of ρ over L. Since we observe I and I_0, the data we gather after sending the X-ray beam through the object along the line L is the quantity

$$\int_L \rho.$$

Since we may initially send the beam in any given direction, we may calculate the above integral for every line in $\mathbb{R}^2$. We define the **X-ray transform** (or **Radon transform** in $\mathbb{R}^2$) of ρ by

$$X(\rho)(L) = \int_L \rho.$$

Note that this transform assigns to each appropriate function ρ on $\mathbb{R}^2$ (for example, $\rho \in \mathcal{S}(\mathbb{R}^2)$) another function $X(\rho)$ whose domain is the set of lines L in $\mathbb{R}^2$.

The unknown is ρ, and since our original interest lies precisely in the composition of the object, the problem now becomes to reconstruct the function ρ from the collected data, that is, its X-ray transform. We therefore pose the following *reconstruction* problem: Find a formula for ρ in terms of $X(\rho)$.

Mathematically, the problem asks for a formula giving the inverse of X. Does such an inverse even exist? As a first step, we pose the following simpler *uniqueness* question: If $X(\rho) = X(\rho')$, can we conclude that $\rho = \rho'$?

There is a reasonable *a priori* expectation that $X(\rho)$ actually determines ρ, as one can see by counting the dimensionality (or degrees of freedom) involved. A function ρ on $\mathbb{R}^2$ depends on two parameters (the x_1 and x_2 coordinates, for example). Similarly, the function $X(\rho)$, which is a function of lines L, is also determined by two parameters (for example, the slope of L and its x_2-intercept). In this sense, ρ and $X(\rho)$

convey an equivalent amount of information, so it is not unreasonable to suppose that $X(\rho)$ determines ρ.

While there is a satisfactory answer to the reconstruction problem, and a positive answer to the uniqueness question in $\mathbb{R}^2$, we shall forego giving them here. (However, see Exercise 13 and Problem 8.) Instead we shall deal with the analogous but simpler situation in $\mathbb{R}^3$.

Let us finally remark that in fact, one can sample the X-ray transform, and determine $X(\rho)(L)$ for only finitely many lines. Therefore, the reconstruction method implemented in practice is based not only on the general theory, but also on sampling procedures, numerical approximations, and computer algorithms. It turns out that a method used in developing effective relevant algorithms is the fast Fourier transform, which incidentally we take up in the next chapter.

5.2 The Radon transform in $\mathbb{R}^3$

The experiment described in the previous section applies in three dimensions as well. If $\mathcal{O}$ is an object in $\mathbb{R}^3$ determined by a function ρ which describes the density and physical characteristics of this object, sending an X-ray beam through $\mathcal{O}$ determines the quantity

$$\int_L \rho,$$

for every line in $\mathbb{R}^3$. In $\mathbb{R}^2$ this knowledge was enough to uniquely determine ρ, but in $\mathbb{R}^3$ we do not need as much data. In fact, by using the heuristic argument above of counting the number of degrees of freedom, we see that for functions ρ in $\mathbb{R}^3$ the number is three, while the number of parameters determining a line L in $\mathbb{R}^3$ is four (for example, two for the intercept in the (x_1, x_2) plane, and two more for the direction of the line). Thus in this sense, the problem is over-determined.

We turn instead to the natural mathematical generalization of the two-dimensional problem. Here we wish to determine the function in $\mathbb{R}^3$ by knowing its integral over all *planes*[3] in $\mathbb{R}^3$. To be precise, when we speak of a plane, we mean a plane not necessarily passing through the origin. If $\mathcal{P}$ is such a plane, we define the Radon transform $\mathcal{R}(f)$ by

$$\mathcal{R}(f)(\mathcal{P}) = \int_{\mathcal{P}} f.$$

To simplify our presentation, we shall follow our practice of assuming that we are dealing with functions in the class $\mathcal{S}(\mathbb{R}^3)$. However, many

[3]Note that the dimensionality associated with points on $\mathbb{R}^3$, and that for planes in $\mathbb{R}^3$, equals three in both cases.

of the results obtained below can be shown to be valid for much larger classes of functions.

First, we explain what we mean by the integral of f over a plane. The description we use for planes in $\mathbb{R}^3$ is the following: given a unit vector $\gamma \in S^2$ and a number $t \in \mathbb{R}$, we define the plane $\mathcal{P}_{t,\gamma}$ by

$$\mathcal{P}_{t,\gamma} = \{x \in \mathbb{R}^3 : x \cdot \gamma = t\}.$$

So we parametrize a plane by a unit vector γ orthogonal to it, and by its "distance" t to the origin (see Figure 5). Note that $\mathcal{P}_{t,\gamma} = \mathcal{P}_{-t,-\gamma}$, and we allow t to take negative values.

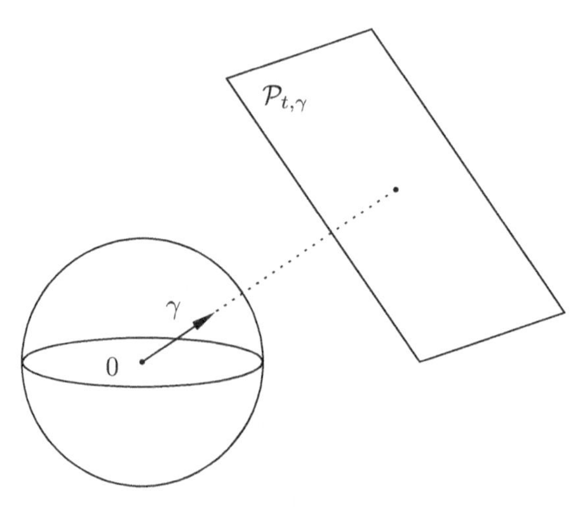

Figure 5. Description of a plane in $\mathbb{R}^3$

Given a function $f \in \mathcal{S}(\mathbb{R}^d)$, we need to make sense of its integral over $\mathcal{P}_{t,\gamma}$. We proceed as follows. Choose unit vectors e_1, e_2 so that e_1, e_2, γ is an orthonormal basis for $\mathbb{R}^3$. Then any $x \in \mathcal{P}_{t,\gamma}$ can be written uniquely as

$$x = t\gamma + u \quad \text{where} \quad u = u_1e_1 + u_2e_2 \quad \text{with } u_1, u_2 \in \mathbb{R}.$$

If $f \in \mathcal{S}(\mathbb{R}^3)$, we define

$$\int_{\mathcal{P}_{t,\gamma}} f = \int_{\mathbb{R}^2} f(t\gamma + u_1e_1 + u_2e_2)\, du_1\, du_2. \tag{12}$$

To be consistent, we must check that this definition is independent of the choice of the vectors e_1, e_2.

Proposition 5.1 *If $f \in \mathcal{S}(\mathbb{R}^3)$, then for each γ the definition of $\int_{\mathcal{P}_{t,\gamma}} f$ is independent of the choice of e_1 and e_2. Moreover*

$$\int_{-\infty}^{\infty} \left(\int_{\mathcal{P}_{t,\gamma}} f \right) dt = \int_{\mathbb{R}^3} f(x)\, dx.$$

Proof. If e_1', e_2' is another choice of basis vectors so that γ, e_1', e_2' is orthonormal, consider the rotation R in $\mathbb{R}^2$ which takes e_1 to e_1' and e_2 to e_2'. Changing variables $u' = R(u)$ in the integral proves that our definition (12) is independent of the choice of basis.

To prove the formula, let R denote the rotation which takes the standard basis of unit vectors[4] in $\mathbb{R}^3$ to γ, e_1, and e_2. Then

$$\begin{aligned} \int_{\mathbb{R}^3} f(x)\, dx &= \int_{\mathbb{R}^3} f(Rx)\, dx \\ &= \int_{\mathbb{R}^3} f(x_1\gamma + x_2 e_1 + x_3 e_2)\, dx_1\, dx_2\, dx_3 \\ &= \int_{-\infty}^{\infty} \left(\int_{\mathcal{P}_{t,\gamma}} f \right) dt. \end{aligned}$$

Remark. We digress to point out that the X-ray transform determines the Radon transform, since two-dimensional integrals can be expressed as iterated one-dimensional integrals. In other words, the knowledge of the integral of a function over all lines determines the integral of that function over any plane.

Having disposed of these preliminary matters, we turn to the study of the original problem. The **Radon transform** of a function $f \in \mathcal{S}(\mathbb{R}^3)$ is defined by

$$\mathcal{R}(f)(t, \gamma) = \int_{\mathcal{P}_{t,\gamma}} f.$$

In particular, we see that the Radon transform is a function on the set of planes in $\mathbb{R}^3$. From the parametrization given for a plane, we may equivalently think of $\mathcal{R}(f)$ as a function on the product $\mathbb{R} \times S^2 = \{(t, \gamma) :\ t \in \mathbb{R},\ \gamma \in S^2\}$, where S^2 denotes the unit sphere in $\mathbb{R}^3$. The relevant class of functions on $\mathbb{R} \times S^2$ consists of those that satisfy the Schwartz condition in t uniformly in γ. In other words, we define $\mathcal{S}(\mathbb{R} \times S^2)$ to be the space of all continuous functions $F(t, \gamma)$ that are indefinitely

[4]Here we are referring to the vectors $(1, 0, 0)$, $(0, 1, 0)$, and $(0, 0, 1)$.

differentiable in t, and that satisfy

$$\sup_{t\in\mathbb{R},\ \gamma\in S^2} |t|^k \left| \frac{d^\ell F}{\partial t^\ell}(t,\gamma) \right| < \infty \quad \text{for all integers } k,\ell \geq 0.$$

Our goal is to solve the following problems.

Uniqueness problem: If $\mathcal{R}(f) = \mathcal{R}(g)$, then $f = g$.

Reconstruction problem: Express f in terms of $\mathcal{R}(f)$.

The solutions will be obtained by using the Fourier transform. In fact, the key point is a very elegant and essential relation between the Radon and Fourier transforms.

Lemma 5.2 *If $f \in \mathcal{S}(\mathbb{R}^3)$, then $\mathcal{R}(f)(t,\gamma) \in \mathcal{S}(\mathbb{R})$ for each fixed γ. Moreover,*

$$\widehat{\mathcal{R}}(f)(s,\gamma) = \hat{f}(s\gamma).$$

To be precise, $\hat{f}$ denotes the (three-dimensional) Fourier transform of f, while $\widehat{\mathcal{R}}(f)(s,\gamma)$ denotes the one-dimensional Fourier transform of $\mathcal{R}(f)(t,\gamma)$ as a function of t, with γ fixed.

Proof. Since $f \in \mathcal{S}(\mathbb{R}^3)$, for every positive integer N there is a constant $A_N < \infty$ so that

$$(1+|t|)^N(1+|u|)^N|f(t\gamma+u)| \leq A_N,$$

if we recall that $x = t\gamma + u$, where γ is orthogonal to u. Therefore, as soon as $N \geq 3$, we find

$$(1+|t|)^N \mathcal{R}(f)(t,\gamma) \leq A_N \int_{\mathbb{R}^2} \frac{du}{(1+|u|)^N} < \infty.$$

A similar argument for the derivatives shows that $\mathcal{R}(f)(t,\gamma) \in \mathcal{S}(\mathbb{R})$ for each fixed γ.

To establish the identity, we first note that

$$\begin{aligned}
\widehat{\mathcal{R}}(f)(s,\gamma) &= \int_{-\infty}^{\infty} \left(\int_{\mathcal{P}_{t,\gamma}} f \right) e^{-2\pi i s t}\, dt \\
&= \int_{-\infty}^{\infty} \int_{\mathbb{R}^2} f(t\gamma + u_1 e_1 + u_2 e_2)\, du_1\, du_2 e^{-2\pi i s t}\, dt.
\end{aligned}$$

However, since $\gamma \cdot u = 0$ and $|\gamma| = 1$, we may write

$$e^{-2\pi i s t} = e^{-2\pi i s \gamma \cdot (t\gamma + u)}.$$

As a result, we find that

$$\begin{aligned} \widehat{\mathcal{R}}(f)(s, \gamma) &= \int_{-\infty}^{\infty} \int_{\mathbb{R}^2} f(t\gamma + u_1 e_1 + u_2 e_2) e^{-2\pi i s \gamma \cdot (t\gamma + u)} \, du_1 \, du_2 \, dt \\ &= \int_{-\infty}^{\infty} \int_{\mathbb{R}^2} f(t\gamma + u) e^{-2\pi i s \gamma \cdot (t\gamma + u)} \, du \, dt. \end{aligned}$$

A final rotation from γ, e_1, e_2 to the standard basis in $\mathbb{R}^3$ proves that $\widehat{\mathcal{R}}(f)(s, \gamma) = \hat{f}(s\gamma)$, as desired.

As a consequence of this identity, we can answer the uniqueness question for the Radon transform in $\mathbb{R}^3$ in the affirmative.

Corollary 5.3 *If $f, g \in \mathcal{S}(\mathbb{R}^3)$ and $\mathcal{R}(f) = \mathcal{R}(g)$, then $f = g$.*

The proof of the corollary follows from an application of the lemma to the difference $f - g$ and use of the Fourier inversion theorem.

Our final task is to give the formula that allows us to recover f from its Radon transform. Since $\mathcal{R}(f)$ is a function on the set of planes in $\mathbb{R}^3$, and f is a function of the space variables $x \in \mathbb{R}^3$, to recover f we introduce the dual Radon transform, which passes from functions defined on planes to functions in $\mathbb{R}^3$.

Given a function F on $\mathbb{R} \times S^2$, we define its **dual Radon transform** by

$$\mathcal{R}^*(F)(x) = \int_{S^2} F(x \cdot \gamma, \gamma) \, d\sigma(\gamma). \tag{13}$$

Observe that a point x belongs to $\mathcal{P}_{t,\gamma}$ if and only if $x \cdot \gamma = t$, so the idea here is that given $x \in \mathbb{R}^3$, we obtain $\mathcal{R}^*(F)(x)$ by integrating F over the subset of all planes passing through x, that is,

$$\mathcal{R}^*(F)(x) = \int_{\{\mathcal{P}_{t,\gamma} \text{ such that } x \in \mathcal{P}_{t,\gamma}\}} F,$$

where the integral on the right is given the precise meaning in (13). We use the terminology "dual" because of the following observation. If $V_1 = \mathcal{S}(\mathbb{R}^3)$ with the usual Hermitian inner product

$$(f, g)_1 = \int_{\mathbb{R}^3} f(x) \overline{g(x)} \, dx,$$

and $V_2 = \mathcal{S}(\mathbb{R} \times S^2)$ with the Hermitian inner product

$$(F, G)_2 = \int_{\mathbb{R}} \int_{S^2} F(t, \gamma) \overline{G(t, \gamma)} \, d\sigma(\gamma) \, dt,$$

then

$$\mathcal{R} : V_1 \to V_2, \qquad \mathcal{R}^* : V_2 \to V_1,$$

with

$$(\mathcal{R}f, F)_2 = (f, \mathcal{R}^* F)_1. \tag{14}$$

The validity of this identity is not needed in the argument below, and its verification is left as an exercise for the reader.

We can now state the reconstruction theorem.

Theorem 5.4 *If $f \in \mathcal{S}(\mathbb{R}^3)$, then*

$$\triangle(\mathcal{R}^*\mathcal{R}(f)) = -8\pi^2 f.$$

We recall that $\triangle = \frac{\partial^2}{\partial x_1^2} + \frac{\partial^2}{\partial x_2^2} + \frac{\partial^2}{\partial x_3^2}$ is the Laplacian.

Proof. By our previous lemma, we have

$$\mathcal{R}(f)(t, \gamma) = \int_{-\infty}^{\infty} \hat{f}(s\gamma) e^{2\pi i t s} \, ds.$$

Therefore

$$\mathcal{R}^*\mathcal{R}(f)(x) = \int_{S^2} \int_{-\infty}^{\infty} \hat{f}(s\gamma) e^{2\pi i x \cdot \gamma s} \, ds \, d\sigma(\gamma),$$

hence

$$\begin{aligned}
\triangle(\mathcal{R}^*\mathcal{R}(f))(x) &= \int_{S^2} \int_{-\infty}^{\infty} \hat{f}(s\gamma)(-4\pi^2 s^2) e^{2\pi i x \cdot \gamma s} \, ds \, d\sigma(\gamma) \\
&= -4\pi^2 \int_{S^2} \int_{-\infty}^{\infty} \hat{f}(s\gamma) e^{2\pi i x \cdot \gamma s} s^2 \, ds \, d\sigma(\gamma) \\
&= -4\pi^2 \int_{S^2} \int_{-\infty}^{0} \hat{f}(s\gamma) e^{2\pi i x \cdot \gamma s} s^2 \, ds \, d\sigma(\gamma) \\
&\qquad\qquad - 4\pi^2 \int_{S^2} \int_{0}^{\infty} \hat{f}(s\gamma) e^{2\pi i x \cdot \gamma s} s^2 \, ds \, d\sigma(\gamma) \\
&= -8\pi^2 \int_{S^2} \int_{0}^{\infty} \hat{f}(s\gamma) e^{2\pi i x \cdot \gamma s} s^2 \, ds \, d\sigma(\gamma) \\
&= -8\pi^2 f(x).
\end{aligned}$$

In the first line, we have differentiated under the integral sign and used the fact $\triangle(e^{2\pi ix\cdot\gamma s}) = (-4\pi^2 s^2)e^{2\pi ix\cdot\gamma s}$, since $|\gamma| = 1$. The last step follows from the formula for polar coordinates in $\mathbb{R}^3$ and the Fourier inversion theorem.

5.3 A note about plane waves

We conclude this chapter by briefly mentioning a nice connection between the Radon transform and solutions of the wave equation. This comes about in the following way. Recall that when $d = 1$, the solution of the wave equation can be expressed as the sum of traveling waves (see Chapter 1), and it is natural to ask if an analogue of such traveling waves exists in higher dimensions. The answer is as follows. Let F be a function of one variable, which we assume is sufficiently smooth (say C^2), and consider $u(x,t)$ defined by

$$u(x,t) = F((x\cdot\gamma) - t),$$

where $x \in \mathbb{R}^d$ and γ is a unit vector in $\mathbb{R}^d$. It is easy to verify directly that u is a solution of the wave equation in $\mathbb{R}^d$ (with $c = 1$). Such a solution is called a **plane wave**; indeed, notice that u is constant on every plane perpendicular to the direction γ, and as time t increases, the wave travels in the γ direction. (It should be remarked that plane waves are never functions in $\mathcal{S}(\mathbb{R}^d)$ when $d > 1$ because they are constant in directions perpendicular to γ).[5]

The basic fact is that when $d > 1$, the solution of the wave equation can be written as an integral (as opposed to sum, when $d = 1$) of plane waves; this can in fact be done via the Radon transform of the initial data f and g. For the relevant formulas when $d = 3$, see Problem 6.

6 Exercises

1. Suppose that R is a rotation in the plane $\mathbb{R}^2$, and let

$$R = \begin{pmatrix} a & b \\ c & d \end{pmatrix}$$

denote its matrix with respect to the standard basis vectors $e_1 = (1,0)$ and $e_2 = (0,1)$.

(a) Write the conditions $R^t = R^{-1}$ and $\det(R) = \pm 1$ in terms of equations in a, b, c, d.

[5]Incidentally, this observation is further indication that a fuller treatment of the wave equation requires lifting the restriction that functions belong to $\mathcal{S}(\mathbb{R}^d)$.

(b) Show that there exists $\varphi \in \mathbb{R}$ such that $a + ib = e^{i\varphi}$.

(c) Conclude that if R is proper, then it can be expressed as $z \mapsto ze^{i\varphi}$, and if R is improper, then it takes the form $z \mapsto \overline{z}e^{i\varphi}$, where $\overline{z} = x - iy$.

2. Suppose that $R : \mathbb{R}^3 \to \mathbb{R}^3$ is a proper rotation.

(a) Show that $p(t) = \det(R - tI)$ is a polynomial of degree 3, and prove that there exists $\gamma \in S^2$ (where S^2 denotes the unit sphere in $\mathbb{R}^3$) with

$$R(\gamma) = \gamma.$$

[Hint: Use the fact that $p(0) > 0$ to see that there is $\lambda > 0$ with $p(\lambda) = 0$. Then $R - \lambda I$ is singular, so its kernel is non-trivial.]

(b) If $\mathcal{P}$ denotes the plane perpendicular to γ and passing through the origin, show that

$$R : \mathcal{P} \to \mathcal{P},$$

and that this linear map is a rotation.

3. Recall the formula

$$\int_{\mathbb{R}^d} F(x)\,dx = \int_{S^{d-1}} \int_0^\infty F(r\gamma)r^{d-1}\,dr\,d\sigma(\gamma).$$

Apply this to the special case when $F(x) = g(r)f(\gamma)$, where $x = r\gamma$, to prove that for any rotation R, one has

$$\int_{S^{d-1}} f(R(\gamma))\,d\sigma(\gamma) = \int_{S^{d-1}} f(\gamma)\,d\sigma(\gamma),$$

whenever f is a continuous function on the sphere S^{d-1}.

4. Let A_d and V_d denote the area and volume of the unit sphere and unit ball in $\mathbb{R}^d$, respectively.

(a) Prove the formula

$$A_d = \frac{2\pi^{d/2}}{\Gamma(d/2)}$$

so that $A_2 = 2\pi$, $A_3 = 4\pi$, $A_4 = 2\pi^2, \ldots$. Here $\Gamma(x) = \int_0^\infty e^{-t}t^{x-1}\,dt$ is the Gamma function. [Hint: Use polar coordinates and the fact that $\int_{\mathbb{R}^d} e^{-\pi|x|^2}\,dx = 1$.]

(b) Show that $dV_d = A_d$, hence

$$V_d = \frac{\pi^{d/2}}{\Gamma(d/2+1)}.$$

In particular $V_2 = \pi$, $V_3 = 4\pi/3, \ldots$.

5. Let A be a $d \times d$ positive definite symmetric matrix with real coefficients. Show that

$$\int_{\mathbb{R}^d} e^{-\pi(x, A(x))}\, dx = (\det(A))^{-1/2}.$$

This generalizes the fact that $\int_{\mathbb{R}^d} e^{-\pi|x|^2}\, dx = 1$, which corresponds to the case where A is the identity.

[Hint: Apply the spectral theorem to write $A = RDR^{-1}$ where R is a rotation and, D is diagonal with entries $\lambda_1, \ldots, \lambda_d$, where $\{\lambda_i\}$ are the eigenvalues of A.]

6. Suppose $\psi \in \mathcal{S}(\mathbb{R}^d)$ satisfies $\int |\psi(x)|^2\, dx = 1$. Show that

$$\left(\int_{\mathbb{R}^d} |x|^2 |\psi(x)|^2\, dx\right)\left(\int_{\mathbb{R}^d} |\xi|^2 |\hat{\psi}(\xi)|^2\, d\xi\right) \geq \frac{d^2}{16\pi^2}.$$

This is the statement of the Heisenberg uncertainty principle in d dimensions.

7. Consider the time-dependent heat equation in $\mathbb{R}^d$:

$$\frac{\partial u}{\partial t} = \frac{\partial^2 u}{\partial x_1^2} + \cdots + \frac{\partial^2 u}{\partial x_d^2}, \qquad \text{where } t > 0, \tag{15}$$

with boundary values $u(x, 0) = f(x) \in \mathcal{S}(\mathbb{R}^d)$. If

$$\mathcal{H}_t^{(d)}(x) = \frac{1}{(4\pi t)^{d/2}} e^{-|x|^2/4t} = \int_{\mathbb{R}^d} e^{-4\pi^2 t|\xi|^2} e^{2\pi i x\cdot\xi}\, d\xi$$

is the d-dimensional **heat kernel**, show that the convolution

$$u(x, t) = (f * \mathcal{H}_t^{(d)})(x)$$

is indefinitely differentiable when $x \in \mathbb{R}^d$ and $t > 0$. Moreover, u solves (15), and is continuous up to the boundary $t = 0$ with $u(x, 0) = f(x)$.

The reader may also wish to formulate the d-dimensional analogues of Theorem 2.1 and 2.3 in Chapter 5.

8. In Chapter 5, we found that a solution to the steady-state heat equation in the upper half-plane with boundary values f is given by the convolution $u = f * \mathcal{P}_y$ where the Poisson kernel is

$$\mathcal{P}_y(x) = \frac{1}{\pi}\frac{y}{x^2 + y^2} \qquad \text{where } x \in \mathbb{R} \text{ and } y > 0.$$

More generally, one can calculate the d-dimensional Poisson kernel using the Fourier transform as follows.

(a) The **subordination principle** allows one to write expressions involving the function e^{-x} in terms of corresponding expressions involving the function e^{-x^2}. One form of this is the identity

$$e^{-\beta} = \int_0^\infty \frac{e^{-u}}{\sqrt{\pi u}} e^{-\beta^2/4u}\, du$$

when $\beta \geq 0$. Prove this identity with $\beta = 2\pi|x|$ by taking the Fourier transform of both sides.

(b) Consider the steady-state heat equation in the upper half-space $\{(x, y) : x \in \mathbb{R}^d,\ y > 0\}$

$$\sum_{j=1}^{d} \frac{\partial^2 u}{\partial x_j^2} + \frac{\partial^2 u}{\partial y^2} = 0$$

with the Dirichlet boundary condition $u(x, 0) = f(x)$. A solution to this problem is given by the convolution $u(x, y) = (f * P_y^{(d)})(x)$ where $P_y^{(d)}(x)$ is the d-dimensional Poisson kernel

$$P_y^{(d)}(x) = \int_{\mathbb{R}^d} e^{2\pi i x\cdot\xi} e^{-2\pi|\xi| y}\, d\xi.$$

Compute $P_y^{(d)}(x)$ by using the subordination principle and the d-dimensional heat kernel. (See Exercise 7.) Show that

$$P_y^{(d)}(x) = \frac{\Gamma((d+1)/2)}{\pi^{(d+1)/2}} \frac{y}{(|x|^2 + y^2)^{(d+1)/2}}.$$

9. A **spherical wave** is a solution $u(x, t)$ of the Cauchy problem for the wave equation in $\mathbb{R}^d$, which as a function of x is radial. Prove that u is a spherical wave if and only if the initial data $f, g \in \mathcal{S}$ are both radial.

10. Let $u(x, t)$ be a solution of the wave equation, and let $E(t)$ denote the energy of this wave

$$E(t) = \int_{\mathbb{R}^d} \left| \frac{\partial u}{\partial t}(x, t) \right|^2 + \sum_{j=1}^{d} \int_{\mathbb{R}^d} \left| \frac{\partial u}{\partial x_j}(x, t) \right|^2 dx.$$

We have seen that $E(t)$ is constant using Plancherel's formula. Give an alternate proof of this fact by differentiating the integral with respect to t and showing that

$$\frac{dE}{dt} = 0.$$

[Hint: Integrate by parts.]

11. Show that the solution of the wave equation

$$\frac{\partial^2 u}{\partial t^2} = \frac{\partial^2 u}{\partial x_1^2} + \frac{\partial^2 u}{\partial x_2^2} + \frac{\partial^2 u}{\partial x_3^2}$$

subject to $u(x,0) = f(x)$ and $\frac{\partial u}{\partial t}(x,0) = g(x)$, where $f, g \in \mathcal{S}(\mathbb{R}^3)$, is given by

$$u(x,t) = \frac{1}{|S(x,t)|} \int_{S(x,t)} [tg(y) + f(y) + \nabla f(y) \cdot (y - x)] \, d\sigma(y),$$

where $S(x,t)$ denotes the sphere of center x and radius t, and $|S(x,t)|$ its area. This is an alternate expression for the solution of the wave equation given in Theorem 3.6. It is sometimes called **Kirchhoff's** formula.

12. Establish the identity (14) about the dual transform given in the text. In other words, prove that

$$\int_{\mathbb{R}} \int_{S^2} \mathcal{R}(f)(t,\gamma)\overline{F(t,\gamma)} d\sigma(\gamma) \, dt = \int_{\mathbb{R}^3} f(x)\overline{\mathcal{R}^*(F)(x)} \, dx \tag{16}$$

where $f \in \mathcal{S}(\mathbb{R}^3)$, $F \in \mathcal{S}(\mathbb{R} \times S^2)$, and

$$\mathcal{R}(f) = \int_{\mathcal{P}_{t,\gamma}} f \quad \text{and} \quad \mathcal{R}^*(F)(x) = \int_{S^2} F(x \cdot \gamma, \gamma) \, d\sigma(\gamma).$$

[Hint: Consider the integral

$$\int\int\int f(t\gamma + u_1 e_2 + u_2 e_2)\overline{F(t,\gamma)} \, dt \, d\sigma(\gamma) \, du_1 \, du_2.$$

Integrating first in u gives the left-hand side of (16), while integrating in u and t and setting $x = t\gamma + u_1 e_2 + u_2 e_2$ gives the right-hand side.]

13. For each (t,θ) with $t \in \mathbb{R}$ and $|\theta| \leq \pi$, let $L = L_{t,\theta}$ denote the line in the (x,y)-plane given by

$$x \cos\theta + y \sin\theta = t.$$

This is the line perpendicular to the direction $(\cos\theta, \sin\theta)$ at "distance" t from the origin (we allow negative t). For $f \in \mathcal{S}(\mathbb{R}^2)$ the X-ray transform or two-dimensional Radon transform of f is defined by

$$X(f)(t,\theta) = \int_{L_{t,\theta}} f = \int_{-\infty}^{\infty} f(t\cos\theta + u\sin\theta, t\sin\theta - u\cos\theta) \, du.$$

Calculate the X-ray transform of the function $f(x, y) = e^{-\pi(x^2+y^2)}$.

14. Let X be the X-ray transform. Show that if $f \in \mathcal{S}$ and $X(f) = 0$, then $f = 0$, by taking the Fourier transform in one variable.

15. For $F \in \mathcal{S}(\mathbb{R} \times S^1)$, define the **dual X-ray transform** $X^*(F)$ by integrating F over all lines that pass through the point (x, y) (that is, those lines $L_{t,\theta}$ with $x \cos\theta + y \sin\theta = t$):

$$X^*(F)(x, y) = \int F(x \cos\theta + y \sin\theta, \theta)\, d\theta.$$

Check that in this case, if $f \in \mathcal{S}(\mathbb{R}^2)$ and $F \in \mathcal{S}(\mathbb{R} \times S^1)$, then

$$\int\!\!\int X(f)(t, \theta)\overline{F(t, \theta)}\, dt\, d\theta = \int\!\!\int f(x, y)\overline{X^*(F)(x, y)}\, dx\, dy.$$

7 Problems

1. Let J_n denote the nth order Bessel function, for $n \in \mathbb{Z}$. Prove that

(a) $J_n(\rho)$ is real for all real ρ.

(b) $J_{-n}(\rho) = (-1)^n J_n(\rho)$.

(c) $2J_n'(\rho) = J_{n-1}(\rho) - J_{n+1}(\rho)$.

(d) $\left(\frac{2n}{\rho}\right) J_n(\rho) = J_{n-1}(\rho) + J_{n+1}(\rho)$.

(e) $(\rho^{-n} J_n(\rho))' = -\rho^{-n} J_{n+1}(\rho)$.

(f) $(\rho^n J_n(\rho))' = \rho^n J_{n-1}(\rho)$.

(g) $J_n(\rho)$ satisfies the second order differential equation

$$J_n''(\rho) + \rho^{-1} J_n'(\rho) + (1 - n^2/\rho^2) J_n(\rho) = 0.$$

(h) Show that

$$J_n(\rho) = \left(\frac{\rho}{2}\right)^n \sum_{m=0}^{\infty} (-1)^m \frac{\rho^{2m}}{2^{2m} m!(n+m)!}.$$

(i) Show that for all integers n and all real numbers a and b we have

$$J_n(a + b) = \sum_{\ell \in \mathbb{Z}} J_\ell(a) J_{n-\ell}(b).$$

2. Another formula for $J_n(\rho)$ that allows one to define Bessel functions for non-integral values of n, $(n > -1/2)$ is

$$J_n(\rho) = \frac{(\rho/2)^n}{\Gamma(n+1/2)\sqrt{\pi}} \int_{-1}^{1} e^{i\rho t}(1-t^2)^{n-(1/2)}\, dt.$$

(a) Check that the above formula agrees with the definition of $J_n(\rho)$ for integral $n \geq 0$. [Hint: Verify it for $n = 0$ and then check that both sides satisfy the recursion formula (e) in Problem 1.]

(b) Note that $J_{1/2}(\rho) = \sqrt{\frac{2}{\pi}}\rho^{-1/2}\sin\rho$.

(c) Prove that

$$\lim_{n\to -1/2} J_n(\rho) = \sqrt{\frac{2}{\pi}}\rho^{-1/2}\cos\rho.$$

(d) Observe that the formulas we have proved in the text giving F_0 in terms of f_0 (when describing the Fourier transform of a radial function) take the form

$$F_0(\rho) = 2\pi\rho^{-(d/2)+1}\int_0^\infty J_{(d/2)-1}(2\pi\rho r)f_0(r)r^{d/2}\, dr, \tag{17}$$

for $d = 1, 2$, and 3, if one uses the formulas above with the understanding that $J_{-1/2}(\rho) = \lim_{n\to -1/2} J_n(\rho)$. It turns out that the relation between F_0 and f_0 given by (17) is valid in all dimensions d.

3. We observed that the solution $u(x,t)$ of the Cauchy problem for the wave equation given by formula (3) depends only on the initial data on the base on the backward light cone. It is natural to ask if this property is shared by *any* solution of the wave equation. An affirmative answer would imply uniqueness of the solution.

Let $B(x_0, r_0)$ denote the closed ball in the hyperplane $t = 0$ centered at x_0 and of radius r_0. The **backward light cone** with base $B(x_0, r_0)$ is defined by

$$\mathcal{L}_{B(x_0,r_0)} = \{(x,t) \in \mathbb{R}^d \times \mathbb{R} : |x - x_0| \leq r_0 - t,\quad 0 \leq t \leq r_0\}.$$

Theorem *Suppose that $u(x,t)$ is a C^2 function on the closed upper half-plane $\{(x,t) :\ x \in \mathbb{R}^d, t \geq 0\}$ that solves the wave equation*

$$\frac{\partial^2 u}{\partial t^2} = \triangle u.$$

If $u(x,0) = \frac{\partial u}{\partial t}(x,0) = 0$ for all $x \in B(x_0, r_0)$, then $u(x,t) = 0$ for all $(x,t) \in \mathcal{L}_{B(x_0,r_0)}$.

In words, if the initial data of the Cauchy problem for the wave equation vanishes on a ball B, then *any* solution u of the problem vanishes in the backward light cone with base B. The following steps outline a proof of the theorem.

(a) Assume that u is real. For $0 \leq t \leq r_0$ let $B_t(x_0, r_0) = \{x : |x - x_0| \leq r_0 - t\}$, and also define

$$\nabla u(x,t) = \left(\frac{\partial u}{\partial x_1}, \ldots, \frac{\partial u}{\partial x_d}, \frac{\partial u}{\partial t}\right).$$

Now consider the energy integral

$$\begin{aligned} E(t) &= \frac{1}{2}\int_{B_t(x_0,r_0)} |\nabla u|^2\, dx \\ &= \frac{1}{2}\int_{B_t(x_0,r_0)} \left(\frac{\partial u}{\partial t}\right)^2 + \sum_{j=1}^{d}\left(\frac{\partial u}{\partial x_j}\right)^2 dx. \end{aligned}$$

Observe that $E(t) \geq 0$ and $E(0) = 0$. Prove that

$$E'(t) = \int_{B_t(x_0,r_0)} \frac{\partial u}{\partial t}\frac{\partial^2 u}{\partial t^2} + \sum_{j=1}^{d} \frac{\partial u}{\partial x_j}\frac{\partial^2 u}{\partial x_j \partial t}\, dx - \frac{1}{2}\int_{\partial B_t(x_0,r_0)} |\nabla u|^2\, d\sigma(\gamma).$$

(b) Show that

$$\frac{\partial}{\partial x_j}\left[\frac{\partial u}{\partial x_j}\frac{\partial u}{\partial t}\right] = \frac{\partial u}{\partial x_j}\frac{\partial^2 u}{\partial x_j \partial t} + \frac{\partial^2 u}{\partial x_j^2}\frac{\partial u}{\partial t}.$$

(c) Use the last identity, the divergence theorem, and the fact that u solves the wave equation to prove that

$$E'(t) = \int_{\partial B_t(x_0,r_0)} \sum_{j=1}^{d} \frac{\partial u}{\partial x_j}\frac{\partial u}{\partial t}\nu_j\, d\sigma(\gamma) - \frac{1}{2}\int_{\partial B_t(x_0,r_0)} |\nabla u|^2\, d\sigma(\gamma),$$

where ν_j denotes the jth coordinate of the outward normal to $B_t(x_0, r_0)$.

(d) Use the Cauchy-Schwarz inequality to conclude that

$$\sum_{j=1}^{d} \frac{\partial u}{\partial x_j}\frac{\partial u}{\partial t}\nu_j \leq \frac{1}{2}|\nabla u|^2,$$

and as a result, $E'(t) \leq 0$. Deduce from this that $E(t) = 0$ and $u = 0$.

4.* There exist formulas for the solution of the Cauchy problem for the wave equation

$$\frac{\partial^2 u}{\partial t^2} = \frac{\partial^2 u}{\partial x_1^2} + \cdots + \frac{\partial^2 u}{\partial x_d^2} \quad \text{with } u(x,0) = f(x) \text{ and } \frac{\partial u}{\partial t}(x,0) = g(x)$$

in $\mathbb{R}^d \times \mathbb{R}$ in terms of spherical means which generalize the formula given in the text for $d = 3$. In fact, the solution for even dimensions is deduced from that for odd dimensions, so we discuss this case first.

Suppose that $d > 1$ is odd and let $h \in \mathcal{S}(\mathbb{R}^d)$. The spherical mean of h on the ball centered at x of radius t is defined by

$$M_r h(x) = Mh(x,r) = \frac{1}{A_d} \int_{S^{d-1}} h(x - r\gamma)\, d\sigma(\gamma),$$

where A_d denotes the area of the unit sphere S^{d-1} in $\mathbb{R}^d$.

(a) Show that

$$\triangle_x Mh(x,r) = \left[\partial_r^2 + \frac{d-1}{r}\right] Mh(x,r),$$

where $\triangle_x$ denotes the Laplacian in the space variables x, and $\partial_r = \partial/\partial r$.

(b) Show that a twice differentiable function $u(x,t)$ satisfies the wave equation if and only if

$$\left[\partial_r^2 + \frac{d-1}{r}\right] Mu(x,r,t) = \partial_t^2 Mu(x,r,t),$$

where $Mu(x,r,t)$ denote the spherical means of the function $u(x,t)$.

(c) If $d = 2k+1$, define $T\varphi(r) = (r^{-1}\partial_r)^{k-1}[r^{2k-1}\varphi(r)]$, and let $\tilde{u} = TMu$. Then this function solves the one-dimensional wave equation for each fixed x:

$$\partial_t^2 \tilde{u}(x,r,t) = \partial_r^2 \tilde{u}(x,r,t).$$

One can then use d'Alembert's formula to find the solution $\tilde{u}(x,r,t)$ of this problem expressed in terms of the initial data.

(d) Now show that

$$u(x,t) = Mu(x,0,t) = \lim_{r \to 0} \frac{\tilde{u}(x,r,t)}{\alpha r}$$

where $\alpha = 1 \cdot 3 \cdots (d-2)$.

(e) Conclude that the solution of the Cauchy problem for the d-dimensional wave equation, when $d > 1$ is odd, is

$$u(x,t) = \frac{1}{1 \cdot 3 \cdots (d-2)} \left[\partial_t (t^{-1}\partial_t)^{(d-3)/2} \left(t^{d-2} M_t f(x) \right) + \right.$$

$$\left. (t^{-1}\partial_t)^{(d-3)/2} \left(t^{d-2} M_t g(x) \right) \right].$$

5.* The method of descent can be used to prove that the solution of the Cauchy problem for the wave equation in the case when d is even is given by the formula

$$u(x,t) = \frac{1}{1 \cdot 3 \cdots (d-2)} \left[\partial_t (t^{-1}\partial_t)^{(d-3)/2} \left(t^{d-2} \widetilde{M}_t f(x) \right) + \right.$$

$$\left. (t^{-1}\partial_t)^{(d-3)/2} \left(t^{d-2} \widetilde{M}_t g(x) \right) \right],$$

where $\widetilde{M}_t$ denotes the modified spherical means defined by

$$\widetilde{M}_t h(x) = \frac{2}{A_{d+1}} \int_{B^d} \frac{f(x+ty)}{\sqrt{1-|y|^2}}\, dy.$$

6.* Given initial data f and g of the form

$$f(x) = F(x \cdot \gamma) \qquad \text{and} \qquad g(x) = G(x \cdot \gamma),$$

check that the plane wave given by

$$u(x,t) = \frac{F(x\cdot\gamma + t) + F(x \cdot \gamma - t)}{2} + \frac{1}{2}\int_{x\cdot\gamma - t}^{x\cdot\gamma+t} G(s)\, ds$$

is a solution of the Cauchy problem for the d-dimensional wave equation.

In general, the solution is given as a superposition of plane waves. For the case $d = 3$, this can be expressed in terms of the Radon transform as follows. Let

$$\tilde{\mathcal{R}}(f)(t,\gamma) = -\frac{1}{8\pi^2}\left(\frac{d}{dt}\right)^2 R(f)(t,\gamma).$$

Then $u(x,t) =$

$$\frac{1}{2}\int_{S^2} \left[\tilde{\mathcal{R}}(f)(x\cdot\gamma - t, \gamma) + \tilde{\mathcal{R}}(f)(x\cdot\gamma + t, \gamma) + \int_{x\cdot\gamma - t}^{x\cdot\gamma + t} \tilde{\mathcal{R}}(g)(s,\gamma)\, ds \right] d\sigma(\gamma).$$

7. For every real number $a > 0$, define the operator $(-\triangle)^a$ by the formula

$$(-\triangle)^a f(x) = \int_{\mathbb{R}^d} (2\pi|\xi|)^{2a} \hat{f}(\xi) e^{2\pi i \xi \cdot x} \, d\xi$$

whenever $f \in \mathcal{S}(\mathbb{R}^d)$.

(a) Check that $(-\triangle)^a$ agrees with the usual definition of the ath power of $-\triangle$ (that is, a compositions of minus the Laplacian) when a is a positive integer.

(b) Verify that $(-\triangle)^a(f)$ is indefinitely differentiable.

(c) Prove that if a is not an integer, then in general $(-\triangle)^a(f)$ is not rapidly decreasing.

(d) Let $u(x, y)$ be the solution of the steady-state heat equation

$$\frac{\partial^2 u}{\partial y^2} + \sum_{j=1}^{d} \frac{\partial^2 u}{\partial x_j^2} = 0, \qquad \text{with } u(x, 0) = f(x),$$

given by convolving f with the Poisson kernel (see Exercise 8). Check that

$$(-\triangle)^{1/2} f(x) = -\lim_{y \to 0} \frac{\partial u}{\partial y}(x, y),$$

and more generally that

$$(-\triangle)^{k/2} f(x) = (-1)^k \lim_{y \to 0} \frac{\partial^k u}{\partial y^k}(x, y)$$

for any positive integer k.

8.* The reconstruction formula for the Radon transform in $\mathbb{R}^d$ is as follows:

(a) When $d = 2$,

$$\frac{(-\triangle)^{1/2}}{4\pi} \mathcal{R}^*(\mathcal{R}(f)) = f,$$

where $(-\triangle)^{1/2}$ is defined in Problem 7.

(b) If the Radon transform and its dual are defined by analogy to the cases $d = 2$ and $d = 3$, then for general d,

$$\frac{(2\pi)^{1-d}}{2} (-\triangle)^{(d-1)/2} \mathcal{R}^*(\mathcal{R}(f)) = f.$$

7 Finite Fourier Analysis

> This past year has seen the birth, or rather the rebirth, of an exciting revolution in computing Fourier transforms. A class of algorithms known as the fast Fourier transform or FFT, is forcing a complete reassessment of many computational paths, not only in frequency analysis, but in any fields where problems can be reduced to Fourier transforms and/or convolutions...
>
> *C. Bingham and J. W. Tukey,* 1966

In the previous chapters we studied the Fourier series of functions on the circle and the Fourier transform of functions defined on the Euclidean space $\mathbb{R}^d$. The goal here is to introduce another version of Fourier analysis, now for functions defined on finite sets, and more precisely, on finite abelian groups. This theory is particularly elegant and simple since infinite sums and integrals are replaced by finite sums, and thus questions of convergence disappear.

In turning our attention to finite Fourier analysis, we begin with the simplest example, $\mathbb{Z}(N)$, where the underlying space is the (multiplicative) group of Nth roots of unity on the circle. This group can also be realized in additive form, as $\mathbb{Z}/N\mathbb{Z}$, the equivalence classes of integers modulo N. The group $\mathbb{Z}(N)$ arises as the natural approximation to the circle (as N tends to infinity) since in the first picture the points of $\mathbb{Z}(N)$ correspond to N points on the circle which are uniformly distributed. For this reason, in practical applications, the group $\mathbb{Z}(N)$ becomes a natural candidate for the storage of information of a function on the circle, and for the resulting numerical computations involving Fourier series. The situation is particularly nice when N is large and of the form $N = 2^n$. The computations of the Fourier coefficients now lead to the "fast Fourier transform," which exploits the fact that an induction in n requires only about $\log N$ steps to go from $N = 1$ to $N = 2^n$. This yields a substantial saving in time in practical applications.

In the second part of the chapter we undertake the more general theory of Fourier analysis on finite abelian groups. Here the fundamental example is the multiplicative group $\mathbb{Z}^*(q)$. The Fourier inversion formula for $\mathbb{Z}^*(q)$ will be seen to be a key step in the proof of Dirichlet's theorem on primes in arithmetic progression, which we will take up in the next chapter.

1 Fourier analysis on $\mathbb{Z}(N)$

We turn to the group of Nth roots of unity. This group arises naturally as the simplest finite abelian group. It also gives a uniform partition of the circle, and is therefore a good choice if one wishes to sample appropriate functions on the circle. Moreover, this partition gets finer as N tends to infinity, and one might expect that the discrete Fourier theory that we discuss here tends to the continuous theory of Fourier series on the circle. In a broad sense, this is the case, although this aspect of the problem is not one that we develop.

1.1 The group $\mathbb{Z}(N)$

Let N be a positive integer. A complex number z is an Nth **root of unity** if $z^N = 1$. The set of Nth roots of unity is precisely

$$\left\{1, e^{2\pi i/N}, e^{2\pi i 2/N}, \ldots, e^{2\pi i(N-1)/N}\right\}.$$

Indeed, suppose that $z^N = 1$ with $z = re^{i\theta}$. Then we must have $r^N e^{iN\theta} = 1$, and taking absolute values yields $r = 1$. Therefore $e^{iN\theta} = 1$, and this means that $N\theta = 2\pi k$ where $k \in \mathbb{Z}$. So if $\zeta = e^{2\pi i/N}$ we find that ζ^k exhausts all the Nth roots of unity. However, notice that $\zeta^N = 1$ so if n and m differ by an integer multiple of N, then $\zeta^n = \zeta^m$. In fact, it is clear that

$$\zeta^n = \zeta^m \quad \text{if and only if} \quad n - m \text{ is divisible by } N.$$

We denote the set of all N^{th} roots of unity by $\mathbb{Z}(N)$. The fact that this set gives a uniform partition of the circle is clear from its definition. Note that the set $\mathbb{Z}(N)$ satisfies the following properties:

(i) If $z, w \in \mathbb{Z}(N)$, then $zw \in \mathbb{Z}(N)$ and $zw = wz$.

(ii) $1 \in \mathbb{Z}(N)$.

(iii) If $z \in \mathbb{Z}(N)$, then $z^{-1} = 1/z \in \mathbb{Z}(N)$ and of course $zz^{-1} = 1$.

As a result we can conclude that $\mathbb{Z}(N)$ is an abelian group under complex multiplication. The appropriate definitions are set out in detail later in Section 2.1.

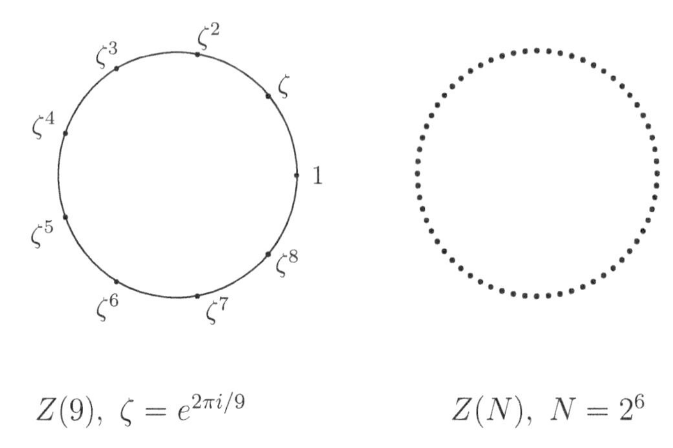

Figure 1. The group of Nth roots of unity when $N = 9$ and $N = 2^6 = 64$

There is another way to visualize the group $\mathbb{Z}(N)$. This consists of choosing the integer power of ζ that determines each root of unity. We observed above that this integer is not unique since $\zeta^n = \zeta^m$ whenever n and m differ by an integer multiple of N. Naturally, we might select the integer which satisfies $0 \leq n \leq N-1$. Although this choice is perfectly reasonable in terms of "sets," we ask what happens when we multiply roots of unity. Clearly, we must add the corresponding integers since $\zeta^n \zeta^m = \zeta^{n+m}$ but nothing guarantees that $0 \leq n+m \leq N-1$. In fact, if $\zeta^n \zeta^m = \zeta^k$ with $0 \leq k \leq N-1$, then $n+m$ and k differ by an integer multiple of N. So, to find the integer in $[0, N-1]$ corresponding to the root of unity $\zeta^n \zeta^m$, we see that after adding the integers n and m we must reduce modulo N, that is, find the unique integer $0 \leq k \leq N-1$ so that $(n+m) - k$ is an integer multiple of N.

An equivalent approach is to associate to each root of unity ω the class of integers n so that $\zeta^n = \omega$. Doing so for each root of unity we obtain a partition of the integers in N disjoint infinite classes. To add two of these classes, choose any integer in each one of them, say n and m, respectively, and define the sum of the classes to be the class which contains the integer $n+m$.

We formalize the above notions. Two integers x and y are **congruent modulo** N if the difference $x - y$ is divisible by N, and we write $x \equiv y \mod N$. In other words, this means that x and y differ by an integer multiple of N. It is an easy exercise to check the following three properties:

- $x \equiv x \mod N$ for all integers x.
- If $x \equiv y \mod N$, then $y \equiv x \mod N$.
- If $x \equiv y \mod N$ and $y \equiv z \mod N$, then $x \equiv z \mod N$.

The above defines an equivalence relation on $\mathbb{Z}$. Let $R(x)$ denote the equivalence class, or residue class, of the integer x. Any integer of the form $x + kN$ with $k \in \mathbb{Z}$ is an element (or "representative") of $R(x)$. In fact, there are precisely N equivalence classes, and each class has a unique representative between 0 and $N - 1$. We may now add equivalence classes by defining

$$R(x) + R(y) = R(x + y).$$

This definition is of course independent of the representatives x and y because if $x' \in R(x)$ and $y' \in R(y)$, then one checks easily that $x' + y' \in R(x + y)$. This turns the set of equivalence classes into an abelian group called the **group of integers modulo** N, which is sometimes denoted by $\mathbb{Z}/N\mathbb{Z}$. The association

$$R(k) \longleftrightarrow e^{2\pi ik/N}$$

gives a correspondence between the two abelian groups, $\mathbb{Z}/N\mathbb{Z}$ and $\mathbb{Z}(N)$. Since the operations are respected, in the sense that addition of integers modulo N becomes multiplication of complex numbers, we shall also denote the group of integers modulo N by $\mathbb{Z}(N)$. Observe that $0 \in \mathbb{Z}/N\mathbb{Z}$ corresponds to 1 on the unit circle.

Let V and W denote the vector spaces of complex-valued functions on the group of integers modulo N and the Nth roots of unity, respectively. Then, the identification given above carries over to V and W as follows:

$$F(k) \longleftrightarrow f(e^{2\pi ik/N}),$$

where F is a function on the integers modulo N and f is a function on the Nth roots of unity.

From now on, we write $\mathbb{Z}(N)$ but think of either the group of integers modulo N or the group of Nth roots of unity.

1.2 Fourier inversion theorem and Plancherel identity on $\mathbb{Z}(N)$

The first and most crucial step in developing Fourier analysis on $\mathbb{Z}(N)$ is to find the functions which correspond to the exponentials $e_n(x) = e^{2\pi inx}$ in the case of the circle. Some important properties of these exponentials are:

(i) $\{e_n\}_{n\in\mathbb{Z}}$ is an orthonormal set for the inner product (1) (in Chapter 3) on the space of Riemann integrable functions on the circle.

(ii) Finite linear combinations of the e_n's (the trigonometric polynomials) are dense in the space of continuous functions on the circle.

(iii) $e_n(x+y) = e_n(x)e_n(y)$.

On $\mathbb{Z}(N)$, the appropriate analogues are the N functions $e_0, \ldots, e_{N-1}$ defined by

$$e_\ell(k) = \zeta^{\ell k} = e^{2\pi i \ell k/N} \quad \text{for } \ell = 0, \ldots, N-1 \text{ and } k = 0, \ldots, N-1,$$

where $\zeta = e^{2\pi i/N}$. To understand the parallel with (i) and (ii), we can think of the complex-valued functions on $\mathbb{Z}(N)$ as a vector space V, endowed with the Hermitian inner product

$$(F, G) = \sum_{k=0}^{N-1} F(k)\overline{G(k)}$$

and associated norm

$$\|F\|^2 = \sum_{k=0}^{N-1} |F(k)|^2.$$

Lemma 1.1 *The family $\{e_0, \ldots, e_{N-1}\}$ is orthogonal. In fact,*

$$(e_m, e_\ell) = \begin{cases} N & \text{if } m = \ell, \\ 0 & \text{if } m \neq \ell. \end{cases}$$

Proof. We have

$$(e_m, e_\ell) = \sum_{k=0}^{N-1} \zeta^{mk}\zeta^{-\ell k} = \sum_{k=0}^{N-1} \zeta^{(m-\ell)k}.$$

If $m = \ell$, each term in the sum is equal to 1, and the sum equals N. If $m \neq \ell$, then $q = \zeta^{m-\ell}$ is not equal to 1, and the usual formula

$$1 + q + q^2 + \cdots + q^{N-1} = \frac{1-q^N}{1-q}$$

shows that $(e_m, e_\ell) = 0$, because $q^N = 1$.

Since the N functions $e_0, \ldots, e_{N-1}$ are orthogonal, they must be linearly independent, and since the vector space V is N-dimensional, we conclude that $\{e_0, \ldots, e_{N-1}\}$ is an orthogonal basis for V. Clearly, property (iii) also holds, that is, $e_\ell(k+m) = e_\ell(k)e_\ell(m)$ for all ℓ, and all $k, m \in \mathbb{Z}(N)$.

By the lemma each vector e_ℓ has norm $\sqrt{N}$, so if we define

$$e_\ell^* = \frac{1}{\sqrt{N}}\, e_\ell,$$

then $\{e_0^*, \ldots, e_{N-1}^*\}$ is an orthonormal basis for V. Hence for any $F \in V$ we have

$$F = \sum_{n=0}^{N-1} (F, e_n^*) e_n^* \quad \text{as well as} \quad \|F\|^2 = \sum_{n=0}^{N-1} |(F, e_n^*)|^2. \tag{1}$$

If we define the nth **Fourier coefficient** of F by

$$a_n = \frac{1}{N} \sum_{k=0}^{N-1} F(k) e^{-2\pi i k n/N},$$

the above observations give the following fundamental theorem which is the $\mathbb{Z}(N)$ version of the Fourier inversion and the Parseval-Plancherel formulas.

Theorem 1.2 *If F is a function on $\mathbb{Z}(N)$, then*

$$F(k) = \sum_{n=0}^{N-1} a_n e^{2\pi i n k/N}.$$

Moreover,

$$\sum_{n=0}^{N-1} |a_n|^2 = \frac{1}{N} \sum_{k=0}^{N-1} |F(k)|^2.$$

The proof follows directly from (1) once we observe that

$$a_n = \frac{1}{N}\,(F, e_n) = \frac{1}{\sqrt{N}}\,(F, e_n^*).$$

Remark. It is possible to recover the Fourier inversion on the circle for sufficiently smooth functions (say C^2) by letting $N \to \infty$ in the finite model $\mathbb{Z}(N)$ (see Exercise 3).

1.3 The fast Fourier transform

The fast Fourier transform is a method that was developed as a means of calculating efficiently the Fourier coefficients of a function F on $\mathbb{Z}(N)$.

The problem, which arises naturally in numerical analysis, is to determine an algorithm that minimizes the amount of time it takes a computer to calculate the Fourier coefficients of a given function on $\mathbb{Z}(N)$. Since this amount of time is roughly proportional to the number of operations the computer must perform, our problem becomes that of minimizing the number of operations necessary to obtain all the Fourier coefficients $\{a_n\}$ given the values of F on $\mathbb{Z}(N)$. By operations we mean either an addition or a multiplication of complex numbers.

We begin with a naive approach to the problem. Fix N, and suppose that we are given $F(0), \ldots, F(N-1)$ and $\omega_N = e^{-2\pi i/N}$. If we denote by $a_k^N(F)$ the kth Fourier coefficient of F on $\mathbb{Z}(N)$, then by definition

$$a_k^N(F) = \frac{1}{N} \sum_{r=0}^{N-1} F(r)\omega_N^{kr},$$

and crude estimates show that the number of operations needed to calculate all Fourier coefficients is $\leq 2N^2 + N$. Indeed, it takes at most $N-2$ multiplications to determine $\omega_N^2, \ldots, \omega_N^{N-1}$, and each coefficient a_k^N requires $N+1$ multiplications and $N-1$ additions.

We now present the **fast Fourier transform**, an algorithm that improves the bound $O(N^2)$ obtained above. Such an improvement is possible if, for example, we restrict ourselves to the case where the partition of the circle is dyadic, that is, $N = 2^n$. (See also Exercise 9.)

Theorem 1.3 *Given $\omega_N = e^{-2\pi i/N}$ with $N = 2^n$, it is possible to calculate the Fourier coefficients of a function on $\mathbb{Z}(N)$ with at most*

$$4 \cdot 2^n n = 4N \log_2(N) = O(N \log N)$$

operations.

The proof of the theorem consists of using the calculations for M division points, to obtain the Fourier coefficients for $2M$ division points. Since we choose $N = 2^n$, we obtain the desired formula as a consequence of a recurrence which involves $n = O(\log N)$ steps.

Let $\#(M)$ denote the minimum number of operations needed to calculate all the Fourier coefficients of any function on $\mathbb{Z}(M)$. The key to the proof of the theorem is contained in the following recursion step.

Lemma 1.4 *If we are given* $\omega_{2M} = e^{-2\pi i/(2M)}$*, then*

$$\#(2M) \leq 2\#(M) + 8M.$$

Proof. The calculation of $\omega_{2M}, \ldots, \omega_{2M}^{2M}$ requires no more than $2M$ operations. Note that in particular we get $\omega_M = e^{-2\pi i/M} = \omega_{2M}^2$. The main idea is that for any given function F on $\mathbb{Z}(2M)$, we consider two functions F_0 and F_1 on $\mathbb{Z}(M)$ defined by

$$F_0(r) = F(2r) \quad \text{and} \quad F_1(r) = F(2r+1).$$

We assume that it is possible to calculate the Fourier coefficients of F_0 and F_1 in no more than $\#(M)$ operations each. If we denote the Fourier coefficients corresponding to the groups $\mathbb{Z}(2M)$ and $\mathbb{Z}(M)$ by a_k^{2M} and a_k^M, respectively, then we have

$$a_k^{2M}(F) = \frac{1}{2}\left(a_k^M(F_0) + a_k^M(F_1)\omega_{2M}^k\right).$$

To prove this, we sum over odd and even integers in the definition of the Fourier coefficient $a_k^{2M}(F)$, and find

$$\begin{aligned}
a_k^{2M}(F) &= \frac{1}{2M}\sum_{r=0}^{2M-1} F(r)\omega_{2M}^{kr} \\
&= \frac{1}{2}\left(\frac{1}{M}\sum_{\ell=0}^{M-1} F(2\ell)\omega_{2M}^{k(2\ell)} + \frac{1}{M}\sum_{m=0}^{M-1} F(2m+1)\omega_{2M}^{k(2m+1)}\right) \\
&= \frac{1}{2}\left(\frac{1}{M}\sum_{\ell=0}^{M-1} F_0(\ell)\omega_M^{k\ell} + \frac{1}{M}\sum_{m=0}^{M-1} F_1(m)\omega_M^{km}\omega_{2M}^k\right),
\end{aligned}$$

which establishes our assertion.

As a result, knowing $a_k^M(F_0)$, $a_k^M(F_1)$, and ω_{2M}^k, we see that each $a_k^{2M}(F)$ can be computed using no more than three operations (one addition and two multiplications). So

$$\#(2M) \leq 2M + 2\#(M) + 3 \times 2M = 2\#(M) + 8M,$$

and the proof of the lemma is complete.

An induction on n, where $N = 2^n$, will conclude the proof of the theorem. The initial step $n = 1$ is easy, since $N = 2$ and the two Fourier coefficients are

$$a_0^N(F) = \frac{1}{2}(F(1) + F(-1)) \quad \text{and} \quad a_1^N(F) = \frac{1}{2}(F(1) + (-1)F(-1)).$$

Calculating these Fourier coefficients requires no more than five operations, which is less than $4 \times 2 = 8$. Suppose the theorem is true up to $N = 2^{n-1}$ so that $\#(N) \leq 4 \cdot 2^{n-1}(n-1)$. By the lemma we must have

$$\#(2N) \leq 2 \cdot 4 \cdot 2^{n-1}(n-1) + 8 \cdot 2^{n-1} = 4 \cdot 2^n n,$$

which concludes the inductive step and the proof of the theorem.

2 Fourier analysis on finite abelian groups

The main goal in the rest of this chapter is to generalize the results about Fourier series expansions obtained in the special case of $\mathbb{Z}(N)$.

After a brief introduction to some notions related to finite abelian groups, we turn to the important concept of a character. In our setting, we find that characters play the same role as the exponentials $e_0, \ldots, e_{N-1}$ on the group $\mathbb{Z}(N)$, and thus provide the key ingredient in the development of the theory on arbitrary finite abelian groups. In fact, it suffices to prove that a finite abelian group has "enough" characters, and this leads automatically to the desired Fourier theory.

2.1 Abelian groups

An **abelian group** (or commutative group) is a set G together with a binary operation on pairs of elements of G, $(a, b) \mapsto a \cdot b$, that satisfies the following conditions:

(i) *Associativity*: $a \cdot (b \cdot c) = (a \cdot b) \cdot c$ for all $a, b, c \in G$.

(ii) *Identity*: There exists an element $u \in G$ (often written as either 1 or 0) such that $a \cdot u = u \cdot a = a$ for all $a \in G$.

(iii) *Inverses*: For every $a \in G$, there exists an element $a^{-1} \in G$ such that $a \cdot a^{-1} = a^{-1} \cdot a = u$.

(iv) *Commutativity*: For $a, b \in G$, we have $a \cdot b = b \cdot a$.

We leave as simple verifications the facts that the identity element and inverses are unique.

Warning. In the definition of an abelian group, we used the "multiplicative" notation for the operation in G. Sometimes, one uses the "additive" notation $a + b$ and $-a$, instead of $a \cdot b$ and a^{-1}. There are times when one notation may be more appropriate than the other, and the examples below illustrate this point. The same group may have different interpretations, one where the multiplicative notation is more suggestive, and another where it is natural to view the group with addition, as the operation.

Examples of abelian groups

- The set of real numbers $\mathbb{R}$ with the usual addition. The identity is 0 and the inverse of x is $-x$.

  Also, $\mathbb{R} - \{0\}$ and $\mathbb{R}^+ = \{x \in \mathbb{R} : x > 0\}$ equipped, with the standard multiplication, are abelian groups. In both cases the unit is 1 and the inverse of x is $1/x$.

- With the usual addition, the set of integers $\mathbb{Z}$ is an abelian group. However, $\mathbb{Z} - \{0\}$ is not an abelian group with the standard multiplication, since, for example, 2 does not have a multiplicative inverse in $\mathbb{Z}$. In contrast, $\mathbb{Q} - \{0\}$ is an abelian group with the standard multiplication.

- The unit circle S^1 in the complex plane. If we view the circle as the set of points $\{e^{i\theta} : \theta \in \mathbb{R}\}$, the group operation is the standard multiplication of complex numbers. However, if we identify points on S^1 with their angle θ, then S^1 becomes $\mathbb{R}$ modulo 2π, where the operation is addition modulo 2π.

- $\mathbb{Z}(N)$ is an abelian group. Viewed as the Nth roots of unity on the circle, $\mathbb{Z}(N)$ is a group under multiplication of complex numbers. However, if $\mathbb{Z}(N)$ is interpreted as $\mathbb{Z}/N\mathbb{Z}$, the integers modulo N, then it is an abelian group where the operation is addition modulo N.

- The last example consists of $\mathbb{Z}^*(q)$. This group is defined as the set of all integers modulo q that have *multiplicative* inverses, with the group operation being multiplication modulo q. This important example is discussed in more detail below.

A **homomorphism** between two abelian groups G and H is a map $f : G \to H$ which satisfies the property

$$f(a \cdot b) = f(a) \cdot f(b),$$

where the dot on the left-hand side is the operation in G, and the dot on the right-hand side the operation in H.

We say that two groups G and H are **isomorphic**, and write $G \approx H$, if there is a bijective homomorphism from G to H. Equivalently, G and H are isomorphic if there exists another homomorphism $\tilde{f} : H \to G$, so that for all $a \in G$ and $b \in H$

$$(\tilde{f} \circ f)(a) = a \quad \text{and} \quad (f \circ \tilde{f})(b) = b.$$

Roughly speaking, isomorphic groups describe the "same" object because they have the same underlying group structure (which is really all that matters); however, their particular notational representations might be different.

EXAMPLE 1. A pair of isomorphic abelian groups arose already when we considered the group $\mathbb{Z}(N)$. In one representation it was given as the multiplicative group of Nth roots of unity in $\mathbb{C}$. In a second representation it was the additive group $\mathbb{Z}/N\mathbb{Z}$ of residue classes of integers modulo N. The mapping $n \mapsto R(n)$, which associates to a root of unity $z = e^{2\pi i n/N} = \zeta^n$ the residue class in $\mathbb{Z}/N\mathbb{Z}$ determined by n, provides an isomorphism between the two different representations.

EXAMPLE 2. In parallel with the previous example, we see that the circle (with multiplication) is isomorphic to the real numbers modulo 2π (with addition).

EXAMPLE 3. The properties of the exponential and logarithm guarantee that

$$\exp : \mathbb{R} \to \mathbb{R}^+ \quad \text{and} \quad \log : \mathbb{R}^+ \to \mathbb{R}$$

are two homomorphisms that are inverses of each other. Thus $\mathbb{R}$ (with addition) and $\mathbb{R}^+$ (with multiplication) are isomorphic.

In what follows, we are primarily interested in abelian groups that are finite. In this case, we denote by $|G|$ the number of elements in G, and call $|G|$ the **order** of the group. For example, the order of $\mathbb{Z}(N)$ is N.

A few additional remarks are in order:

- If G_1 and G_2 are two finite abelian groups, their **direct product** $G_1 \times G_2$ is the group whose elements are pairs (g_1, g_2) with $g_1 \in G_1$ and $g_2 \in G_2$. The operation in $G_1 \times G_2$ is then defined by

$$(g_1, g_2) \cdot (g_1', g_2') = (g_1 \cdot g_1', g_2 \cdot g_2').$$

  Clearly, if G_1 and G_2 are finite abelian groups, then so is $G_1 \times G_2$. The definition of direct product generalizes immediately to the case of finitely many factors $G_1 \times G_2 \times \cdots \times G_n$.

- The structure theorem for finite abelian groups states that such a group is isomorphic to a direct product of groups of the type $\mathbb{Z}(N)$; see Problem 2. This is a nice result which gives us an overview of the class of all finite abelian groups. However, since we shall not use this theorem below, we omit its proof.

We now discuss briefly the examples of abelian groups that play a central role in the proof of Dirichlet's theorem in the next chapter.

The group $\mathbb{Z}^*(q)$

Let q be a positive integer. We see that multiplication in $\mathbb{Z}(q)$ can be unambiguously defined, because if n is congruent to n' and m is congruent to m' (both modulo q), then nm is congruent to $n'm'$ modulo q. An integer $n \in \mathbb{Z}(q)$ is a **unit** if there exists an integer $m \in \mathbb{Z}(q)$ so that

$$nm \equiv 1 \mod q.$$

The set of all units in $\mathbb{Z}(q)$ is denoted by $\mathbb{Z}^*(q)$, and it is clear from our definition that $\mathbb{Z}^*(q)$ is an abelian group under *multiplication* modulo q. Thus within the additive group $\mathbb{Z}(q)$ lies a set $\mathbb{Z}^*(q)$ that is a group under multiplication. An alternative characterization of $\mathbb{Z}^*(q)$ will be given in the next chapter, as those elements in $\mathbb{Z}(q)$ that are relatively prime to q.

EXAMPLE 4. The group of units in $\mathbb{Z}(4) = \{0, 1, 2, 3\}$ is

$$\mathbb{Z}^*(4) = \{1, 3\}.$$

This reflects the fact that odd integers are divided into two classes depending on whether they are of the form $4k + 1$ or $4k + 3$. In fact, $\mathbb{Z}^*(4)$ is isomorphic to $\mathbb{Z}(2)$. Indeed, we can make the following association:

$$\begin{array}{ccc} \mathbb{Z}^*(4) & & \mathbb{Z}(2) \\ 1 & \longleftrightarrow & 0 \\ 3 & \longleftrightarrow & 1 \end{array}$$

and then notice that multiplication in $\mathbb{Z}^*(4)$ corresponds to addition in $\mathbb{Z}(2)$.

EXAMPLE 5. The units in $\mathbb{Z}(5)$ are

$$\mathbb{Z}^*(5) = \{1, 2, 3, 4\}.$$

Moreover, $\mathbb{Z}^*(5)$ is isomorphic to $\mathbb{Z}(4)$ with the following identification:

$$\begin{array}{ccc} \mathbb{Z}^*(5) & & \mathbb{Z}(4) \\ 1 & \longleftrightarrow & 0 \\ 2 & \longleftrightarrow & 1 \\ 3 & \longleftrightarrow & 3 \\ 4 & \longleftrightarrow & 2 \end{array}$$

EXAMPLE 6. The units in $\mathbb{Z}(8) = \{0, 1, 2, 3, 4, 5, 6, 7\}$ are given by

$$\mathbb{Z}^*(8) = \{1, 3, 5, 7\}.$$

In fact, $\mathbb{Z}^*(8)$ is isomorphic to the direct product $\mathbb{Z}(2) \times \mathbb{Z}(2)$. In this case, an isomorphism between the groups is given by the identification

$$\begin{array}{ccc} \mathbb{Z}^*(8) & & \mathbb{Z}(2) \times \mathbb{Z}(2) \\ 1 & \longleftrightarrow & (0,0) \\ 3 & \longleftrightarrow & (1,0) \\ 5 & \longleftrightarrow & (0,1) \\ 7 & \longleftrightarrow & (1,1) \end{array}$$

2.2 Characters

Let G be a finite abelian group (with the multiplicative notation) and S^1 the unit circle in the complex plane. A **character** on G is a complex-valued function $e : G \to S^1$ which satisfies the following condition:

$$e(a \cdot b) = e(a)e(b) \quad \text{for all } a, b \in G. \tag{2}$$

In other words, a character is a homomorphism from G to the circle group. The **trivial** or **unit character** is defined by $e(a) = 1$ for all $a \in G$.

Characters play an important role in the context of finite Fourier series, primarily because the multiplicative property (2) generalizes the analogous identity for the exponential functions on the circle and the law

$$e_\ell(k + m) = e_\ell(k)e_\ell(m),$$

which held for the exponentials $e_0, \ldots, e_{N-1}$ used in the Fourier theory on $\mathbb{Z}(N)$. There we had $e_\ell(k) = \zeta^{\ell k} = e^{2\pi i \ell k/N}$, with $0 \leq \ell \leq N - 1$ and $k \in \mathbb{Z}(N)$, and in fact, the functions $e_0, \ldots, e_{N-1}$ are precisely all the characters of the group $\mathbb{Z}(N)$.

If G is a finite abelian group, we denote by $\hat{G}$ the set of all characters of G, and observe next that this set inherits the structure of an abelian group.

Lemma 2.1 *The set $\hat{G}$ is an abelian group under multiplication defined by*

$$(e_1 \cdot e_2)(a) = e_1(a)e_2(a) \quad \textit{for all } a \in G.$$

The proof of this assertion is straightforward if one observes that the trivial character plays the role of the unit. We call $\hat{G}$ the **dual group** of G.

In light of the above analogy between characters for a general abelian group and the exponentials on $\mathbb{Z}(N)$, we gather several more examples of groups and their duals. This provides further evidence of the central role played by characters. (See Exercises 4, 5, and 6.)

EXAMPLE 1. If $G = \mathbb{Z}(N)$, all characters of G take the form $e_\ell(k) = \zeta^{\ell k} = e^{2\pi i \ell k/N}$ for some $0 \leq \ell \leq N-1$, and it is easy to check that $e_\ell \mapsto \ell$ gives an isomorphism from $\widehat{\mathbb{Z}(N)}$ to $\mathbb{Z}(N)$.

EXAMPLE 2. The dual group of the circle[1] is precisely $\{e_n\}_{n\in\mathbb{Z}}$ (where $e_n(x) = e^{2\pi i n x}$). Moreover, $e_n \mapsto n$ gives an isomorphism between $\widehat{S^1}$ and the integers $\mathbb{Z}$.

EXAMPLE 3. Characters on $\mathbb{R}$ are described by

$$e_\xi(x) = e^{2\pi i \xi x} \quad \text{where } \xi \in \mathbb{R}.$$

Thus $e_\xi \mapsto \xi$ is an isomorphism from $\widehat{\mathbb{R}}$ to $\mathbb{R}$.

EXAMPLE 4. Since $\exp : \mathbb{R} \to \mathbb{R}^+$ is an isomorphism, we deduce from the previous example that the characters on $\mathbb{R}^+$ are given by

$$e_\xi(x) = x^{2\pi i \xi} = e^{2\pi i \xi \log x} \quad \text{where } \xi \in \mathbb{R},$$

and $\widehat{\mathbb{R}^+}$ is isomorphic to $\mathbb{R}$ (or $\mathbb{R}^+$).

The following lemma says that a nowhere vanishing multiplicative function is a character, a result that will be useful later.

Lemma 2.2 *Let G be a finite abelian group, and $e : G \to \mathbb{C} - \{0\}$ a multiplicative function, namely $e(a \cdot b) = e(a)e(b)$ for all $a, b \in G$. Then e is a character.*

---

[1] In addition to (2), the definition of a character on an infinite abelian group requires continuity. When G is the circle, $\mathbb{R}$, or $\mathbb{R}^+$, the meaning of "continuous" refers to the standard notion of limit.

Proof. The group G being finite, the absolute value of $e(a)$ is bounded above and below as a ranges over G. Since $|e(b^n)| = |e(b)|^n$, we conclude that $|e(b)| = 1$ for all $b \in G$.

The next step is to verify that the characters form an orthonormal basis of the vector space V of functions over the group G. This fact was obtained directly in the special case $G = \mathbb{Z}(N)$ from the explicit description of the characters $e_0, \ldots, e_{N-1}$.

In the general case, we begin with the orthogonality relations; then we prove that there are "enough" characters by showing that there are as many as the order of the group.

2.3 The orthogonality relations

Let V denote the vector space of complex-valued functions defined on the finite abelian group G. Note that the dimension of V is $|G|$, the order of G. We define a Hermitian inner product on V by

$$(f, g) = \frac{1}{|G|} \sum_{a \in G} f(a)\overline{g(a)}, \qquad \text{whenever } f, g \in V. \tag{3}$$

Here the sum is taken over the group and is therefore finite.

Theorem 2.3 *The characters of G form an orthonormal family with respect to the inner product defined above.*

Since $|e(a)| = 1$ for any character, we find that

$$(e, e) = \frac{1}{|G|} \sum_{a \in G} e(a)\overline{e(a)} = \frac{1}{|G|} \sum_{a \in G} |e(a)|^2 = 1.$$

If $e \neq e'$ and both are characters, we must prove that $(e, e') = 0$; we isolate the key step in a lemma.

Lemma 2.4 *If e is a non-trivial character of the group G, then $\sum_{a \in G} e(a) = 0$.*

Proof. Choose $b \in G$ such that $e(b) \neq 1$. Then we have

$$e(b) \sum_{a \in G} e(a) = \sum_{a \in G} e(b)e(a) = \sum_{a \in G} e(ab) = \sum_{a \in G} e(a).$$

The last equality follows because as a ranges over the group, ab ranges over G as well. Therefore $\sum_{a \in G} e(a) = 0$.

We can now conclude the proof of the theorem. Suppose e' is a character distinct from e. Because $e(e')^{-1}$ is non-trivial, the lemma implies that

$$\sum_{a \in G} e(a)(e'(a))^{-1} = 0.$$

Since $(e'(a))^{-1} = \overline{e'(a)}$, the theorem is proved.

As a consequence of the theorem, we see that distinct characters are linearly independent. Since the dimension of V over $\mathbb{C}$ is $|G|$, we conclude that the order of $\hat{G}$ is finite and $\leq |G|$. The main result to which we now turn is that, in fact, $|\hat{G}| = |G|$.

2.4 Characters as a total family

The following completes the analogy between characters and the complex exponentials.

Theorem 2.5 *The characters of a finite abelian group G form a basis for the vector space of functions on G.*

There are several proofs of this theorem. One consists of using the structure theorem for finite abelian groups we have mentioned earlier, which states that any such group is the direct product of cyclic groups, that is, groups of the type $\mathbb{Z}(N)$. Since cyclic groups are self-dual, using this fact we would conclude that $|\hat{G}| = |G|$, and therefore the characters form a basis for G. (See Problem 3.)

Here we shall prove the theorem directly without these considerations.

Suppose V is a vector space of dimension d with inner product $(\cdot, \cdot)$. A linear transformation $T : V \to V$ is **unitary** if it preserves the inner product, $(Tv, Tw) = (v, w)$ for all $v, w \in V$. The spectral theorem from linear algebra asserts that any unitary transformation is diagonalizable. In other words, there exists a basis $\{v_1, \ldots, v_d\}$ (eigenvectors) of V such that $T(v_i) = \lambda_i v_i$, where $\lambda_i \in \mathbb{C}$ is the eigenvalue attached to v_i.

The proof of Theorem 2.5 is based on the following extension of the spectral theorem.

Lemma 2.6 *Suppose $\{T_1, \ldots, T_k\}$ is a commuting family of unitary transformations on the finite-dimensional inner product space V; that is,*

$$T_i T_j = T_j T_i \quad \textit{for all } i, j.$$

Then $T_1, \ldots, T_k$ are simultaneously diagonalizable. In other words, there exists a basis for V which consists of eigenvectors for every T_i, $i = 1, \ldots, k$.

Proof. We use induction on k. The case $k = 1$ is simply the spectral theorem. Suppose that the lemma is true for any family of $k-1$ commuting unitary transformations. The spectral theorem applied to T_k says that V is the direct sum of its eigenspaces

$$V = V_{\lambda_1} \oplus \cdots \oplus V_{\lambda_s},$$

where V_{λ_i} denotes the subspace of all eigenvectors with eigenvalue λ_i. We claim that each one of the $T_1, \ldots, T_{k-1}$ maps each eigenspace V_{λ_i} to itself. Indeed, if $v \in V_{\lambda_i}$ and $1 \leq j \leq k-1$, then

$$T_k T_j(v) = T_j T_k(v) = T_j(\lambda_i v) = \lambda_i T_j(v)$$

so $T_j(v) \in V_{\lambda_i}$, and the claim is proved.

Since the restrictions to V_{λ_i} of $T_1, \ldots, T_{k-1}$ form a family of commuting unitary linear transformations, the induction hypothesis guarantees that these are simultaneously diagonalizable on each subspace V_{λ_i}. This diagonalization provides us with the desired basis for each V_{λ_i}, and thus for V.

We can now prove Theorem 2.5. Recall that the vector space V of complex-valued functions defined on G has dimension $|G|$. For each $a \in G$ we define a linear transformation $T_a : V \to V$ by

$$(T_a f)(x) = f(a \cdot x) \quad \text{for } x \in G.$$

Since G is abelian it is clear that $T_a T_b = T_b T_a$ for all $a, b \in G$, and one checks easily that T_a is unitary for the Hermitian inner product (3) defined on V. By Lemma 2.6 the family $\{T_a\}_{a \in G}$ is simultaneously diagonalizable. This means there is a basis $\{v_b(x)\}_{b \in G}$ for V such that each $v_b(x)$ is an eigenfunction for T_a, for every a. Let v be one of these basis elements and 1 the unit element in G. We must have $v(1) \neq 0$ for otherwise

$$v(a) = v(a \cdot 1) = (T_a v)(1) = \lambda_a v(1) = 0,$$

where λ_a is the eigenvalue of v for T_a. Hence $v = 0$, and this is a contradiction. We claim that the function defined by $w(x) = \lambda_x = v(x)/v(1)$ is a character of G. Arguing as above we find that $w(x) \neq 0$ for every x, and

$$w(a \cdot b) = \frac{v(a \cdot b)}{v(1)} = \frac{\lambda_a v(b)}{v(1)} = \lambda_a \lambda_b \frac{v(1)}{v(1)} = \lambda_a \lambda_b = w(a) w(b).$$

We now invoke Lemma 2.2 to conclude the proof.

2.5 Fourier inversion and Plancherel formula

We now put together the results obtained in the previous sections to discuss the Fourier expansion of a function on a finite abelian group G. Given a function f on G and character e of G, we define the **Fourier coefficient** of f with respect to e, by

$$\hat{f}(e) = (f, e) = \frac{1}{|G|} \sum_{a \in G} f(a)\overline{e(a)},$$

and the **Fourier series** of f as

$$f \sim \sum_{e \in \hat{G}} \hat{f}(e) e.$$

Since the characters form a basis, we know that

$$f = \sum_{e \in \hat{G}} c_e e$$

for some set of constants c_e. By the orthogonality relations satisfied by the characters, we find that

$$(f, e) = c_e,$$

so f is indeed equal to its Fourier series, namely,

$$f = \sum_{e \in \hat{G}} \hat{f}(e) e.$$

We summarize our results.

Theorem 2.7 *Let G be a finite abelian group. The characters of G form an orthonormal basis for the vector space V of functions on G equipped with the inner product*

$$(f, g) = \frac{1}{|G|} \sum_{a \in G} f(a)\overline{g(a)}.$$

In particular, any function f on G is equal to its Fourier series

$$f = \sum_{e \in \hat{G}} \hat{f}(e) e.$$

Finally, we have the Parseval-Plancherel formula for finite abelian groups.

Theorem 2.8 *If f is a function on G, then* $\|f\|^2 = \sum_{e\in\hat{G}} |\hat{f}(e)|^2$.

Proof. Since the characters of G form an orthonormal basis for the vector space V, and $(f, e) = \hat{f}(e)$, we have that

$$\|f\|^2 = (f, f) = \sum_{e\in\hat{G}} (f, e)\overline{\hat{f}(e)} = \sum_{e\in\hat{G}} |\hat{f}(e)|^2.$$

The apparent difference of this statement with that of Theorem 1.2 is due to the different normalizations of the Fourier coefficients that are used.

3 Exercises

1. Let f be a function on the circle. For each $N \geq 1$ the discrete Fourier coefficients of f are defined by

$$a_N(n) = \frac{1}{N}\sum_{k=1}^{N} f(e^{2\pi ik/N})e^{-2\pi ikn/N}, \qquad \text{for } n \in \mathbb{Z}.$$

We also let

$$a(n) = \int_0^1 f(e^{2\pi ix})e^{-2\pi inx}\, dx$$

denote the ordinary Fourier coefficients of f.

(a) Show that $a_N(n) = a_N(n + N)$.

(b) Prove that if f is continuous, then $a_N(n) \to a(n)$ as $N \to \infty$.

2. If f is a C^1 function on the circle, prove that $|a_N(n)| \leq c/|n|$ whenever $0 < |n| \leq N/2$.
[Hint: Write

$$a_N(n)[1 - e^{2\pi i\ell n/N}] = \frac{1}{N}\sum_{k=1}^{N}[f(e^{2\pi ik/N}) - f(e^{2\pi i(k+\ell)/N})]e^{-2\pi ikn/N},$$

and choose ℓ so that $\ell n/N$ is nearly $1/2$.]

3. By a similar method, show that if f is a C^2 function on the circle, then

$$|a_N(n)| \leq c/|n|^2, \qquad \text{whenever } 0 < |n| \leq N/2.$$

As a result, prove the inversion formula for $f \in C^2$,

$$f(e^{2\pi ix}) = \sum_{n=-\infty}^{\infty} a(n)e^{2\pi inx}$$

from its finite version.

[Hint: For the first part, use the second symmetric difference

$$f(e^{2\pi i(k+\ell)/N}) + f(e^{2\pi i(k-\ell)/N}) - 2f(e^{2\pi ik/N}).$$

For the second part, if N is odd (say), write the inversion formula as

$$f(e^{2\pi ik/N}) = \sum_{|n|<N/2} a_N(n)e^{2\pi ikn/N}.]$$

4. Let e be a character on $G = \mathbb{Z}(N)$, the additive group of integers modulo N. Show that there exists a unique $0 \leq \ell \leq N-1$ so that

$$e(k) = e_\ell(k) = e^{2\pi i\ell k/N} \qquad \text{for all } k \in \mathbb{Z}(N).$$

Conversely, every function of this type is a character on $\mathbb{Z}(N)$. Deduce that $e_\ell \mapsto \ell$ defines an isomorphism from $\hat{G}$ to G.

[Hint: Show that $e(1)$ is an Nth root of unity.]

5. Show that all characters on S^1 are given by

$$e_n(x) = e^{2\pi inx} \qquad \text{with } n \in \mathbb{Z},$$

and check that $e_n \mapsto n$ defines an isomorphism from $\widehat{S^1}$ to $\mathbb{Z}$.

[Hint: If F is continuous and $F(x+y) = F(x)F(y)$, then F is differentiable. To see this, note that if $F(0) \neq 0$, then for appropriate δ, $c = \int_0^\delta F(y)\,dy \neq 0$, and $cF(x) = \int_x^{\delta+x} F(y)\,dy$. Differentiate to conclude that $F(x) = e^{Ax}$ for some A.]

6. Prove that all characters on $\mathbb{R}$ take the form

$$e_\xi(x) = e^{2\pi i\xi x} \qquad \text{with } \xi \in \mathbb{R},$$

and that $e_\xi \mapsto \xi$ defines an isomorphism from $\widehat{\mathbb{R}}$ to $\mathbb{R}$. The argument in Exercise 5 applies here as well.

7. Let $\zeta = e^{2\pi i/N}$. Define the $N \times N$ matrix $M = (a_{jk})_{1 \leq j,k \leq N}$ by $a_{jk} = N^{-1/2}\zeta^{jk}$.

(a) Show that M is unitary.

(b) Interpret the identity $(Mu, Mv) = (u, v)$ and the fact that $M^* = M^{-1}$ in terms of Fourier series on $\mathbb{Z}(N)$.

8. Suppose that $P(x) = \sum_{n=1}^{N} a_n e^{2\pi inx}$.

(a) Show by using the Parseval identities for the circle and $\mathbb{Z}(N)$, that

$$\int_0^1 |P(x)|^2\, dx = \frac{1}{N}\sum_{j=1}^{N} |P(j/N)|^2.$$

(b) Prove the reconstruction formula

$$P(x) = \sum_{j=1}^{N} P(j/N)K(x - (j/N))$$

where

$$K(x) = \frac{e^{2\pi ix}}{N}\frac{1 - e^{2\pi iNx}}{1 - e^{2\pi ix}} = \frac{1}{N}(e^{2\pi ix} + e^{2\pi i2x} + \cdots + e^{2\pi iNx}).$$

Observe that P is completely determined by the values $P(j/N)$ for $1 \leq j \leq N$. Note also that $K(0) = 1$, and $K(j/N) = 0$ whenever j is not congruent to 0 modulo N.

9. To prove the following assertions, modify the argument given in the text.

(a) Show that one can compute the Fourier coefficients of a function on $\mathbb{Z}(N)$ when $N = 3^n$ with at most $6N \log_3 N$ operations.

(b) Generalize this to $N = \alpha^n$ where α is an integer > 1.

10. A group G is **cyclic** if there exists $g \in G$ that generates all of G, that is, if any element in G can be written as g^n for some $n \in \mathbb{Z}$. Prove that a finite abelian group is cyclic if and only if it is isomorphic to $\mathbb{Z}(N)$ for some N.

11. Write down the multiplicative tables for the groups $\mathbb{Z}^*(3)$, $\mathbb{Z}^*(4)$, $\mathbb{Z}^*(5)$, $\mathbb{Z}^*(6)$, $\mathbb{Z}^*(8)$, and $\mathbb{Z}^*(9)$. Which of these groups are cyclic?

12. Suppose that G is a finite abelian group and $e : G \to \mathbb{C}$ is a function that satisfies $e(x \cdot y) = e(x)e(y)$ for all $x, y \in G$. Prove that either e is identically 0, or e never vanishes. In the second case, show that for each x, $e(x) = e^{2\pi ir}$ for some $r \in \mathbb{Q}$ of the form $r = p/q$, where $q = |G|$.

13. In analogy with ordinary Fourier series, one may interpret finite Fourier expansions using convolutions as follows. Suppose G is a finite abelian group, 1_G its unit, and V the vector space of complex-valued functions on G.

(a) The convolution of two functions f and g in V is defined for each $a \in G$ by

$$(f * g)(a) = \frac{1}{|G|} \sum_{b \in G} f(b) g(a \cdot b^{-1}).$$

Show that for all $e \in \hat{G}$ one has $\widehat{(f * g)}(e) = \hat{f}(e)\hat{g}(e)$.

(b) Use Theorem 2.5 to show that if e is a character on G, then

$$\sum_{e \in \hat{G}} e(c) = 0 \qquad \text{whenever } c \in G \text{ and } c \neq 1_G.$$

(c) As a result of (b), show that the Fourier series $Sf(a) = \sum_{e \in \hat{G}} \hat{f}(e) e(a)$ of a function $f \in V$ takes the form

$$Sf = f * D,$$

where D is defined by

$$D(c) = \sum_{e \in \hat{G}} e(c) = \begin{cases} |G| & \text{if } c = 1_G, \\ 0 & \text{otherwise.} \end{cases} \tag{4}$$

Since $f * D = f$, we recover the fact that $Sf = f$. Loosely speaking, D corresponds to a "Dirac delta function"; it has unit mass

$$\frac{1}{|G|} \sum_{c \in G} D(c) = 1,$$

and (4) says that this mass is concentrated at the unit element in G. Thus D has the same interpretation as the "limit" of a family of good kernels. (See Section 4, Chapter 2.)

Note. The function D reappears in the next chapter as $\delta_1(n)$.

4 Problems

1. Prove that if n and m are two positive integers that are relatively prime, then

$$\mathbb{Z}(nm) \approx \mathbb{Z}(n) \times \mathbb{Z}(m).$$

[Hint: Consider the map $\mathbb{Z}(nm) \to \mathbb{Z}(n) \times \mathbb{Z}(m)$ given by $k \mapsto (k \bmod n, k \bmod m)$, and use the fact that there exist integers x and y such that $xn + ym = 1$.]

2.* Every finite abelian group G is isomorphic to a direct product of cyclic groups. Here are two more precise formulations of this theorem.

- If $p_1, \ldots, p_s$ are the distinct primes appearing in the factorization of the order of G, then

$$G \approx G(p_1) \times \cdots \times G(p_s),$$

  where each $G(p)$ is of the form $G(p) = \mathbb{Z}(p^{r_1}) \times \cdots \times \mathbb{Z}(p^{r_\ell})$, with $0 \leq r_1 \leq \cdots \leq r_\ell$ (this sequence of integers depends on p of course). This decomposition is unique.

- There exist unique integers $d_1, \ldots, d_k$ such that

$$d_1|d_2, \quad d_2|d_3, \quad \cdots, \quad d_{k-1}|d_k$$

  and

$$G \approx \mathbb{Z}(d_1) \times \cdots \times \mathbb{Z}(d_k).$$

Deduce the second formulation from the first.

3. Let $\hat{G}$ denote the collection of distinct characters of the finite abelian group G.

(a) Note that if $G = \mathbb{Z}(N)$, then $\hat{G}$ is isomorphic to G.

(b) Prove that $\widehat{G_1 \times G_2} = \hat{G}_1 \times \hat{G}_2$.

(c) Prove using Problem 2 that if G is a finite abelian group, then $\hat{G}$ is isomorphic to G.

4.* When p is prime the group $\mathbb{Z}^*(p)$ is cyclic, and $\mathbb{Z}^*(p) \approx \mathbb{Z}(p-1)$.

8 Dirichlet's Theorem

> Dirichlet, Gustav Lejeune (Düren 1805-Göttingen 1859), German mathematician. He was a number theorist at heart. But, while studying in Paris, being a very likeable person, he was befriended by Fourier and other like-minded mathematicians, and he learned analysis from them. Thus equipped, he was able to lay the foundation for the application of Fourier analysis to (analytic) theory of numbers.
>
> *S. Bochner*, 1966

As a striking application of the theory of finite Fourier series, we now prove Dirichlet's theorem on primes in arithmetic progression. This theorem states that if q and ℓ are positive integers with no common factor, then the progression

$$\ell,\ \ell+q,\ \ell+2q,\ \ell+3q,\ldots,\ell+kq,\ldots$$

contains infinitely many prime numbers. This change of subject matter that we undertake illustrates the wide applicability of ideas from Fourier analysis to various areas outside its seemingly narrower confines. In this particular case, it is the theory of Fourier series on the finite abelian group $\mathbb{Z}^*(q)$ that plays a key role in the solution of the problem.

1 A little elementary number theory

We begin by introducing the requisite background. This involves elementary ideas of divisibility of integers, and in particular properties regarding prime numbers. Here the basic fact, called the fundamental theorem of arithmetic, is that every integer is the product of primes in an essentially unique way.

1.1 The fundamental theorem of arithmetic

The following theorem is a mathematical formulation of long division.

Theorem 1.1 (Euclid's algorithm) *For any integers a and b with $b > 0$, there exist unique integers q and r with $0 \leq r < b$ such that*

$$a = qb + r.$$

Here q denotes the quotient of a by b, and r is the remainder, which is smaller than b.

Proof. First we prove the existence of q and r. Let S denote the set of all non-negative integers of the form $a - qb$ with $q \in \mathbb{Z}$. This set is non-empty and in fact S contains arbitrarily large positive integers since $b \neq 0$. Let r denote the smallest element in S, so that

$$r = a - qb$$

for some integer q. By construction $0 \leq r$, and we claim that $r < b$. If not, we may write $r = b + s$ with $0 \leq s < r$, so $b + s = a - qb$, which then implies

$$s = a - (q+1)b.$$

Hence $s \in S$ with $s < r$, and this contradicts the minimality of r. So $r < b$, hence q and r satisfy the conditions of the theorem.

To prove uniqueness, suppose we also had $a = q_1 b + r_1$ where $0 \leq r_1 < b$. By subtraction we find

$$(q - q_1)b = r_1 - r.$$

The left-hand side has absolute value 0 or $\geq b$, while the right-hand side has absolute value $< b$. Hence both sides of the equation must be 0, which gives $q = q_1$ and $r = r_1$.

An integer a **divides** b if there exists another integer c such that $ac = b$; we then write $a|b$ and say that a is a **divisor** of b. Note that in particular 1 divides every integer, and $a|a$ for all integers a. A **prime number** is a positive integer greater than 1 that has no positive divisors besides 1 and itself. The main theorem in this section says that any positive integer can be written uniquely as the product of prime numbers.

The **greatest common divisor** of two positive integers a and b is the largest integer that divides both a and b. We usually denote the greatest common divisor by $\gcd(a, b)$. Two positive integers are **relatively prime** if their greatest common divisor is 1. In other words, 1 is the only positive divisor common to both a and b.

Theorem 1.2 *If* $\gcd(a, b) = d$*, then there exist integers* x *and* y *such that*

$$ax + by = d.$$

Proof. Consider the set S of all positive integers of the form $ax + by$ where $x, y \in \mathbb{Z}$, and let s be the smallest element in S. We claim that $s = d$. By construction, there exist integers x and y such that

$$ax + by = s.$$

Clearly, any divisor of a and b divides s, so we must have $d \leq s$. The proof will be complete if we can show that $s|a$ and $s|b$. By Euclid's algorithm, we can write $a = qs + r$ with $0 \leq r < s$. Multiplying the above by q we find $qax + qby = qs$, and therefore

$$qax + qby = a - r.$$

Hence $r = a(1 - qx) + b(-qy)$. Since s was minimal in S and $0 \leq r < s$, we conclude that $r = 0$, therefore s divides a. A similar argument shows that s divides b, hence $s = d$ as desired.

In particular we record the following three consequences of the theorem.

Corollary 1.3 *Two positive integers* a *and* b *are relatively prime if and only if there exist integers* x *and* y *such that* $ax + by = 1$.

Proof. If a and b are relatively prime, two integers x and y with the desired property exist by Theorem 1.2. Conversely, if $ax + by = 1$ holds and d is positive and divides both a and b, then d divides 1, hence $d = 1$.

Corollary 1.4 *If* a *and* c *are relatively prime and* c *divides* ab*, then* c *divides* b*. In particular, if* p *is a prime that does not divide* a *and* p *divides* ab*, then* p *divides* b.

Proof. We can write $1 = ax + cy$, so multiplying by b we find $b = abx + cby$. Hence $c|b$.

Corollary 1.5 *If* p *is prime and* p *divides the product* $a_1 \cdots a_r$*, then* p *divides* a_i *for some* i.

Proof. By the previous corollary, if p does not divide a_1, then p divides $a_2 \cdots a_r$, so eventually $p|a_i$.

We can now prove the main result of this section.

Theorem 1.6 *Every positive integer greater than 1 can be factored uniquely into a product of primes.*

Proof. First, we show that such a factorization is possible. We do so by proving that the set S of positive integers > 1 which do not have a factorization into primes is empty. Arguing by contradiction, we assume that $S \neq \emptyset$. Let n be the smallest element of S. Since n cannot be a prime, there exist integers $a > 1$ and $b > 1$ such that $ab = n$. But then $a < n$ and $b < n$, so $a \notin S$ as well as $b \notin S$. Hence both a and b have prime factorizations and so does their product n. This implies $n \notin S$, therefore S is empty, as desired.

We now turn our attention to the uniqueness of the factorization. Suppose that n has two factorizations into primes

$$\begin{aligned} n &= p_1 p_2 \cdots p_r \\ &= q_1 q_2 \cdots q_s. \end{aligned}$$

So p_1 divides $q_1 q_2 \cdots q_s$, and we can apply Corollary 1.5 to conclude that $p_1 | q_i$ for some i. Since q_i is prime, we must have $p_1 = q_i$. Continuing with this argument we find that the two factorizations of n are equal up to a permutation of the factors.

We briefly digress to give an alternate definition of the group $\mathbb{Z}^*(q)$ which appeared in the previous chapter. According to our initial definition, $\mathbb{Z}^*(q)$ is the multiplicative group of units in $\mathbb{Z}(q)$: those $n \in \mathbb{Z}(q)$ for which there exists an integer m so that

$$nm \equiv 1 \mod q. \tag{1}$$

Equivalently, $\mathbb{Z}^*(q)$ is the group under multiplication of all integers in $\mathbb{Z}(q)$ that are relatively prime to q. Indeed, notice that if (1) is satisfied, then automatically n and q are relatively prime. Conversely, suppose we assume that n and q are relatively prime. Then, if we put $a = n$ and $b = q$ in Corollary 1.3, we find

$$nx + qy = 1.$$

Hence $nx \equiv 1 \mod q$, and we can take $m = x$ to establish the equivalence.

1.2 The infinitude of primes

The study of prime numbers has always been a central topic in arithmetic, and the first fundamental problem that arose was to determine whether there are infinitely many primes or not. This problem was solved in Euclid's *Elements* with a simple and very elegant argument.

Theorem 1.7 *There are infinitely many primes.*

Proof. Suppose not, and denote by $p_1, \ldots, p_n$ the complete set of primes. Define

$$N = p_1 p_2 \cdots p_n + 1.$$

Since N is larger than any p_i, the integer N cannot be prime. Therefore, N is divisible by a prime that belongs to our list. But this is also an absurdity since every prime divides the product, yet no prime divides 1. This contradiction concludes the proof.

Euclid's argument actually can be modified to deduce finer results about the infinitude of primes. To see this, consider the following problem. Prime numbers (except for 2) can be divided into two classes depending on whether they are of the form $4k + 1$ or $4k + 3$, and the above theorem says that at least one of these classes has to be infinite. A natural question is to ask whether both classes are infinite, and if not, which one is? In the case of primes of the form $4k + 3$, the fact that the class is infinite has a proof that is similar to Euclid's, but with a twist. If there are only finitely many such primes, enumerate them in increasing order omitting 3,

$$p_1 = 7, \quad p_2 = 11, \quad \ldots, \quad p_n,$$

and let

$$N = 4 p_1 p_2 \cdots p_n + 3.$$

Clearly, N is of the form $4k + 3$ and cannot be prime since $N > p_n$. Since the product of two numbers of the form $4m + 1$ is again of the form $4m + 1$, one of the prime divisors of N, say p, must be of the form $4k + 3$. We must have $p \neq 3$, since 3 does not divide the product in the definition of N. Also, p cannot be one of the other primes of the form $4k + 3$, that is, $p \neq p_i$ for $i = 1, \ldots n$, because then p divides the product $p_1 \cdots p_n$ but does not divide 3.

It remains to determine if the class of primes of the form $4k + 1$ is infinite. A simple-minded modification of the above argument does not work since the product of two numbers of the form $4m + 3$ is never of the form $4m + 3$. More generally, in an attempt to prove the law of quadratic reciprocity, Legendre formulated the following statement:

If q and ℓ are relatively prime, then the sequence

$$\ell + kq, \qquad k \in \mathbb{Z}$$

contains infinitely many primes (hence at least one prime!).

Of course, the condition that q and ℓ be relatively prime is necessary, for otherwise $\ell + kq$ is never prime. In other words, this hypothesis says that any arithmetic progression that could contain primes necessarily contains infinitely many of them.

Legendre's assertion was proved by Dirichlet. The key idea in his proof is Euler's analytical approach to prime numbers involving his product formula, which gives a strengthened version of Theorem 1.7. This insight of Euler led to a deep connection between the theory of primes and analysis.

The zeta function and its Euler product

We begin with a rapid review of infinite products. If $\{A_n\}_{n=1}^{\infty}$ is a sequence of real numbers, we define

$$\prod_{n=1}^{\infty} A_n = \lim_{N\to\infty} \prod_{n=1}^{N} A_n$$

if the limit exists, in which case we say that the product converges. The natural approach is to take logarithms and transform products into sums. We gather in a lemma the properties we shall need of the function $\log x$, defined for positive real numbers.

Lemma 1.8 *The exponential and logarithm functions satisfy the following properties:*

(i) $e^{\log x} = x$.

(ii) $\log(1+x) = x + E(x)$ *where* $|E(x)| \leq x^2$ *if* $|x| < 1/2$.

(iii) *If* $\log(1+x) = y$ *and* $|x| < 1/2$, *then* $|y| \leq 2|x|$.

In terms of the O notation, property (ii) will be recorded as $\log(1+x) = x + O(x^2)$.

Proof. Property (i) is standard. To prove property (ii) we use the power series expansion of $\log(1+x)$ for $|x| < 1$, that is,

$$\log(1+x) = \sum_{n=1}^{\infty} \frac{(-1)^{n+1}}{n} x^n. \tag{2}$$

Then we have

$$E(x) = \log(1+x) - x = -\frac{x^2}{2} + \frac{x^3}{3} - \frac{x^4}{4} + \cdots,$$

and the triangle inequality implies

$$|E(x)| \le \frac{x^2}{2}\left(1 + |x| + |x|^2 + \cdots\right).$$

Therefore, if $|x| \le 1/2$ we can sum the geometric series on the right-hand side to find that

$$\begin{aligned} |E(x)| &\le \frac{x^2}{2}\left(1 + \frac{1}{2} + \frac{1}{2^2} + \cdots\right) \\ &\le \frac{x^2}{2}\left(\frac{1}{1-1/2}\right) \\ &\le x^2. \end{aligned}$$

The proof of property (iii) is now immediate; if $x \ne 0$ and $|x| \le 1/2$, then

$$\begin{aligned} \left|\frac{\log(1+x)}{x}\right| &\le 1 + \left|\frac{E(x)}{x}\right| \\ &\le 1 + |x| \\ &\le 2, \end{aligned}$$

and if $x = 0$, (iii) is clearly also true.

We can now prove the main result on infinite products of real numbers.

Proposition 1.9 *If $A_n = 1 + a_n$ and $\sum |a_n|$ converges, then the product $\prod_n A_n$ converges, and this product vanishes if and only if one of its factors A_n vanishes. Also, if $a_n \ne 1$ for all n, then $\prod_n 1/(1-a_n)$ converges.*

Proof. If $\sum |a_n|$ converges, then for all large n we must have $|a_n| < 1/2$. Disregarding finitely many terms if necessary, we may assume that this inequality holds for all n. Then we may write the partial products as follows:

$$\prod_{n=1}^{N} A_n = \prod_{n=1}^{N} e^{\log(1+a_n)} = e^{B_N},$$

where $B_N = \sum_{n=1}^{N} b_n$ with $b_n = \log(1+a_n)$. By the lemma, we know that $|b_n| \le 2|a_n|$, so that B_N converges to a real number, say B. Since the exponential function is continuous, we conclude that e^{B_N} converges to e^B as N goes to infinity, proving the first assertion of the proposition. Observe also that if $1+a_n \neq 0$ for all n, the product converges to a non-zero limit since it is expressed as e^B.

Finally observe that the partial products of $\prod_n 1/(1-a_n)$ are $1/\prod_{n=1}^N (1-a_n)$, so the same argument as above proves that the product in the denominator converges to a non-zero limit.

With these preliminaries behind us, we can now return to the heart of the matter. For s a real number (strictly) greater than 1, we define the **zeta function** by

$$\zeta(s) = \sum_{n=1}^{\infty} \frac{1}{n^s}.$$

To see that the series defining ζ converges, we use the principle that whenever f is a decreasing function one can compare $\sum f(n)$ with $\int f(x)\,dx$, as is suggested by Figure 1. Note also that a similar technique was used in Chapter 3, that time bounding a sum from below by an integral.

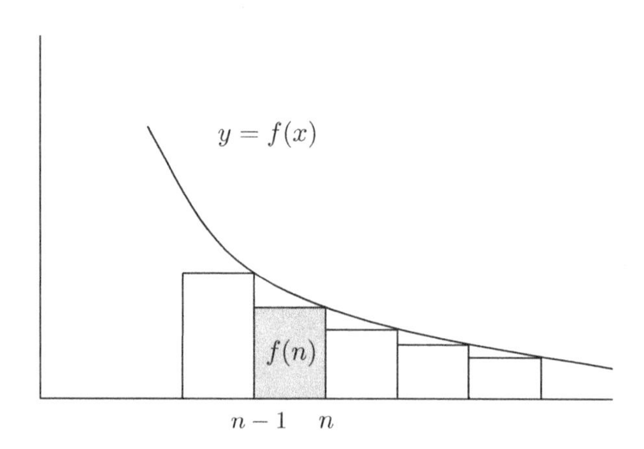

Figure 1. Comparing sums with integrals

Here we take $f(x) = 1/x^s$ to see that

$$\sum_{n=1}^{\infty} \frac{1}{n^s} \leq 1 + \sum_{n=2}^{\infty} \int_{n-1}^{n} \frac{dx}{x^s} = 1 + \int_1^{\infty} \frac{dx}{x^s},$$

and therefore,

$$\zeta(s) \leq 1 + \frac{1}{s-1}. \tag{3}$$

Clearly, the series defining ζ converges uniformly on each half-line $s > s_0 > 1$, hence ζ is continuous when $s > 1$. The zeta function was already mentioned earlier in the discussion of the Poisson summation formula and the theta function.

The key result is Euler's product formula.

Theorem 1.10 *For every $s > 1$, we have*

$$\zeta(s) = \prod_p \frac{1}{1 - 1/p^s},$$

where the product is taken over all primes.

It is important to remark that this identity is an analytic expression of the fundamental theorem of arithmetic. In fact, each factor of the product $1/(1-p^{-s})$ can be written as a convergent geometric series

$$1 + \frac{1}{p^s} + \frac{1}{p^{2s}} + \cdots + \frac{1}{p^{Ms}} + \cdots .$$

So we consider

$$\prod_{p_j} \left(1 + \frac{1}{p_j^s} + \frac{1}{p_j^{2s}} + \cdots + \frac{1}{p_j^{Ms}} + \cdots \right),$$

where the product is taken over all primes, which we order in increasing order $p_1 < p_2 < \cdots$. Proceeding formally (these manipulations will be justified below), we calculate the product as a sum of terms, each term originating by picking out a term $1/p_j^{ks}$ (in the sum corresponding to p_j) with a k, which of course will depend on j, and with $k = 0$ for j sufficiently large. The product obtained this way is

$$\frac{1}{(p_1^{k_1} p_2^{k_2} \cdots p_m^{k_m})^s} = \frac{1}{n^s},$$

where the integer n is written as a product of primes $n = p_1^{k_1} p_2^{k_2} \cdots p_m^{k_m}$. By the fundamental theorem of arithmetic, each integer ≥ 1 occurs in this way uniquely, hence the product equals

$$\sum_{n=1}^{\infty} \frac{1}{n^s}.$$

We now justify this heuristic argument.

Proof. Suppose M and N are positive integers with $M > N$. Observe now that any positive integer $n \leq N$ can be written uniquely as a product of primes, and that each prime must be less than or equal to N and repeated less than M times. Therefore

$$\begin{aligned}\sum_{n=1}^{N} \frac{1}{n^s} &\leq \prod_{p \leq N} \left(1 + \frac{1}{p^s} + \frac{1}{p^{2s}} + \cdots + \frac{1}{p^{Ms}}\right) \\ &\leq \prod_{p \leq N} \left(\frac{1}{1 - p^{-s}}\right) \\ &\leq \prod_{p} \left(\frac{1}{1 - p^{-s}}\right).\end{aligned}$$

Letting N tend to infinity now yields

$$\sum_{n=1}^{\infty} \frac{1}{n^s} \leq \prod_{p} \left(\frac{1}{1 - p^{-s}}\right).$$

For the reverse inequality, we argue as follows. Again, by the fundamental theorem of arithmetic, we find that

$$\prod_{p \leq N} \left(1 + \frac{1}{p^s} + \frac{1}{p^{2s}} + \cdots + \frac{1}{p^{Ms}}\right) \leq \sum_{n=1}^{\infty} \frac{1}{n^s}.$$

Letting M tend to infinity gives

$$\prod_{p \leq N} \left(\frac{1}{1 - p^{-s}}\right) \leq \sum_{n=1}^{\infty} \frac{1}{n^s}.$$

Hence

$$\prod_{p} \left(\frac{1}{1 - p^{-s}}\right) \leq \sum_{n=1}^{\infty} \frac{1}{n^s},$$

and the proof of the product formula is complete.

We now come to Euler's version of Theorem 1.7, which inspired Dirichlet's approach to the general problem of primes in arithmetic progression. The point is the following proposition.

Proposition 1.11 *The series*

$$\sum_p 1/p$$

diverges, when the sum is taken over all primes p.

Of course, if there were only finitely many primes the series would converge automatically.

Proof. We take logarithms of both sides of the Euler formula. Since $\log x$ is continuous, we may write the logarithm of the infinite product as the sum of the logarithms. Therefore, we obtain for $s > 1$

$$-\sum_p \log(1 - 1/p^s) = \log \zeta(s).$$

Since $\log(1 + x) = x + O(|x|^2)$ whenever $|x| \leq 1/2$, we get

$$-\sum_p \left[-1/p^s + O(1/p^{2s})\right] = \log \zeta(s),$$

which gives

$$\sum_p 1/p^s + O(1) = \log \zeta(s).$$

The term $O(1)$ appears because $\sum_p 1/p^{2s} \leq \sum_{n=1}^{\infty} 1/n^2$. Now we let s tend to 1 from above, namely $s \to 1^+$, and note that $\zeta(s) \to \infty$ since $\sum_{n=1}^{\infty} 1/n^s \geq \sum_{n=1}^{M} 1/n^s$, and therefore

$$\liminf_{s \to 1^+} \sum_{n=1}^{\infty} 1/n^s \geq \sum_{n=1}^{M} 1/n \qquad \text{for every } M.$$

We conclude that $\sum_p 1/p^s \to \infty$ as $s \to 1^+$, and since $1/p > 1/p^s$ for all $s > 1$, we finally have that

$$\sum_p 1/p = \infty.$$

In the rest of this chapter we see how Dirichlet adapted Euler's insight.

2 Dirichlet's theorem

We remind the reader of our goal:

Theorem 2.1 *If q and ℓ are relatively prime positive integers, then there are infinitely many primes of the form $\ell + kq$ with $k \in \mathbb{Z}$.*

Following Euler's argument, Dirichlet proved this theorem by showing that the series

$$\sum_{p \equiv \ell \mod q} \frac{1}{p}$$

diverges, where the sum is over all primes congruent to ℓ modulo q. Once q is fixed and no confusion is possible, we write $p \equiv \ell$ to denote a prime congruent to ℓ modulo q. The proof consists of several steps, one of which requires Fourier analysis on the group $\mathbb{Z}^*(q)$. Before proceeding with the theorem in its complete generality, we outline the solution to the particular problem raised earlier: are there infinitely many primes of the form $4k+1$? This example, which consists of the special case $q = 4$ and $\ell = 1$, illustrates all the important steps in the proof of Dirichlet's theorem.

We begin with the character on $\mathbb{Z}^*(4)$ defined by $\chi(1) = 1$ and $\chi(3) = -1$. We extend this character to all of $\mathbb{Z}$ as follows:

$$\chi(n) = \begin{cases} 0 & \text{if } n \text{ is even,} \\ 1 & \text{if } n = 4k+1, \\ -1 & \text{if } n = 4k+3. \end{cases}$$

Note that this function is multiplicative, that is, $\chi(nm) = \chi(n)\chi(m)$ on *all* of $\mathbb{Z}$. Let $L(s, \chi) = \sum_{n=1}^{\infty} \chi(n)/n^s$, so that

$$L(s, \chi) = 1 - \frac{1}{3^s} + \frac{1}{5^s} - \frac{1}{7^s} + \cdots.$$

Then $L(1, \chi)$ is the convergent series given by

$$1 - \frac{1}{3} + \frac{1}{5} - \frac{1}{7} + \cdots.$$

Since the terms in the series are alternating and their absolute values decrease to zero we have $L(1, \chi) \neq 0$. Because χ is multiplicative, the Euler product generalizes (as we will prove later) to give

$$\sum_{n=1}^{\infty} \frac{\chi(n)}{n^s} = \prod_p \frac{1}{1 - \chi(p)/p^s}.$$

Taking the logarithm of both sides, we find that

$$\log L(s,\chi) = \sum_p \frac{\chi(p)}{p^s} + O(1).$$

Letting $s \to 1^+$, the observation that $L(1,\chi) \neq 0$ shows that $\sum_p \chi(p)/p^s$ remains bounded. Hence

$$\sum_{p\equiv 1} \frac{1}{p^s} - \sum_{p\equiv 3} \frac{1}{p^s}$$

is bounded as $s \to 1^+$. However, we know from Proposition 1.11 that

$$\sum_p \frac{1}{p^s}$$

is unbounded as $s \to 1^+$, so putting these two facts together, we find that

$$2\sum_{p\equiv 1} \frac{1}{p^s}$$

is unbounded as $s \to 1^+$. Hence $\sum_{p\equiv 1} 1/p$ diverges, and as a consequence there are infinitely many primes of the form $4k+1$.

We digress briefly to show that in fact $L(1,\chi) = \pi/4$. To see this, we integrate the identity

$$\frac{1}{1+x^2} = 1 - x^2 + x^4 - x^6 + \cdots,$$

and get

$$\int_0^y \frac{dx}{1+x^2} = y - \frac{y^3}{3} + \frac{y^5}{5} - \cdots, \qquad 0 < y < 1.$$

We then let y tend to 1. The integral can be calculated as

$$\int_0^1 \frac{dx}{1+x^2} = \arctan u\big|_0^1 = \frac{\pi}{4},$$

so this proves that the series $1 - 1/3 + 1/5 - \cdots$ is Abel summable to $\pi/4$. Since we know the series converges, its limit is the same as its Abel limit, hence $1 - 1/3 + 1/5 - \cdots = \pi/4$.

The rest of this chapter gives the full proof of Dirichlet's theorem. We begin with the Fourier analysis (which is actually the last step in the example given above), and reduce the theorem to the non-vanishing of L-functions.

2.1 Fourier analysis, Dirichlet characters, and reduction of the theorem

In what follows we take the abelian group G to be $\mathbb{Z}^*(q)$. Our formulas below involve the order of G, which is the number of integers $0 \le n < q$ that are relatively prime to q; this number defines the **Euler phi-function** $\varphi(q)$, and $|G| = \varphi(q)$.

Consider the function δ_ℓ on G, which we think of as the characteristic function of ℓ; if $n \in \mathbb{Z}^*(q)$, then

$$\delta_\ell(n) = \begin{cases} 1 & \text{if } n \equiv \ell \mod q, \\ 0 & \text{otherwise.} \end{cases}$$

We can expand this function in a Fourier series as follows:

$$\delta_\ell(n) = \sum_{e \in \hat{G}} \widehat{\delta_\ell}(e) e(n),$$

where

$$\widehat{\delta_\ell}(e) = \frac{1}{|G|} \sum_{m \in G} \delta_\ell(m) \overline{e(m)} = \frac{1}{|G|} \overline{e(\ell)}.$$

Hence

$$\delta_\ell(n) = \frac{1}{|G|} \sum_{e \in \hat{G}} \overline{e(\ell)} e(n).$$

We can extend the function δ_ℓ to all of $\mathbb{Z}$ by setting $\delta_\ell(m) = 0$ whenever m and q are not relatively prime. Similarly, the extensions of the characters $e \in \hat{G}$ to all of $\mathbb{Z}$ which are given by the recipe

$$\chi(m) = \begin{cases} e(m) & \text{if } m \text{ and } q \text{ are relatively prime} \\ 0 & \text{otherwise,} \end{cases}$$

are called the **Dirichlet characters** modulo q. We shall denote the extension to $\mathbb{Z}$ of the trivial character of G by χ_0, so that $\chi_0(m) = 1$ if m and q are relatively prime, and 0 otherwise. Note that the Dirichlet characters modulo q are multiplicative on all of $\mathbb{Z}$, in the sense that

$$\chi(nm) = \chi(n)\chi(m) \quad \text{for all } n, m \in \mathbb{Z}.$$

Since the integer q is fixed, we may without fear of confusion, speak of "Dirichlet characters" omitting reference to q.[1]

With $|G| = \varphi(q)$, we may restate the above results as follows:

---

[1]We use the notation χ instead of e to distinguish the Dirichlet characters (defined on $\mathbb{Z}$) from the characters e (defined on $\mathbb{Z}^*(q)$).

Lemma 2.2 *The Dirichlet characters are multiplicative. Moreover,*

$$\delta_\ell(m) = \frac{1}{\varphi(q)} \sum_\chi \overline{\chi(\ell)}\chi(m),$$

where the sum is over all Dirichlet characters.

With the above lemma we have taken our first step towards a proof of the theorem, since this lemma shows that

$$\begin{aligned}
\sum_{p\equiv\ell} \frac{1}{p^s} &= \sum_p \frac{\delta_\ell(p)}{p^s} \\
&= \frac{1}{\varphi(q)} \sum_\chi \overline{\chi(\ell)} \sum_p \frac{\chi(p)}{p^s}.
\end{aligned}$$

Thus it suffices to understand the behavior of $\sum_p \chi(p)p^{-s}$ as $s \to 1^+$. In fact, we divide the above sum in two parts depending on whether or not χ is trivial. So we have

$$\begin{aligned}
\sum_{p\equiv\ell} \frac{1}{p^s} &= \frac{1}{\varphi(q)} \sum_p \frac{\chi_0(p)}{p^s} + \frac{1}{\varphi(q)} \sum_{\chi\neq\chi_0} \overline{\chi(\ell)} \sum_p \frac{\chi(p)}{p^s} \\
&= \frac{1}{\varphi(q)} \sum_{p \text{ not dividing } q} \frac{1}{p^s} + \frac{1}{\varphi(q)} \sum_{\chi\neq\chi_0} \overline{\chi(\ell)} \sum_p \frac{\chi(p)}{p^s}. \qquad (4)
\end{aligned}$$

Since there are only finitely many primes dividing q, Euler's theorem (Proposition 1.11) implies that the first sum on the right-hand side diverges when s tends to 1. These observations show that Dirichlet's theorem is a consequence of the following assertion.

Theorem 2.3 *If χ is a nontrivial Dirichlet character, then the sum*

$$\sum_p \frac{\chi(p)}{p^s}$$

remains bounded as $s \to 1^+$.

The proof of Theorem 2.3 requires the introduction of the L-functions, to which we now turn.

2.2 Dirichlet L-functions

We proved earlier that the zeta function $\zeta(s) = \sum_n 1/n^s$ could be expressed as a product, namely

$$\sum_{n=1}^{\infty} \frac{1}{n^s} = \prod_p \frac{1}{(1 - p^{-s})}.$$

Dirichlet observed an analogue of this formula for the so-called **L-functions** defined for $s > 1$ by

$$L(s, \chi) = \sum_{n=1}^{\infty} \frac{\chi(n)}{n^s},$$

where χ is a Dirichlet character.

Theorem 2.4 *If $s > 1$, then*

$$\sum_{n=1}^{\infty} \frac{\chi(n)}{n^s} = \prod_{p} \frac{1}{(1 - \chi(p)p^{-s})},$$

where the product is over all primes.

Assuming this theorem for now, we can follow Euler's argument formally: taking the logarithm of the product and using the fact that $\log(1 + x) = x + O(x^2)$ whenever x is small, we would get

$$\begin{aligned}\log L(s, \chi) &= -\sum_{p} \log(1 - \chi(p)/p^s) \\ &= -\sum_{p} \left[-\frac{\chi(p)}{p^s} + O\left(\frac{1}{p^{2s}}\right) \right] \\ &= \sum_{p} \frac{\chi(p)}{p^s} + O(1).\end{aligned}$$

If $L(1, \chi)$ is finite and non-zero, then $\log L(s, \chi)$ is bounded as $s \to 1^+$, and we can conclude that the sum

$$\sum_{p} \frac{\chi(p)}{p^s}$$

is bounded as $s \to 1^+$. We now make several observations about the above formal argument.

First, we must prove the product formula in Theorem 2.4. Since the Dirichlet characters χ can be complex-valued we will extend the logarithm to complex numbers w of the form $w = 1/(1 - z)$ with $|z| < 1$. (This will be done in terms of a power series.) Then we show that with this definition of the logarithm, the proof of Euler's product formula given earlier carries over to L-functions.

Second, we must make sense of taking the logarithm of both sides of the product formula. If the Dirichlet characters are real, this argument works and is precisely the one given in the example corresponding to primes of the form $4k + 1$. In general, the difficulty lies in the fact that $\chi(p)$ is a complex number, and the complex logarithm is not single valued; in particular, the logarithm of a product is not the sum of the logarithms.

Third, it remains to prove that whenever $\chi \neq \chi_0$, then $\log L(s, \chi)$ is bounded as $s \to 1^+$. If (as we shall see) $L(s, \chi)$ is continuous at $s = 1$, then it suffices to show that

$$L(1, \chi) \neq 0.$$

This is the non-vanishing we mentioned earlier, which corresponds to the alternating series being non-zero in the previous example. The fact that $L(1, \chi) \neq 0$ is the most difficult part of the argument.

So we will focus on three points:

1. Complex logarithms and infinite products.

2. Study of $L(s, \chi)$.

3. Proof that $L(1, \chi) \neq 0$ if χ is non-trivial.

However, before we enter further into the details, we pause briefly to discuss some historical facts surrounding Dirichlet's theorem.

Historical digression

In the following list, we have gathered the names of those mathematicians whose work dealt most closely with the series of achievements related to Dirichlet's theorem. To give a better perspective, we attach the years in which they reached the age of 35:

Euler 1742
Legendre 1787
Gauss 1812
Dirichlet 1840
Riemann 1861

As we mentioned earlier, Euler's discovery of the product formula for the zeta function is the starting point in Dirichlet's argument. Legendre in effect conjectured the theorem because he needed it in his proof of the law of quadratic reciprocity. However, this goal was first accomplished by Gauss who, while not knowing how to establish the theorem about primes in arithmetic progression, nevertheless found a number of different proofs of quadratic reciprocity. Later, Riemann extended the study of the zeta function to the complex plane and indicated how properties related to the non-vanishing of that function were central in the further understanding of the distribution of prime numbers.

Dirichlet proved his theorem in 1837. It should be noted that Fourier, who had befriended Dirichlet when the latter was a young mathematician visiting Paris, had died several years before. Besides the great activity in mathematics, that period was also a very fertile time in the arts, and in particular music. The era of Beethoven had ended only ten years earlier, and Schumann was now reaching the heights of his creativity. But the musician whose career was closest to Dirichlet was Felix Mendelssohn (four years his junior). It so happens that the latter began composing his famous violin concerto the year after Dirichlet succeeded in proving his theorem.

3 Proof of the theorem

We return to the proof of Dirichlet's theorem and to the three difficulties mentioned above.

3.1 Logarithms

The device to deal with the first point is to define two logarithms, one for complex numbers of the form $1/(1-z)$ with $|z|<1$ which we denote by $\log_1$, and one for the function $L(s,\chi)$ which we will denote by $\log_2$.

For the first logarithm, we define

$$\log_1\left(\frac{1}{1-z}\right) = \sum_{k=1}^{\infty}\frac{z^k}{k} \quad \text{for } |z|<1.$$

Note that $\log_1 w$ is then defined if $\mathrm{Re}(w) > 1/2$, and because of equation (2), $\log_1 w$ gives an extension of the usual $\log x$ when x is a real number $> 1/2$.

Proposition 3.1 *The logarithm function* $\log_1$ *satisfies the following properties:*

(i) *If* $|z|<1$, *then*

$$e^{\log_1\left(\frac{1}{1-z}\right)} = \frac{1}{1-z}.$$

(ii) *If* $|z|<1$, *then*

$$\log_1\left(\frac{1}{1-z}\right) = z + E_1(z),$$

where the error E_1 *satisfies* $|E_1(z)| \leq |z|^2$ *if* $|z|<1/2$.

(iii) *If* $|z| < 1/2$, *then*

$$\left|\log_1\left(\frac{1}{1-z}\right)\right| \leq 2|z|.$$

Proof. To establish the first property, let $z = re^{i\theta}$ with $0 \leq r < 1$, and observe that it suffices to show that

$$(1 - re^{i\theta})\, e^{\sum_{k=1}^{\infty}(re^{i\theta})^k/k} = 1. \tag{5}$$

To do so, we differentiate the left-hand side with respect to r, and this gives

$$\left[-e^{i\theta} + (1 - re^{i\theta})\left(\sum_{k=1}^{\infty}(re^{i\theta})^k/k\right)'\right] e^{\sum_{k=1}^{\infty}(re^{i\theta})^k/k}.$$

The term in brackets equals

$$-e^{i\theta} + (1 - re^{i\theta})e^{i\theta}\left(\sum_{k=1}^{\infty}(re^{i\theta})^{k-1}\right) = -e^{i\theta} + (1 - re^{i\theta})e^{i\theta}\frac{1}{1 - re^{i\theta}} = 0.$$

Having found that the left-hand side of the equation (5) is constant, we set $r = 0$ and get the desired result.

The proofs of the second and third properties are the same as their real counterparts given in Lemma 1.8.

Using these results we can state a sufficient condition guaranteeing the convergence of infinite products of complex numbers. Its proof is the same as in the real case, except that we now use the logarithm $\log_1$.

Proposition 3.2 *If* $\sum |a_n|$ *converges, and* $a_n \neq 1$ *for all* n, *then*

$$\prod_{n=1}^{\infty}\left(\frac{1}{1-a_n}\right)$$

converges. Moreover, this product is non-zero.

Proof. For n large enough, $|a_n| < 1/2$, so we may assume without loss of generality that this inequality holds for all $n \geq 1$. Then

$$\prod_{n=1}^{N}\left(\frac{1}{1-a_n}\right) = \prod_{n=1}^{N} e^{\log_1\left(\frac{1}{1-a_n}\right)} = e^{\sum_{n=1}^{N}\log_1\left(\frac{1}{1-a_n}\right)}.$$

But we know from the previous proposition that

$$\left|\log_1\left(\frac{1}{1-z}\right)\right| \leq 2|z|,$$

so the fact that the series $\sum |a_n|$ converges, immediately implies that the limit

$$\lim_{N\to\infty}\sum_{n=1}^{N}\log_1\left(\frac{1}{1-a_n}\right) = A$$

exists. Since the exponential function is continuous, we conclude that the product converges to e^A, which is clearly non-zero.

We may now prove the promised Dirichlet product formula

$$\sum_n \frac{\chi(n)}{n^s} = \prod_p \frac{1}{(1-\chi(p)p^{-s})}.$$

For simplicity of notation, let L denote the left-hand side of the above equation. Define

$$S_N = \sum_{n\leq N}\chi(n)n^{-s} \quad \text{and} \quad \Pi_N = \prod_{p\leq N}\left(\frac{1}{1-\chi(p)p^{-s}}\right).$$

The infinite product $\Pi = \lim_{N\to\infty}\Pi_N = \prod_p \left(\frac{1}{1-\chi(p)p^{-s}}\right)$ converges by the previous proposition. Indeed, if we set $a_n = \chi(p_n)p_n^{-s}$, where p_n is the nth prime, we note that if $s > 1$, then $\sum |a_n| < \infty$.

Also, define

$$\Pi_{N,M} = \prod_{p\leq N}\left(1 + \frac{\chi(p)}{p^s} + \cdots + \frac{\chi(p^M)}{p^{Ms}}\right).$$

Now fix $\epsilon > 0$ and choose N so large that

$$|S_N - L| < \epsilon \quad \text{and} \quad |\Pi_N - \Pi| < \epsilon.$$

We can next select M large enough so that

$$|S_N - \Pi_{N,M}| < \epsilon \quad \text{and} \quad |\Pi_{N,M} - \Pi_N| < \epsilon.$$

To see the first inequality, one uses the fundamental theorem of arithmetic and the fact that the Dirichlet characters are multiplicative. The second inequality follows merely because each series $\sum_{n=1}^{\infty} \frac{\chi(p^n)}{p^{ns}}$ converges.

Therefore

$$|L - \Pi| \leq |L - S_N| + |S_N - \Pi_{N,M}| + |\Pi_{N,M} - \Pi_N| + |\Pi_N - \Pi| < 4\epsilon,$$

as was to be shown.

3.2 L-functions

The next step is a better understanding of the L-functions. Their behavior as functions of s (especially near $s = 1$) depends on whether or not χ is trivial. In the first case, $L(s, \chi_0)$ is up to some simple factors just the zeta function.

Proposition 3.3 *Suppose χ_0 is the trivial Dirichlet character,*

$$\chi_0(n) = \begin{cases} 1 & \text{if } n \text{ and } q \text{ are relatively prime,} \\ 0 & \text{otherwise,} \end{cases}$$

and $q = p_1^{a_1} \cdots p_N^{a_N}$ is the prime factorization of q. Then

$$L(s, \chi_0) = (1 - p_1^{-s})(1 - p_2^{-s}) \cdots (1 - p_N^{-s})\zeta(s).$$

Therefore $L(s, \chi_0) \to \infty$ as $s \to 1^+$.

Proof. The identity follows at once on comparing the Dirichlet and Euler product formulas. The final statement holds because $\zeta(s) \to \infty$ as $s \to 1^+$.

The behavior of the remaining L-functions, those for which $\chi \neq \chi_0$, is more subtle. A remarkable property is that these functions are now defined and continuous for $s > 0$. In fact, more is true.

Proposition 3.4 *If χ is a non-trivial Dirichlet character, then the series*

$$\sum_{n=1}^{\infty} \chi(n)/n^s$$

converges for $s > 0$, and we denote its sum by $L(s, \chi)$. Moreover:

(i) *The function $L(s, \chi)$ is continuously differentiable for $0 < s < \infty$.*

(ii) *There exists constants $c, c' > 0$ so that*

$$L(s, \chi) = 1 + O(e^{-cs}) \quad \text{as } s \to \infty, \text{ and}$$

$$L'(s, \chi) = O(e^{-c's}) \qquad \text{as } s \to \infty.$$

We first isolate the key cancellation property that non-trivial Dirichlet characters possess, which accounts for the behavior of the L-function described in the proposition.

Lemma 3.5 *If χ is a non-trivial Dirichlet character, then*

$$\left| \sum_{n=1}^{k} \chi(n) \right| \leq q, \qquad \textit{for any } k.$$

Proof. First, we recall that

$$\sum_{n=1}^{q} \chi(n) = 0.$$

In fact, if S denotes the sum and $a \in \mathbb{Z}^*(q)$, then the multiplicative property of the Dirichlet character χ gives

$$\chi(a)S = \sum \chi(a)\chi(n) = \sum \chi(an) = \sum \chi(n) = S.$$

Since χ is non-trivial, $\chi(a) \neq 1$ for some a, hence $S = 0$. We now write $k = aq + b$ with $0 \leq b < q$, and note that

$$\sum_{n=1}^{k} \chi(n) = \sum_{n=1}^{aq} \chi(n) + \sum_{aq<n\leq aq+b} \chi(n) = \sum_{aq<n\leq aq+b} \chi(n),$$

and there are no more than q terms in the last sum. The proof is complete once we recall that $|\chi(n)| \leq 1$.

We can now prove the proposition. Let $s_k = \sum_{n=1}^{k} \chi(n)$, and $s_0 = 0$. We know that $L(s, \chi)$ is defined for $s > 1$ by the series

$$\sum_{n=1}^{\infty} \frac{\chi(n)}{n^s}$$

which converges absolutely and uniformly for $s > \delta > 1$. Moreover, the differentiated series also converges absolutely and uniformly for $s > \delta > 1$, which shows that $L(s, \chi)$ is continuously differentiable for $s > 1$. We sum by parts[2] to extend this result to $s > 0$. Indeed, we have

$$\begin{aligned}\sum_{k=1}^{N} \frac{\chi(k)}{k^s} &= \sum_{k=1}^{N} \frac{s_k - s_{k-1}}{k^s} \\ &= \sum_{k=1}^{N-1} s_k \left[\frac{1}{k^s} - \frac{1}{(k+1)^s} \right] + \frac{s_N}{N^s} \\ &= \sum_{k=1}^{N-1} f_k(s) + \frac{s_N}{N^s},\end{aligned}$$

where $f_k(s) = s_k\left[k^{-s} - (k+1)^{-s}\right]$. If $g(x) = x^{-s}$, then $g'(x) = -sx^{-s-1}$, so applying the mean-value theorem between $x = k$ and $x = k+1$, and the fact that $|s_k| \le q$, we find that

$$|f_k(s)| \le qsk^{-s-1}.$$

Therefore, the series $\sum f_k(s)$ converges absolutely and uniformly for $s > \delta > 0$, and this proves that $L(s, \chi)$ is continuous for $s > 0$. To prove that it is also continuously differentiable, we differentiate the series term by term, obtaining

$$\sum (\log n) \frac{\chi(n)}{n^s}.$$

Again, we rewrite this series using summation by parts as

$$\sum s_k \left[-k^{-s} \log k + (k+1)^{-s} \log(k+1) \right],$$

and an application of the mean-value theorem to the function $g(x) = x^{-s} \log x$ shows that the terms are $O(k^{-\delta/2-1})$, thus proving that the differentiated series converges uniformly for $s > \delta > 0$. Hence $L(s, \chi)$ is continuously differentiable for $s > 0$.

Now, observe that for all s large,

$$\begin{aligned}|L(s,\chi) - 1| &\le 2q \sum_{n=2}^{\infty} n^{-s} \\ &\le 2^{-s} O(1),\end{aligned}$$

and we can take $c = \log 2$, to see that $L(s, \chi) = 1 + O(e^{-cs})$ as $s \to \infty$. A similar argument also shows that $L'(s, \chi) = O(e^{-c's})$ as $s \to \infty$ with in fact $c' = c$, and the proof of the proposition is complete.

---

[2] For the formula of summation by parts, see Exercise 7 in Chapter 2.

With the facts gathered so far about $L(s, \chi)$ we are in a position to define the logarithm of the L-functions. This is done by integrating its logarithmic derivative. In other words, if χ is a non-trivial Dirichlet character and $s > 1$ we define[3]

$$\log_2 L(s, \chi) = -\int_s^\infty \frac{L'(t, \chi)}{L(t, \chi)}\, dt.$$

We know that $L(t, \chi) \neq 0$ for every $t > 1$ since it is given by a product (Proposition 3.2), and the integral is convergent because

$$\frac{L'(t, \chi)}{L(t, \chi)} = O(e^{-ct}),$$

which follows from the behavior at infinity of $L(t, \chi)$ and $L'(t, \chi)$ recorded earlier.

The following links the two logarithms.

Proposition 3.6 *If $s > 1$, then*

$$e^{\log_2 L(s,\chi)} = L(s, \chi).$$

Moreover

$$\log_2 L(s, \chi) = \sum_p \log_1 \left(\frac{1}{1 - \chi(p)/p^s} \right).$$

Proof. Differentiating $e^{-\log_2 L(s,\chi)} L(s, \chi)$ with respect to s gives

$$-\frac{L'(s, \chi)}{L(s, \chi)} e^{-\log_2 L(s,\chi)} L(s, \chi) + e^{-\log_2 L(s,\chi)} L'(s, \chi) = 0.$$

So $e^{-\log_2 L(s,\chi)} L(s, \chi)$ is constant, and this constant can be seen to be 1 by letting s tend to infinity. This proves the first conclusion.

To prove the equality between the logarithms, we fix s and take the exponential of both sides. The left-hand side becomes $e^{\log_2 L(s,\chi)} = L(s, \chi)$, and the right-hand side becomes

$$e^{\sum_p \log_1 \left(\frac{1}{1-\chi(p)/p^s} \right)} = \prod_p e^{\log_1 \left(\frac{1}{1-\chi(p)/p^s} \right)} = \prod_p \left(\frac{1}{1 - \chi(p)/p^s} \right) = L(s, \chi),$$

---

[3]The notation $\log_2$ used in this context should not be confused with the logarithm to the base 2.

by (i) in Proposition 3.1 and the Dirichlet product formula. Therefore, for each s there exists an integer $M(s)$ so that

$$\log_2 L(s,\chi) - \sum_p \log_1\left(\frac{1}{1-\chi(p)/p^s}\right) = 2\pi i M(s).$$

As the reader may verify, the left-hand side is continuous in s, and this implies the continuity of the function $M(s)$. But $M(s)$ is integer-valued so we conclude that $M(s)$ is constant, and this constant can be seen to be 0 by letting s go to infinity.

Putting together the work we have done so far gives rigorous meaning to the formal argument presented earlier. Indeed, the properties of $\log_1$ show that

$$\begin{aligned}\sum_p \log_1\left(\frac{1}{1-\chi(p)/p^s}\right) &= \sum_p \frac{\chi(p)}{p^s} + O\left(\sum_p \frac{1}{p^{2s}}\right)\\ &= \sum_p \frac{\chi(p)}{p^s} + O(1).\end{aligned}$$

Now if $L(1,\chi) \neq 0$ for a non-trivial Dirichlet character, then by its integral representation $\log_2 L(s,\chi)$ remains bounded as $s \to 1^+$. Thus the identity between the logarithms implies that $\sum_p \chi(p)p^{-s}$ remains bounded as $s \to 1^+$, which is the desired result. Therefore, to finish the proof of Dirichlet's theorem, we need to see that $L(1,\chi) \neq 0$ when χ is non-trivial.

3.3 Non-vanishing of the L-function

We now turn to a proof of the following deep result:

Theorem 3.7 *If $\chi \neq \chi_0$, then $L(1,\chi) \neq 0$.*

There are several proofs of this fact, some involving algebraic number theory (among them Dirichlet's original argument), and others involving complex analysis. Here we opt for a more elementary argument that requires no special knowledge of either of these areas. The proof splits in two cases, depending on whether χ is complex or real. A Dirichlet character is said to be **real** if it takes on only real values (that is, $+1$, -1, or 0) and **complex** otherwise. In other words, χ is real if and only if $\chi(n) = \overline{\chi(n)}$ for all integers n.

Case I: complex Dirichlet characters

This is the easier of the two cases. The proof is by contradiction, and we use two lemmas.

Lemma 3.8 *If $s > 1$, then*

$$\prod_{\chi} L(s, \chi) \geq 1,$$

where the product is taken over all Dirichlet characters. In particular the product is real-valued.

Proof. We have shown earlier that for $s > 1$

$$L(s, \chi) = \exp\left(\sum_{p} \log_1\left(\frac{1}{1 - \chi(p)p^{-s}}\right)\right).$$

Hence,

$$\begin{aligned}
\prod_{\chi} L(s, \chi) &= \exp\left(\sum_{\chi}\sum_{p} \log_1\left(\frac{1}{1 - \chi(p)p^{-s}}\right)\right) \\
&= \exp\left(\sum_{\chi}\sum_{p}\sum_{k=1}^{\infty} \frac{1}{k}\frac{\chi(p^k)}{p^{ks}}\right) \\
&= \exp\left(\sum_{p}\sum_{k=1}^{\infty}\sum_{\chi} \frac{1}{k}\frac{\chi(p^k)}{p^{ks}}\right).
\end{aligned}$$

Because of Lemma 2.2 (with $\ell = 1$) we have $\sum_{\chi} \chi(p^k) = \varphi(q)\delta_1(p^k)$, and hence

$$\prod_{\chi} L(s, \chi) = \exp\left(\varphi(q)\sum_{p}\sum_{k=1}^{\infty} \frac{1}{k}\frac{\delta_1(p^k)}{p^{ks}}\right) \geq 1,$$

since the term in the exponential is non-negative.

Lemma 3.9 *The following three properties hold:*

(i) *If $L(1, \chi) = 0$, then $L(1, \overline{\chi}) = 0$.*

(ii) *If χ is non-trivial and $L(1, \chi) = 0$, then*

$$|L(s, \chi)| \leq C|s - 1| \quad \textit{when } 1 \leq s \leq 2.$$

(iii) *For the trivial Dirichlet character* χ_0, *we have*

$$|L(s,\chi_0)| \leq \frac{C}{|s-1|} \quad \textit{when } 1 < s \leq 2.$$

Proof. The first statement is immediate because $L(1,\overline{\chi}) = \overline{L(1,\chi)}$. The second statement follows from the mean-value theorem since $L(s,\chi)$ is continuously differentiable for $s > 0$ when χ is non-trivial. Finally, the last statement follows because by Proposition 3.3

$$L(s,\chi_0) = (1 - p_1^{-s})(1 - p_2^{-s}) \cdots (1 - p_N^{-s})\zeta(s),$$

and ζ satisfies the similar estimate (3).

We can now conclude the proof that $L(1,\chi) \neq 0$ for χ a non-trivial complex Dirichlet character. If not, say $L(1,\chi) = 0$, then we also have $L(1,\overline{\chi}) = 0$. Since $\chi \neq \overline{\chi}$, there are at least two terms in the product

$$\prod_{\chi} L(s,\chi),$$

that vanish like $|s-1|$ as $s \to 1^+$. Since only the trivial character contributes a term that grows, and this growth is no worse than $O(1/|s-1|)$, we find that the product goes to 0 as $s \to 1^+$, contradicting the fact that it is ≥ 1 by Lemma 3.8.

Case II: real Dirichlet characters

The proof that $L(1,\chi) \neq 0$ when χ is a non-trivial real Dirichlet character is very different from the earlier complex case. The method we shall exploit involves summation along hyperbolas. It is a curious fact that this method was introduced by Dirichlet himself, twelve years after the proof of his theorem on arithmetic progressions, to establish another famous result of his: the average order of the divisor function. However, he made no connection between the proofs of these two theorems. We will instead proceed by proving first Dirichlet's divisor theorem, as a simple example of the method of summation along hyperbolas. Then, we shall adapt these ideas to prove the fact that $L(1,\chi) \neq 0$. As a preliminary matter, we need to deal with some simple sums, and their corresponding integral analogues.

Sums vs. Integrals

Here we use the idea of comparing a sum with its corresponding integral, which already occurred in the estimate (3) for the zeta function.

Proposition 3.10 *If N is a positive integer, then:*

(i) $\displaystyle\sum_{1\le n\le N}\frac{1}{n}=\int_1^N\frac{dx}{x}+O(1)=\log N+O(1).$

(ii) *More precisely, there exists a real number γ, called Euler's constant, so that*

$$\sum_{1\le n\le N}\frac{1}{n}=\log N+\gamma+O(1/N).$$

Proof. It suffices to establish the more refined estimate given in part (ii). Let

$$\gamma_n=\frac{1}{n}-\int_n^{n+1}\frac{dx}{x}.$$

Since $1/x$ is decreasing, we clearly have

$$0\le\gamma_n\le\frac{1}{n}-\frac{1}{n+1}\le\frac{1}{n^2},$$

so the series $\sum_{n=1}^{\infty}\gamma_n$ converges to a limit which we denote by γ. Moreover, if we estimate $\sum f(n)$ by $\int f(x)\,dx$, where $f(x)=1/x^2$, we find

$$\sum_{n=N+1}^{\infty}\gamma_n\le\sum_{n=N+1}^{\infty}\frac{1}{n^2}\le\int_N^{\infty}\frac{dx}{x^2}=O(1/N).$$

Therefore

$$\sum_{n=1}^{N}\frac{1}{n}-\int_1^N\frac{dx}{x}=\gamma-\sum_{n=N+1}^{\infty}\gamma_n+\int_N^{N+1}\frac{dx}{x},$$

and this last integral is $O(1/N)$ as $N\to\infty$.

Proposition 3.11 *If N is a positive integer, then*

$$\begin{aligned}\sum_{1\le n\le N}\frac{1}{n^{1/2}}&=\int_1^N\frac{dx}{x^{1/2}}+c'+O(1/N^{1/2})\\&=2N^{1/2}+c+O(1/N^{1/2}).\end{aligned}$$

The proof is essentially a repetition of the proof of the previous proposition, this time using the fact that

$$\left|\frac{1}{n^{1/2}}-\frac{1}{(n+1)^{1/2}}\right|\le\frac{C}{n^{3/2}}.$$

This last inequality follows from the mean-value theorem applied to $f(x) = x^{-1/2}$, between $x = n$ and $x = n + 1$.

Hyperbolic sums

If F is a function defined on pairs of positive integers, there are three ways to calculate

$$S_N = \sum\sum F(m,n),$$

where the sum is taken over all pairs of positive integers (m, n) which satisfy $mn \leq N$.

We may carry out the summation in any one of the following three ways. (See Figure 2.)

(a) Along hyperbolas:

$$S_N = \sum_{1\leq k\leq N} \left(\sum_{nm=k} F(m,n) \right)$$

(b) Vertically:

$$S_N = \sum_{1\leq m\leq N} \left(\sum_{1\leq n\leq N/m} F(m,n) \right)$$

(c) Horizontally:

$$S_N = \sum_{1\leq n\leq N} \left(\sum_{1\leq m\leq N/n} F(m,n) \right)$$

It is a remarkable fact that one can obtain interesting conclusions from the obvious fact that these three methods of summation give the same sum. We apply this idea first in the study of the divisor problem.

Intermezzo: the divisor problem

For a positive integer k, let $d(k)$ denote the number of positive divisors of k. For example,

k	1	2	3	4	5	6	7	8	9	10	11	12	13	14	15	16	17
$d(k)$	1	2	2	3	2	4	2	4	3	4	2	6	2	4	4	5	2

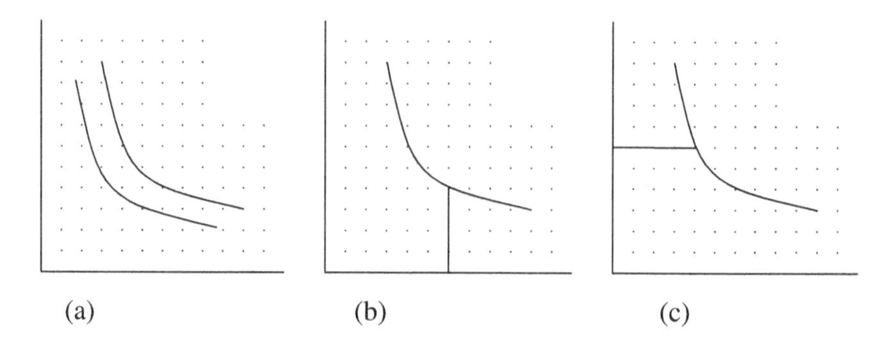

Figure 2. The three methods of summation

One observes that the behavior of $d(k)$ as k tends to infinity is rather irregular, and in fact, it does not seem possible to approximate $d(k)$ by a simple analytic expression in k. However, it is natural to inquire about the average size of $d(k)$. In other words, one might ask, what is the behavior of

$$\frac{1}{N}\sum_{k=1}^{N} d(k) \qquad \text{as } N \to \infty?$$

The answer was provided by Dirichlet, who made use of hyperbolic sums. Indeed, we observe that

$$d(k) = \sum_{nm=k,\ 1\leq n,m} 1.$$

Theorem 3.12 *If k is a positive integer, then*

$$\frac{1}{N}\sum_{k=1}^{N} d(k) = \log N + O(1).$$

More precisely,

$$\frac{1}{N}\sum_{k=1}^{N} d(k) = \log N + (2\gamma - 1) + O(1/N^{1/2}),$$

where γ is Euler's constant.

Proof. Let $S_N = \sum_{k=1}^{N} d(k)$. We observed that summing $F = 1$ along hyperbolas gives S_N. Summing vertically, we find

$$S_N = \sum_{1\leq m\leq N}\ \sum_{1\leq n\leq N/m} 1.$$

But $\sum_{1\leq n\leq N/m} 1 = [N/m] = N/m + O(1)$, where $[x]$ denote the greatest integer $\leq x$. Therefore

$$S_N = \sum_{1\leq m\leq N} (N/m + O(1)) = N\left(\sum_{1\leq m\leq N} 1/m\right) + O(N).$$

Hence, by part (i) of Proposition 3.10,

$$\frac{S_N}{N} = \log N + O(1)$$

which gives the first conclusion.

For the more refined estimate we proceed as follows. Consider the three regions I, II, and III shown in Figure 3. These are defined by

$$\begin{aligned} I &= \{1 \leq m < N^{1/2},\ N^{1/2} < n \leq N/m\}, \\ II &= \{1 \leq m \leq N^{1/2},\ 1 \leq n \leq N^{1/2}\}, \\ III &= \{N^{1/2} < m \leq N/n,\ 1 \leq n < N^{1/2}\}. \end{aligned}$$

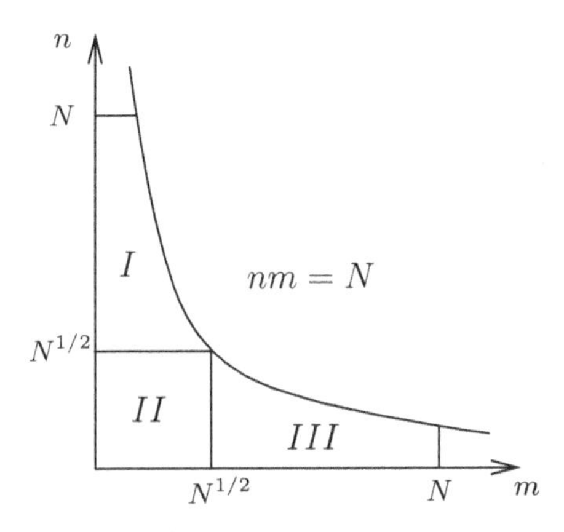

Figure 3. The three regions I, II, and III

If S_I, S_{II}, and S_{III} denote the sums taken over the regions I, II, and III, respectively, then

$$\begin{aligned} S_N &= S_I + S_{II} + S_{III} \\ &= 2(S_I + S_{II}) - S_{II}, \end{aligned}$$

since by symmetry $S_I = S_{III}$. Now we sum vertically, and use (ii) of Proposition 3.10 to obtain

$$\begin{aligned} S_I + S_{II} &= \sum_{1 \leq m \leq N^{1/2}} \left(\sum_{1 \leq n \leq N/m} 1 \right) \\ &= \sum_{1 \leq m \leq N^{1/2}} [N/m] \\ &= \sum_{1 \leq m \leq N^{1/2}} (N/m + O(1)) \\ &= N \left(\sum_{1 \leq m \leq N^{1/2}} 1/m \right) + O(N^{1/2}) \\ &= N \log N^{1/2} + N\gamma + O(N^{1/2}). \end{aligned}$$

Finally, S_{II} corresponds to a square so

$$S_{II} = \sum_{1 \leq m \leq N^{1/2}} \sum_{1 \leq n \leq N^{1/2}} 1 = [N^{1/2}]^2 = N + O(N^{1/2}).$$

Putting these estimates together and dividing by N yields the more refined statement in the theorem.

Non-vanishing of the L-function

Our essential application of summation along hyperbolas is to the main point of this section, namely that $L(1, \chi) \neq 0$ for a non-trivial real Dirichlet character χ.

Given such a character, let

$$F(m, n) = \frac{\chi(n)}{(nm)^{1/2}},$$

and define

$$S_N = \sum \sum F(m, n),$$

where the sum is over all integers $m, n \geq 1$ that satisfy $mn \leq N$.

Proposition 3.13 *The following statements are true:*

(i) $S_N \geq c \log N$ *for some constant* $c > 0$.

(ii) $S_N = 2N^{1/2} L(1, \chi) + O(1)$.

It suffices to prove the proposition, since the assumption $L(1, \chi) = 0$ would give an immediate contradiction.

We first sum along hyperbolas. Observe that

$$\sum_{nm=k} \frac{\chi(n)}{(nm)^{1/2}} = \frac{1}{k^{1/2}} \sum_{n|k} \chi(n).$$

For conclusion (i) it will be enough to show the following lemma.

Lemma 3.14 $\displaystyle\sum_{n|k} \chi(n) \geq \begin{cases} 0 & \textit{for all } k \\ 1 & \textit{if } k = \ell^2 \textit{ for some } \ell \in \mathbb{Z}. \end{cases}$

From the lemma, we then get

$$S_N \geq \sum_{k=\ell^2,\ \ell \leq N^{1/2}} \frac{1}{k^{1/2}} \geq c \log N,$$

where the last inequality follows from (i) in Proposition 3.10.

The proof of the lemma is simple. If k is a power of a prime, say $k = p^a$, then the divisors of k are $1, p, p^2, \ldots, p^a$ and

$$\begin{aligned} \sum_{n|k} \chi(n) &= \chi(1) + \chi(p) + \chi(p^2) + \cdots + \chi(p^a) \\ &= 1 + \chi(p) + \chi(p)^2 + \cdots + \chi(p)^a. \end{aligned}$$

So this sum is equal to

$$\begin{cases} a+1 & \text{if } \chi(p) = 1, \\ 1 & \text{if } \chi(p) = -1 \text{ and } a \text{ is even}, \\ 0 & \text{if } \chi(p) = -1 \text{ and } a \text{ is odd}, \\ 1 & \text{if } \chi(p) = 0, \text{ that is } p|q. \end{cases}$$

In general, if $k = p_1^{a_1} \cdots p_N^{a_N}$, then any divisor of k is of the form $p_1^{b_1} \cdots p_N^{b_N}$ where $0 \leq b_j \leq a_j$ for all j. Therefore, the multiplicative property of χ gives

$$\sum_{n|k} \chi(n) = \prod_{j=1}^{N} \left(\chi(1) + \chi(p_j) + \chi(p_j^2) + \cdots + \chi(p_j^{a_j}) \right),$$

and the proof is complete.

To prove the second statement in the proposition, we write

$$S_N = S_I + (S_{II} + S_{III}),$$

where the sums S_I, S_{II}, and S_{III} were defined earlier (see also Figure 3). We evaluate S_I by summing vertically, and $S_{II} + S_{III}$ by summing horizontally. In order to carry this out we need the following simple results.

Lemma 3.15 *For all integers $0 < a < b$ we have*

(i) $\displaystyle\sum_{n=a}^{b} \frac{\chi(n)}{n^{1/2}} = O(a^{-1/2})$,

(ii) $\displaystyle\sum_{n=a}^{b} \frac{\chi(n)}{n} = O(a^{-1})$.

Proof. This argument is similar to the proof of Proposition 3.4; we use summation by parts. Let $s_n = \sum_{1 \leq k \leq n} \chi(k)$, and remember that $|s_n| \leq q$ for all n. Then

$$\begin{aligned}\sum_{n=a}^{b} \frac{\chi(n)}{n^{1/2}} &= \sum_{n=a}^{b-1} s_n \left[n^{-1/2} - (n+1)^{-1/2}\right] + O(a^{-1/2}) \\ &= O\left(\sum_{n=a}^{\infty} n^{-3/2}\right) + O(a^{-1/2}).\end{aligned}$$

By comparing the sum $\sum_{n=a}^{\infty} n^{-3/2}$ with the integral of $f(x) = x^{-3/2}$, we find that the former is also $O(a^{-1/2})$.

A similar argument establishes (ii).

We may now finish the proof of the proposition. Summing vertically we find

$$S_I = \sum_{m < N^{1/2}} \frac{1}{m^{1/2}} \left(\sum_{N^{1/2} < n \leq N/m} \chi(n)/n^{1/2} \right).$$

The lemma together with Proposition 3.11 shows that $S_I = O(1)$. Finally we sum horizontally to get

$$\begin{aligned}
S_{II} + S_{III} &= \sum_{1 \le n \le N^{1/2}} \frac{\chi(n)}{n^{1/2}} \left(\sum_{m \le N/n} 1/m^{1/2} \right) \\
&= \sum_{1 \le n \le N^{1/2}} \frac{\chi(n)}{n^{1/2}} \left\{ 2(N/n)^{1/2} + c + O((n/N)^{1/2}) \right\} \\
&= 2N^{1/2} \sum_{1 \le n \le N^{1/2}} \frac{\chi(n)}{n} + c \sum_{1 \le n \le N^{1/2}} \frac{\chi(n)}{n^{1/2}} \\
&\quad + O\left(\frac{1}{N^{1/2}} \sum_{1 \le n \le N^{1/2}} 1 \right) \\
&= A + B + C.
\end{aligned}$$

Now observe that the lemma, together with the definition of $L(s, \chi)$, implies

$$A = 2N^{1/2} L(1, \chi) + O(N^{1/2} N^{-1/2}).$$

Moreover, part (i) of the lemma gives $B = O(1)$, and we also clearly have $C = O(1)$. Thus $S_N = 2N^{1/2} L(1, \chi) + O(1)$, which is part (ii) in Proposition 3.13.

This completes the proof that $L(1, \chi) \neq 0$, and thus the proof of Dirichlet's theorem.

4 Exercises

1. Prove that there are infinitely many primes by observing that if there were only finitely many, $p_1, \ldots, p_N$, then

$$\prod_{j=1}^{N} \frac{1}{1 - 1/p_j} \ge \sum_{n=1}^{\infty} \frac{1}{n}.$$

2. In the text we showed that there are infinitely many primes of the form $4k + 3$ by a modification of Euclid's original argument. Adapt this technique to prove the similar result for primes of the form $3k + 2$, and for those of the form $6k + 5$.

3. Prove that if p and q are relatively prime, then $\mathbb{Z}^*(p) \times \mathbb{Z}^*(q)$ is isomorphic to $\mathbb{Z}^*(pq)$.

4. Let $\varphi(n)$ denote the number of positive integers $\leq n$ that are relatively prime to n. Use the previous exercise to show that if n and m are relatively prime, then

$$\varphi(nm) = \varphi(n)\varphi(m).$$

One can give a formula for the Euler phi-function as follows:

(a) Calculate $\varphi(p)$ when p is prime by counting the number of elements in $\mathbb{Z}^*(p)$.

(b) Give a formula for $\varphi(p^k)$ when p is prime and $k \geq 1$ by counting the number of elements in $\mathbb{Z}^*(p^k)$.

(c) Show that

$$\varphi(n) = n \prod_i \left(1 - \frac{1}{p_i}\right),$$

where p_i are the primes that divide n.

5. If n is a positive integer, show that

$$n = \sum_{d|n} \varphi(d),$$

where φ is the Euler phi-function.
[Hint: There are precisely $\varphi(n/d)$ integers $1 \leq m \leq n$ with $\gcd(m, n) = d$.]

6. Write down the characters of the groups $\mathbb{Z}^*(3)$, $\mathbb{Z}^*(4)$, $\mathbb{Z}^*(5)$, $\mathbb{Z}^*(6)$, and $\mathbb{Z}^*(8)$.

(a) Which ones are real, or complex?

(b) Which ones are even, or odd? (A character is even if $\chi(-1) = 1$, and odd otherwise).

7. Recall that for $|z| < 1$,

$$\log_1\left(\frac{1}{1-z}\right) = \sum_{k\geq 1} \frac{z^k}{k}.$$

We have seen that

$$e^{\log_1\left(\frac{1}{1-z}\right)} = \frac{1}{1-z}.$$

(a) Show that if $w = 1/(1-z)$, then $|z| < 1$ if and only if $\mathrm{Re}(w) > 1/2$.

(b) Show that if $\mathrm{Re}(w) > 1/2$ and $w = \rho e^{i\varphi}$ with $\rho > 0$, $|\varphi| < \pi$, then

$$\log_1 w = \log \rho + i\varphi.$$

[Hint: If $e^{\zeta} = w$, then the real part of ζ is uniquely determined and its imaginary part is determined modulo 2π.]

8. Let ζ denote the zeta function defined for $s > 1$.

(a) Compare $\zeta(s)$ with $\int_1^\infty x^{-s}\,dx$ to show that

$$\zeta(s) = \frac{1}{s-1} + O(1) \qquad \text{as } s \to 1^+.$$

(b) Prove as a consequence that

$$\sum_p \frac{1}{p^s} = \log\left(\frac{1}{s-1}\right) + O(1) \qquad \text{as } s \to 1^+.$$

9. Let χ_0 denote the trivial Dirichlet character mod q, and $p_1, \ldots, p_k$ the distinct prime divisors of q. Recall that $L(s, \chi_0) = (1 - p_1^{-s}) \cdots (1 - p_k^{-s})\zeta(s)$, and show as a consequence

$$L(s, \chi_0) = \frac{\varphi(q)}{q}\frac{1}{s-1} + O(1) \qquad \text{as } s \to 1^+.$$

[Hint: Use the asymptotics for ζ in Exercise 8.]

10. Show that if ℓ is relatively prime to q, then

$$\sum_{p \equiv \ell} \frac{1}{p^s} = \frac{1}{\varphi(q)} \log\left(\frac{1}{s-1}\right) + O(1) \qquad \text{as } s \to 1^+.$$

This is a quantitative version of Dirichlet's theorem.

[Hint: Recall (4).]

11. Use the characters for $\mathbb{Z}^*(3)$, $\mathbb{Z}^*(4)$, $\mathbb{Z}^*(5)$, and $\mathbb{Z}^*(6)$ to verify directly that $L(1, \chi) \neq 0$ for all non-trivial Dirichlet characters modulo q when $q = 3, 4, 5$, and 6.

[Hint: Consider in each case the appropriate alternating series.]

12. Suppose χ is real and non-trivial; assuming the theorem that $L(1,\chi) \neq 0$, show directly that $L(1,\chi) > 0$.

[Hint: Use the product formula for $L(s,\chi)$.]

13. Let $\{a_n\}_{n=-\infty}^{\infty}$ be a sequence of complex numbers such that $a_n = a_m$ if $n = m \mod q$. Show that the series

$$\sum_{n=1}^{\infty} \frac{a_n}{n}$$

converges if and only if $\sum_{n=1}^{q} a_n = 0$.

[Hint: Sum by parts.]

14. The series

$$F(\theta) = \sum_{|n|\neq 0} \frac{e^{in\theta}}{n}, \qquad \text{for } |\theta| < \pi,$$

converges for every θ and is the Fourier series of the function defined on $[-\pi,\pi]$ by $F(0) = 0$ and

$$F(\theta) = \begin{cases} i(-\pi-\theta) & \text{if } -\pi \leq \theta < 0 \\ i(\pi-\theta) & \text{if } 0 < \theta \leq \pi, \end{cases}$$

and extended by periodicity (period 2π) to all of $\mathbb{R}$ (see Exercise 8 in Chapter 2).

Show also that if $\theta \neq 0 \mod 2\pi$, then the series

$$E(\theta) = \sum_{n=1}^{\infty} \frac{e^{in\theta}}{n}$$

converges, and that

$$E(\theta) = \frac{1}{2}\log\left(\frac{1}{2-2\cos\theta}\right) + \frac{i}{2}F(\theta).$$

15. To sum the series $\sum_{n=1}^{\infty} a_n/n$, with $a_n = a_m$ if $n = m \mod q$ and $\sum_{n=1}^{q} a_n = 0$, proceed as follows.

(a) Define

$$A(m) = \sum_{n=1}^{q} a_n \zeta^{-mn} \qquad \text{where } \zeta = e^{2\pi i/q}.$$

Note that $A(q) = 0$. With the notation of the previous exercise, prove that

$$\sum_{n=1}^{\infty} \frac{a_n}{n} = \frac{1}{q} \sum_{m=1}^{q-1} A(m)E(2\pi m/q).$$

[Hint: Use Fourier inversion on $\mathbb{Z}(q)$.]

(b) If $\{a_m\}$ is odd, $(a_{-m} = -a_m)$ for $m \in \mathbb{Z}$, observe that $a_0 = a_q = 0$ and show that

$$A(m) = \sum_{1 \leq n < q/2} a_n(\zeta^{-mn} - \zeta^{mn}).$$

(c) Still assuming that $\{a_m\}$ is odd, show that

$$\sum_{n=1}^{\infty} \frac{a_n}{n} = \frac{1}{2q} \sum_{m=1}^{q-1} A(m)F(2\pi m/q).$$

[Hint: Define $\tilde{A}(m) = \sum_{n=1}^{q} a_n \zeta^{mn}$ and apply the Fourier inversion formula.]

16. Use the previous exercises to show that

$$\frac{\pi}{3\sqrt{3}} = 1 - \frac{1}{2} + \frac{1}{4} - \frac{1}{5} + \frac{1}{7} - \frac{1}{8} + \cdots,$$

which is $L(1, \chi)$ for the non-trivial (odd) Dirichlet character modulo 3.

5 Problems

1.* Here are other series that can be summed by the methods in Exercise 15.

(a) For the non-trivial Dirichlet character modulo 6, $L(1, \chi)$ equals

$$\frac{\pi}{2\sqrt{3}} = 1 - \frac{1}{5} + \frac{1}{7} - \frac{1}{11} + \frac{1}{13} + \cdots.$$

(b) If χ is the odd Dirichlet character modulo 8, then $L(1, \chi)$ equals

$$\frac{\pi}{2\sqrt{2}} = 1 + \frac{1}{3} - \frac{1}{5} - \frac{1}{7} + \frac{1}{9} + \frac{1}{11} \cdots.$$

(c) For an odd Dirichlet character modulo 7, $L(1, \chi)$ equals

$$\frac{\pi}{\sqrt{7}} = 1 + \frac{1}{2} - \frac{1}{3} + \frac{1}{4} - \frac{1}{5} - \frac{1}{6} \cdots.$$

(d) For an even Dirichlet character modulo 8, $L(1, \chi)$ equals

$$\frac{\log(1+\sqrt{2})}{\sqrt{2}} = 1 - \frac{1}{3} - \frac{1}{5} + \frac{1}{7} + \frac{1}{9} - \frac{1}{11} \cdots .$$

(e) For an even Dirichlet character modulo 5, $L(1, \chi)$ equals

$$\frac{2}{\sqrt{5}} \log\left(\frac{1+\sqrt{5}}{2}\right) = 1 - \frac{1}{2} - \frac{1}{3} + \frac{1}{4} + \frac{1}{6} - \frac{1}{7} - \frac{1}{8} + \frac{1}{9} + \frac{1}{11} \cdots .$$

2. Let $d(k)$ denote the number of positive divisors of k.

(a) Show that if $k = p_1^{a_1} \cdots p_n^{a_n}$ is the prime factorization of k, then

$$d(k) = (a_1 + 1) \cdots (a_n + 1).$$

Although Theorem 3.12 shows that on "average" $d(k)$ is of the order of $\log k$, prove the following on the basis of (a):

(b) $d(k) = 2$ for infinitely many k.

(c) For any positive integer N, there is a constant $c > 0$ so that $d(k) \geq c(\log k)^N$ for infinitely many k. [Hint: Let $p_1, \ldots, p_N$ be N distinct primes, and consider k of the form $(p_1 p_2 \cdots p_N)^m$ for $m = 1, 2, \ldots$.]

3. Show that if p is relatively prime to q, then

$$\prod_{\chi}\left(1 - \frac{\chi(p)}{p^s}\right) = \left(\frac{1}{1 - p^{fs}}\right)^g,$$

where $g = \varphi(q)/f$, and f is the order of p in $\mathbb{Z}^*(q)$ (that is, the smallest n for which $p^n \equiv 1 \mod q$). Here the product is taken over all Dirichlet characters modulo q.

4. Prove as a consequence of the previous problem that

$$\prod_{\chi} L(s, \chi) = \sum_{n \geq 1} \frac{a_n}{n^s},$$

where $a_n \geq 0$, and the product is over all Dirichlet characters modulo q.

Appendix : Integration

This appendix is meant as a quick review of the definition and main properties of the Riemann integral on $\mathbb{R}$, and integration of appropriate continuous functions on $\mathbb{R}^d$. Our exposition is brief since we assume that the reader already has some familiarity with this material.

We begin with the theory of Riemann integration on a closed and bounded interval on the real line. Besides the standard results about the integral, we also discuss the notion of sets of measure 0, and give a necessary and sufficient condition on the set of discontinuities of a function that guarantee its integrability.

We also discuss multiple and repeated integrals. In particular, we extend the notion of integration to the entire space $\mathbb{R}^d$ by restricting ourselves to functions that decay fast enough at infinity.

1 Definition of the Riemann integral

Let f be a *bounded* real-valued function defined on the closed interval $[a, b] \subset \mathbb{R}$. By a **partition** P of $[a, b]$ we mean a finite sequence of numbers $x_0, x_1, \ldots, x_N$ with

$$a = x_0 < x_1 < \cdots < x_{N-1} < x_N = b.$$

Given such a partition, we let I_j denote the interval $[x_{j-1}, x_j]$ and write $|I_j|$ for its length, namely $|I_j| = x_j - x_{j-1}$. We define the upper and lower sums of f with respect to P by

$$\mathcal{U}(P, f) = \sum_{j=1}^{N} [\sup_{x \in I_j} f(x)]\, |I_j| \quad \text{and} \quad \mathcal{L}(P, f) = \sum_{j=1}^{N} [\inf_{x \in I_j} f(x)]\, |I_j|.$$

Note that the infimum and supremum exist because by assumption, f is bounded. Clearly $\mathcal{U}(P, f) \geq \mathcal{L}(P, f)$, and the function f is said to be **Riemann integrable**, or simply **integrable**, if for every $\epsilon > 0$ there exists a partition P such that

$$\mathcal{U}(P, f) - \mathcal{L}(P, f) < \epsilon.$$

To define the value of the integral of f, we need to make a simple yet important observation. A partition P' is said to be a **refinement** of the partition P if P' is obtained from P by adding points. Then, adding one point at a time, it is easy to check that

$$\mathcal{U}(P', f) \leq \mathcal{U}(P, f) \quad \text{and} \quad \mathcal{L}(P', f) \geq \mathcal{L}(P, f).$$

From this, we see that if P_1 and P_2 are two partitions of $[a, b]$, then

$$\mathcal{U}(P_1, f) \geq \mathcal{L}(P_2, f),$$

since it is possible to take P' as a common refinement of both P_1 and P_2 to obtain

$$\mathcal{U}(P_1, f) \geq \mathcal{U}(P', f) \geq \mathcal{L}(P', f) \geq \mathcal{L}(P_2, f).$$

Since f is bounded we see that both

$$U = \inf_P \mathcal{U}(P, f) \quad \text{and} \quad L = \sup_P \mathcal{L}(P, f)$$

exist (where the infimum and supremum are taken over all partitions of $[a, b]$), and also that $U \geq L$. Moreover, if f is integrable we must have $U = L$, and we *define* $\int_a^b f(x)\, dx$ to be this common value.

Finally, a bounded complex-valued function $f = u + iv$ is said to be integrable if its real and imaginary parts u and v are integrable, and we define

$$\int_a^b f(x)\, dx = \int_a^b u(x)\, dx + i \int_a^b v(x)\, dx.$$

For example, the constants are integrable functions and it is clear that if $c \in \mathbb{C}$, then $\int_a^b c\, dx = c(b - a)$. Also, continuous functions are integrable. This is because a continuous function on a closed and bounded interval $[a, b]$ is uniformly continuous, that is, given $\epsilon > 0$ there exists δ such that if $|x - y| < \delta$ then $|f(x) - f(y)| < \epsilon$. So if we choose n with $(b - a)/n < \delta$, then the partition P given by

$$a,\ a + \frac{b-a}{n}, \ldots, a + k\frac{b-a}{n}, \ldots, a + (n-1)\frac{b-a}{n},\ b$$

satisfies $\mathcal{U}(P, f) - \mathcal{L}(P, f) \leq \epsilon(b - a)$.

1.1 Basic properties

Proposition 1.1 *If f and g are integrable on $[a, b]$, then:*

(i) *$f + g$ is integrable, and $\int_a^b f(x) + g(x)\, dx = \int_a^b f(x)\, dx + \int_a^b g(x)\, dx$.*

(ii) *If* $c \in \mathbb{C}$, *then* $\int_a^b cf(x)\,dx = c\int_a^b f(x)\,dx$.

(iii) *If* f *and* g *are real-valued and* $f(x) \leq g(x)$, *then* $\int_a^b f(x)\,dx \leq \int_a^b g(x)\,dx$.

(iv) *If* $c \in [a,b]$, *then* $\int_a^b f(x)\,dx = \int_a^c f(x)\,dx + \int_c^b f(x)\,dx$.

Proof. For property (i) we may assume that f and g are real-valued. If P is a partition of $[a,b]$, then

$$\mathcal{U}(P, f+g) \leq \mathcal{U}(P,f) + \mathcal{U}(P,g) \quad \text{and} \quad \mathcal{L}(P, f+g) \geq \mathcal{L}(P,f) + \mathcal{L}(P,g).$$

Given $\epsilon > 0$, there exist partitions P_1 and P_2 such that $\mathcal{U}(P_1, f) - \mathcal{L}(P_1, f) < \epsilon$ and $\mathcal{U}(P_2, g) - \mathcal{L}(P_2, g) < \epsilon$, so that if P_0 is a common refinement of P_1 and P_2, we get

$$\mathcal{U}(P_0, f+g) - \mathcal{L}(P_0, f+g) < 2\epsilon.$$

So $f+g$ is integrable, and if we let $I = \inf_P \mathcal{U}(P, f+g) = \sup_P \mathcal{L}(P, f+g)$, then we see that

$$\begin{aligned} I &\leq \mathcal{U}(P_0, f+g) + 2\epsilon \leq \mathcal{U}(P_0, f) + \mathcal{U}(P_0, g) + 2\epsilon \\ &\leq \int_a^b f(x)\,dx + \int_a^b g(x)\,dx + 4\epsilon. \end{aligned}$$

Similarly $I \geq \int_a^b f(x)\,dx + \int_a^b g(x)\,dx - 4\epsilon$, which proves that $\int_a^b f(x) + g(x)\,dx = \int_a^b f(x)\,dx + \int_a^b g(x)\,dx$. The second and third parts of the proposition are just as easy to prove. For the last property, simply refine partitions of $[a,b]$ by adding the point c.

Another important property we need to prove is that fg is integrable whenever f and g are integrable.

Lemma 1.2 *If* f *is real-valued integrable on* $[a,b]$ *and* φ *is a real-valued continuous function on* $\mathbb{R}$, *then* $\varphi \circ f$ *is also integrable on* $[a,b]$.

Proof. Let $\epsilon > 0$ and remember that f is bounded, say $|f| \leq M$. Since φ is uniformly continuous on $[-M, M]$ we may choose $\delta > 0$ so that if $s, t \in [-M, M]$ and $|s-t| < \delta$, then $|\varphi(s) - \varphi(t)| < \epsilon$. Now choose a partition $P = \{x_0, \ldots, x_N\}$ of $[a,b]$ with $\mathcal{U}(P,f) - \mathcal{L}(P,f) < \delta^2$. Let $I_j = [x_{j-1}, x_j]$ and distinguish two classes: we write $j \in \Lambda$ if $\sup_{x \in I_j} f(x) - \inf_{x \in I_j} f(x) < \delta$ so that by construction

$$\sup_{x \in I_j} \varphi \circ f(x) - \inf_{x \in I_j} \varphi \circ f(x) < \epsilon.$$

Otherwise, we write $j \in \Lambda'$ and note that

$$\delta \sum_{j \in \Lambda'} |I_j| \leq \sum_{j \in \Lambda'} [\sup_{x \in I_j} f(x) - \inf_{x \in I_j} f(x)] \, |I_j| \leq \delta^2$$

so $\sum_{j \in \Lambda'} |I_j| < \delta$. Therefore, separating the cases $j \in \Lambda$ and $j \in \Lambda'$ we find that

$$\mathcal{U}(P, \varphi \circ f) - \mathcal{L}(P, \varphi \circ f) \leq \epsilon(b - a) + 2\mathcal{B}\delta,$$

where $\mathcal{B}$ is a bound for φ on $[-M, M]$. Since we can also choose $\delta < \epsilon$, we see that the proposition is proved.

¿From the lemma we get the following facts:

- If f and g are integrable on $[a, b]$, then the product fg is integrable on $[a, b]$.

This follows from the lemma with $\varphi(t) = t^2$, and the fact that $fg = \frac{1}{4}\left([f + g]^2 - [f - g]^2\right)$.

- If f is integrable on $[a, b]$, then the function $|f|$ is integrable, and $\left|\int_a^b f(x)\, dx\right| \leq \int_a^b |f(x)|\, dx$.

We can take $\varphi(t) = |t|$ to see that $|f|$ is integrable. Moreover, the inequality follows from (iii) in Proposition 1.1.

We record two results that imply integrability.

Proposition 1.3 *A bounded monotonic function f on an interval $[a, b]$ is integrable.*

Proof. We may assume without loss of generality that $a = 0$, $b = 1$, and f is monotonically increasing. Then, for each N, we choose the uniform partition P_N given by $x_j = j/N$ for all $j = 0, \ldots, N$. If $\alpha_j = f(x_j)$, then we have

$$\mathcal{U}(P_N, f) = \frac{1}{N} \sum_{j=1}^{N} \alpha_j \quad \text{and} \quad \mathcal{L}(P_N, f) = \frac{1}{N} \sum_{j=1}^{N} \alpha_{j-1}.$$

Therefore, if $|f(x)| \leq B$ for all x we have

$$\mathcal{U}(P_N, f) - \mathcal{L}(P_N, f) = \frac{\alpha_N - \alpha_0}{N} \leq \frac{2B}{N},$$

and the proposition is proved.

Proposition 1.4 *Let f be a bounded function on the compact interval $[a,b]$. If $c \in (a,b)$, and if for all small $\delta > 0$ the function f is integrable on the intervals $[a, c-\delta]$ and $[c+\delta, b]$, then f is integrable on $[a,b]$.*

Proof. Suppose $|f| \leq M$ and let $\epsilon > 0$. Choose $\delta > 0$ (small) so that $4\delta M \leq \epsilon/3$. Now let P_1 and P_2 be partitions of $[a, c-\delta]$ and $[c+\delta, b]$ so that for each $i = 1, 2$ we have $\mathcal{U}(P_i, f) - \mathcal{L}(P_i, f) < \epsilon/3$. This is possible since f is integrable on each one of the intervals. Then by taking as a partition $P = P_1 \cup \{c-\delta\} \cup \{c+\delta\} \cup P_2$ we immediately see that $\mathcal{U}(P, f) - \mathcal{L}(P, f) < \epsilon$.

We end this section with a useful approximation lemma. Recall that a function on the circle is the same as a 2π-periodic function on $\mathbb{R}$.

Lemma 1.5 *Suppose f is integrable on the circle and f is bounded by B. Then there exists a sequence $\{f_k\}_{k=1}^{\infty}$ of continuous functions on the circle so that*

$$\sup_{x \in [-\pi, \pi]} |f_k(x)| \leq B \quad \textit{for all } k = 1, 2, \ldots,$$

and

$$\int_{-\pi}^{\pi} |f(x) - f_k(x)| \, dx \to 0 \quad \textit{as } k \to \infty.$$

Proof. Assume f is real-valued (in general apply the following argument to the real and imaginary parts separately). Given $\epsilon > 0$, we may choose a partition $-\pi = x_0 < x_1 < \cdots < x_N = \pi$ of the interval $[-\pi, \pi]$ so that the upper and lower sums of f differ by at most ϵ. Denote by f^* the step function defined by

$$f^*(x) = \sup_{x_{j-1} \leq y \leq x_j} f(y) \quad \text{if } x \in [x_{j-1}, x_j) \text{ for } 1 \leq j \leq N.$$

By construction we have $|f^*| \leq B$, and moreover

$$(1) \qquad \int_{-\pi}^{\pi} |f^*(x) - f(x)| \, dx = \int_{-\pi}^{\pi} (f^*(x) - f(x)) \, dx < \epsilon.$$

Now we can modify f^* to make it continuous and periodic yet still approximate f in the sense of the lemma. For small $\delta > 0$, let $\tilde{f}(x) = f^*(x)$ when the distance of x from any of the division points $x_0, \ldots, x_N$ is $\geq \delta$. In the δ-neighborhood of x_j for $j = 1, \ldots, N-1$, define $\tilde{f}(x)$ to be the linear function for which $\tilde{f}(x_j \pm \delta) = f^*(x_j \pm \delta)$. Near $x_0 = -\pi$, $\tilde{f}$

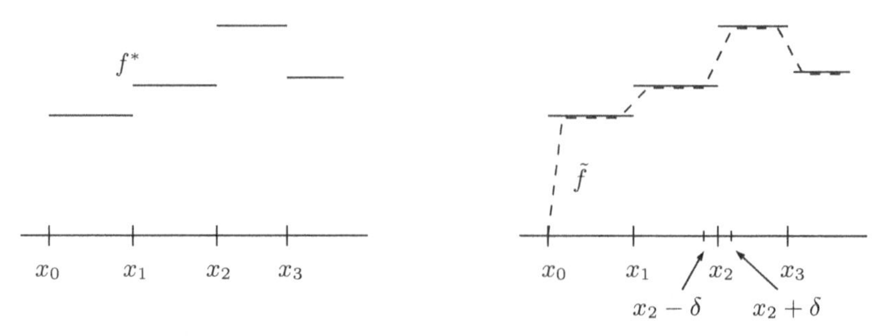

Figure 1. Portions of the functions f^* and $\tilde{f}$

is linear with $\tilde{f}(-\pi) = 0$ and $\tilde{f}(-\pi + \delta) = f^*(-\pi + \delta)$. Similarly, near $x_N = \pi$ the function $\tilde{f}$ is linear with $\tilde{f}(\pi) = 0$ and $\tilde{f}(\pi - \delta) = f^*(\pi - \delta)$. In Figure 1 we illustrate the situation near $x_0 = -\pi$. In the second picture the graph of $\tilde{f}$ is shifted slightly below to clarify the situation.

Then, since $\tilde{f}(-\pi) = \tilde{f}(\pi)$, we may extend $\tilde{f}$ to a continuous and 2π-periodic function on $\mathbb{R}$. The absolute value of this extension is also bounded by B. Moreover, $\tilde{f}$ differs from f^* only in the N intervals of length 2δ surrounding the division points. Thus

$$\int_{-\pi}^{\pi} |f^*(x) - \tilde{f}(x)|\, dx \leq 2BN(2\delta).$$

If we choose δ sufficiently small, we get

$$\int_{-\pi}^{\pi} |f^*(x) - \tilde{f}(x)|\, dx < \epsilon. \tag{2}$$

As a result, equations (1), (2), and the triangle inequality yield

$$\int_{-\pi}^{\pi} |f(x) - \tilde{f}(x)|\, dx < 2\epsilon.$$

Denoting by f_k the $\tilde{f}$ so constructed, when $2\epsilon = 1/k$, we see that the sequence $\{f_k\}$ has the properties required by the lemma.

1.2 Sets of measure zero and discontinuities of integrable functions

We observed that all continuous functions are integrable. By modifying the argument slightly, one can show that all piecewise continuous functions are also integrable. In fact, this is a consequence of Proposition 1.4

applied finitely many times. We now turn to a more careful study of the discontinuities of integrable functions.

We start with a definition[1]: a subset E of $\mathbb{R}$ is said to have **measure** 0 if for every $\epsilon > 0$ there exists a countable family of open intervals $\{I_k\}_{k=1}^{\infty}$ such that

(i) $E \subset \bigcup_{k=1}^{\infty} I_k$,

(ii) $\sum_{k=1}^{\infty} |I_k| < \epsilon$, where $|I_k|$ denotes the length of the interval I_k.

The first condition says that the union of the intervals covers E, and the second that this union is small. The reader will have no difficulty proving that any finite set of points has measure 0. A more subtle argument is needed to prove that a countable set of points has measure 0. In fact, this result is contained in the following lemma.

Lemma 1.6 *The union of countably many sets of measure* 0 *has measure* 0.

Proof. Say $E_1, E_2, \ldots$ are sets of measure 0, and let $E = \cup_{i=1}^{\infty} E_i$. Let $\epsilon > 0$, and for each i choose open interval $I_{i,1}, I_{i,2}, \ldots$ so that

$$E_i \subset \bigcup_{k=1}^{\infty} I_{i,k} \quad \text{and} \quad \sum_{k=1}^{\infty} |I_{i,k}| < \epsilon/2^i.$$

Now clearly we have $E \subset \bigcup_{i,k=1}^{\infty} I_{i,k}$, and

$$\sum_{i=1}^{\infty} \sum_{k=1}^{\infty} |I_{i,k}| \leq \sum_{i=1}^{\infty} \frac{\epsilon}{2^i} \leq \epsilon,$$

as was to be shown.

An important observation is that if E has measure 0 and is *compact*, then it is possible to find a *finite* number of open intervals I_k, $k = 1, \ldots, N$, that satisfy the two conditions (i) and (ii) above.

We can prove the characterization of Riemann integrable functions in terms of their discontinuities.

Theorem 1.7 *A bounded function f on $[a, b]$ is integrable if and only if its set of discontinuities has measure* 0.

[1] A systematic study of the measure of sets arises in the theory of Lebesgue integration, which is taken up in Book III.

We write $J = [a, b]$ and $I(c, r) = (c - r, c + r)$ for the open interval centered at c of radius $r > 0$. Define the **oscillation of f on $I(c, r)$** by

$$\operatorname{osc}(f, c, r) = \sup |f(x) - f(y)|$$

where the supremum is taken over all $x, y \in J \cap I(c, r)$. This quantity exists since f is bounded. Define the **oscillation of f at c** by

$$\operatorname{osc}(f, c) = \lim_{r \to 0} \operatorname{osc}(f, c, r).$$

This limit exists because $\operatorname{osc}(f, c, r)$ is ≥ 0 and a decreasing function of r. The point is that f is continuous at c if and only if $\operatorname{osc}(f, c) = 0$. This is clear from the definitions. For each $\epsilon > 0$ we define a set A_ϵ by

$$A_\epsilon = \{c \in J : \operatorname{osc}(f, c) \geq \epsilon\}.$$

Having done that, we see that the set of points in J where f is discontinuous is simply $\bigcup_{\epsilon > 0} A_\epsilon$. This is an important step in the proof of our theorem.

Lemma 1.8 *If $\epsilon > 0$, then the set A_ϵ is closed and therefore compact.*

Proof. The argument is simple. Suppose $c_n \in A_\epsilon$ converges to c and assume that $c \notin A_\epsilon$. Write $\operatorname{osc}(f, c) = \epsilon - \delta$ where $\delta > 0$. Select r so that $\operatorname{osc}(f, c, r) < \epsilon - \delta/2$, and choose n with $|c_n - c| < r/2$. Then $\operatorname{osc}(f, c_n, r/2) < \epsilon$ which implies $\operatorname{osc}(f, c_n) < \epsilon$, a contradiction.

We are now ready to prove the first part of the theorem. Suppose that the set $\mathcal{D}$ of discontinuities of f has measure 0, and let $\epsilon > 0$. Since $A_\epsilon \subset \mathcal{D}$, we can cover A_ϵ by a finite number of open intervals, say $I_1, \ldots, I_N$, whose total length is $< \epsilon$. The complement of this union I of intervals is compact, and around each point z in this complement we can find an interval F_z with $\sup_{x,y \in F_z} |f(x) - f(y)| \leq \epsilon$, simply because $z \notin A_\epsilon$. We may now choose a finite subcovering of $\cup_{z \in I^c} I_z$, which we denote by $I_{N+1}, \ldots, I_{N'}$. Now, taking all the end points of the intervals $I_1, I_2, \ldots, I_{N'}$ we obtain a partition P of $[a, b]$ with

$$\mathcal{U}(P, f) - \mathcal{L}(P, f) \leq 2M \sum_{j=1}^{N} |I_j| + \epsilon(b - a) \leq C\epsilon.$$

Hence f is integrable on $[a, b]$, as was to be shown.

Conversely, suppose that f is integrable on $[a, b]$, and let $\mathcal{D}$ be its set of discontinuities. Since $\mathcal{D}$ equals $\cup_{n=1}^{\infty} A_{1/n}$, it suffices to prove that each $A_{1/n}$ has measure 0. Let $\epsilon > 0$ and choose a partition $P = \{x_0, x_1, \ldots, x_N\}$ so that $\mathcal{U}(P, f) - \mathcal{L}(P, f) < \epsilon/n$. Then, if $A_{1/n}$ intersects $I_j = (x_{j-1}, x_j)$ we must have $\sup_{x \in I_j} f(x) - \inf_{x \in I_j} f(x) \geq 1/n$, and this shows that

$$\frac{1}{n} \sum_{\{j : I_j \cap A_{1/n} \neq \emptyset\}} |I_j| \leq \mathcal{U}(P, f) - \mathcal{L}(P, f) < \epsilon/n.$$

So by taking intervals intersecting $A_{1/n}$ and making them slightly larger, we can cover $A_{1/n}$ with open intervals of total length $\leq 2\epsilon$. Therefore $A_{1/n}$ has measure 0, and we are done.

Note that incidentally, this gives another proof that fg is integrable whenever f and g are.

2 Multiple integrals

We assume that the reader is familiar with the standard theory of multiple integrals of functions defined on bounded sets. Here, we give a quick review of the main definitions and results of this theory. Then, we describe the notion of "improper" multiple integration where the range of integration is extended to all of $\mathbb{R}^d$. This is relevant to our study of the Fourier transform. In the spirit of Chapters 5 and 6, we shall define the integral of functions that are continuous and satisfy an adequate decay condition at infinity.

Recall that the vector space $\mathbb{R}^d$ consists of all d-tuples of real numbers $x = (x_1, x_2, \ldots, x_d)$ with $x_j \in \mathbb{R}$, where addition and multiplication by scalars are defined componentwise.

2.1 The Riemann integral in $\mathbb{R}^d$

Definitions

The notion of Riemann integration on a rectangle $R \subset \mathbb{R}^d$ is an immediate generalization of the notion of Riemann integration on an interval $[a, b] \subset \mathbb{R}$. We restrict our attention to continuous functions; these are always integrable.

By a **closed rectangle** in $\mathbb{R}^d$, we mean a set of the form

$$R = \{a_j \leq x_j \leq b_j \ : 1 \leq j \leq d\}$$

where $a_j, b_j \in \mathbb{R}$ for $1 \leq j \leq n$. In other words, R is the product of the one-dimensional intervals $[a_j, b_j]$:

$$R = [a_1, b_1] \times \cdots \times [a_d, b_d].$$

If P_j is a partition of the closed interval $[a_j, b_j]$, then we call $P = (P_1, \ldots, P_d)$ a **partition** of R; and if S_j is a subinterval of the partition P_j, then $S = S_1 \times \cdots \times S_d$ is a **subrectangle** of the partition P. The volume $|S|$ of a subrectangle S is naturally given by the product of the length of its sides $|S| = |S_1| \times \cdots \times |S_d|$, where $|S_j|$ denotes the length of the interval S_j.

We are now ready to define the notion of integral over R. Given a bounded real-valued function f defined on R and a partition P, we define the upper and lower sums of f with respect to P by

$$\mathcal{U}(P, f) = \sum [\sup_{x \in S} f(x)] \, |S| \quad \text{and} \quad \mathcal{L}(P, f) = \sum [\inf_{x \in S} f(x)] \, |S|,$$

where the sums are taken over all subrectangles of the partition P. These definitions are direct generalizations of the analogous notions in one dimension.

A partition $P' = (P'_1, \ldots, P'_d)$ is a **refinement** of $P = (P_1, \ldots, P_d)$ if each P'_j is a refinement of P_j. Arguing with these refinements as we did in the one-dimensional case, we see that if we define

$$U = \inf_P \mathcal{U}(P, f) \quad \text{and} \quad L = \sup_P \mathcal{L}(P, f),$$

then both U and L exist, are finite, and $U \geq L$. We say that f is **Riemann integrable** on R if for every $\epsilon > 0$ there exists a partition P so that

$$\mathcal{U}(P, f) - \mathcal{L}(P, f) < \epsilon.$$

This implies that $U = L$, and this common value, which we shall denote by either

$$\int_R f(x_1, \ldots, x_d) \, dx_1 \cdots dx_d, \qquad \int_R f(x) \, dx, \qquad \text{or} \qquad \int_R f,$$

is by definition the integral of f over R. If f is complex-valued, say $f(x) = u(x) + iv(x)$, where u and v are real-valued, we naturally define

$$\int_R f(x) \, dx = \int_R u(x) \, dx + i \int_R v(x) \, dx.$$

In the results that follow, we are primarily interested in continuous functions. Clearly, if f is continuous on a closed rectangle R then f is integrable since it is uniformly continuous on R. Also, we note that if f is continuous on, say, a closed ball B, then we may define its integral over B in the following way: if g is the extension of f defined by $g(x) = 0$ if $x \notin B$, then g is integrable on any rectangle R that contains B, and we may set

$$\int_B f(x)\,dx = \int_R g(x)\,dx.$$

2.2 Repeated integrals

The fundamental theorem of calculus allows us to compute many one dimensional integrals, since it is possible in many instances to find an antiderivative for the integrand. In $\mathbb{R}^d$, this permits the calculation of multiple integrals, since a d-dimensional integral actually reduces to d one-dimensional integrals. A precise statement describing this fact is given by the following.

Theorem 2.1 *Let f be a continuous function defined on a closed rectangle $R \subset \mathbb{R}^d$. Suppose $R = R_1 \times R_2$ where $R_1 \subset \mathbb{R}^{d_1}$ and $R_2 \subset \mathbb{R}^{d_2}$ with $d = d_1 + d_2$. If we write $x = (x_1, x_2)$ with $x_i \in \mathbb{R}^{d_i}$, then $F(x_1) = \int_{R_2} f(x_1, x_2)\,dx_2$ is continuous on R_1, and we have*

$$\int_R f(x)\,dx = \int_{R_1}\left(\int_{R_2} f(x_1, x_2)\,dx_2\right)dx_1.$$

Proof. The continuity of F follows from the uniform continuity of f on R and the fact that

$$|F(x_1) - F(x_1')| \leq \int_{R_2} |f(x_1, x_2) - f(x_1', x_2)|\,dx_2.$$

To prove the identity, let P_1 and P_2 be partitions of R_1 and R_2, respectively. If S and T are subrectangles in P_1 and P_2, respectively, then the key observation is that

$$\sup_{S\times T} f(x_1, x_2) \geq \sup_{x_1\in S}\left(\sup_{x_2\in T} f(x_1, x_2)\right)$$

and

$$\inf_{S\times T} f(x_1, x_2) \leq \inf_{x_1\in S}\left(\inf_{x_2\in T} f(x_1, x_2)\right).$$

Then,

$$\begin{aligned}\mathcal{U}(P,f) &= \sum_{S,T}[\sup_{S\times T} f(x_1,x_2)]\,|S\times T| \\ &\geq \sum_S\sum_T \sup_{x_1\in S}[\sup_{x_2\in T} f(x_1,x_2)]\,|T|\times|S| \\ &\geq \sum_S \sup_{x_1\in S}\left(\int_{R_2} f(x_1,x_2)\,dx_2\right)|S| \\ &\geq \mathcal{U}\left(P_1,\int_{R_2} f(x_1,x_2)\,dx_2\right).\end{aligned}$$

Arguing similarly for the lower sums, we find that

$$\mathcal{L}(P,f)\leq\mathcal{L}(P_1,\int_{R_2} f(x_1,x_2)\,dx_2)\leq\mathcal{U}(P_1,\int_{R_2} f(x_1,x_2)\,dx_2)\leq\mathcal{U}(P,f),$$

and the theorem follows from these inequalities.

Repeating this argument, we find as a corollary that if f is continuous on the rectangle $R\subset\mathbb{R}^d$ given by $R=[a_1,b_1]\times\cdots[a_d,b_d]$, then

$$\int_R f(x)\,dx=\int_{a_1}^{b_1}\left(\int_{a_2}^{b_2}\cdots\left(\int_{a_d}^{b_d} f(x_1,\ldots,x_d)\,dx_d\right)\ldots dx_2\right)dx_1,$$

where the right-hand side denotes d-iterates of one-dimensional integrals. It is also clear from the theorem that we can interchange the order of integration in the repeated integral as desired.

2.3 The change of variables formula

A diffeomorphism of class C^1, $g:A\to B$, is a mapping that is continuously differentiable, invertible, and whose inverse $g^{-1}:B\to A$ is also continuously differentiable. We denote by Dg the Jacobian or derivative of g. Then, the change of variables formula says the following.

Theorem 2.2 *Suppose A and B are compact subsets of $\mathbb{R}^d$ and $g:A\to B$ is a diffeomorphism of class C^1. If f is continuous on B, then*

$$\int_{g(A)} f(x)\,dx=\int_A f(g(y))\,|\det(Dg)(y)|\,dy.$$

The proof of this theorem consists first of an analysis of the special situation when g is a linear transformation L. In this case, if R is a rectangle, then

$$|g(R)|=|\det(L)|\,|R|,$$

which explains the term $|\det(Dg)|$. Indeed, this term corresponds to the new infinitesimal element of volume after the change of variables.

2.4 Spherical coordinates

An important application of the change of variables formula is to the case of polar coordinates in $\mathbb{R}^2$, spherical coordinates in $\mathbb{R}^3$, and their generalization in $\mathbb{R}^d$. These are particularly important when the function, or set we are integrating over, exhibit some rotational (or spherical) symmetries. The cases $d = 2$ and $d = 3$ were given in Chapter 6. More generally, the spherical coordinates system in $\mathbb{R}^d$ is given by $x = g(r, \theta_1, \ldots, \theta_{d-1})$ where

$$\begin{cases} x_1 &= r \sin\theta_1 \sin\theta_2 \cdots \sin\theta_{d-2} \sin\theta_{d-1}, \\ x_2 &= r \sin\theta_1 \sin\theta_2 \cdots \sin\theta_{d-2} \cos\theta_{d-1}, \\ \vdots & \\ x_{d-1} &= r \sin\theta_1 \cos\theta_2, \\ x_d &= r \cos\theta_1, \end{cases}$$

with $0 \leq \theta_i \leq \pi$ for $1 \leq i \leq d-2$ and $0 \leq \theta_{d-1} \leq 2\pi$. The determinant of the Jacobian of this transformation is given by

$$r^{d-1} \sin^{d-2}\theta_1 \sin^{d-3}\theta_2 \cdots \sin\theta_{d-2}.$$

Any point in $x \in \mathbb{R}^d - \{0\}$ can be written uniquely as $r\gamma$ with $\gamma \in S^{d-1}$ the unit sphere in $\mathbb{R}^d$. If we define

$$\int_{S^{d-1}} f(\gamma)\, d\sigma(\gamma) =$$

$$\int_0^{\pi} \int_0^{\pi} \cdots \int_0^{2\pi} f(g(r,\theta)) \sin^{d-2}\theta_1 \sin^{d-3}\theta_2 \cdots \sin\theta_{d-2}\, d\theta_{d-1} \cdots d\theta_1,$$

then we see that if $B(0, N)$ denotes the ball of radius N centered at the origin, then

$$(3) \qquad \int_{B(0,N)} f(x)\, dx = \int_{S^{d-1}} \int_0^N f(r\gamma)\, r^{d-1}\, dr\, d\sigma(\gamma).$$

In fact, we define the **area** of the unit sphere $S^{d-1} \subset \mathbb{R}^d$ as

$$\omega_d = \int_{S^{d-1}} d\sigma(\gamma).$$

An important application of spherical coordinates is to the calculation of the integral $\int_{A(R_1,R_2)} |x|^{\lambda}\, dx$, where $A(R_1, R_2)$ denotes the annulus

$A(R_1, R_2) = \{R_1 \leq |x| \leq R_2\}$ and $\lambda \in \mathbb{R}$. Applying polar coordinates, we find

$$\int_{A(R_1,R_2)} |x|^\lambda \, dx = \int_{S^{d-1}} \int_{R_1}^{R_2} r^{\lambda+d-1} \, dr d\sigma(\gamma).$$

Therefore

$$\int_{A(R_1,R_2)} |x|^\lambda \, dx = \begin{cases} \frac{\omega_d}{\lambda+d}[R_2^{\lambda+d} - R_1^{\lambda+d}] & \text{if } \lambda \neq -d, \\ \omega_d[\log(R_2) - \log(R_1)] & \text{if } \lambda = -d. \end{cases}$$

3 Improper integrals. Integration over $\mathbb{R}^d$

Most of the theorems we just discussed extend to functions integrated over all of $\mathbb{R}^d$ once we impose some decay at infinity on the functions we integrate.

3.1 Integration of functions of moderate decrease

For each fixed $N > 0$ consider the closed cube in $\mathbb{R}^d$ centered at the origin with sides parallel to the axis, and of side length N: $Q_N = \{|x_j| \leq N/2 : 1 \leq j \leq d\}$. Let f be a continuous function on $\mathbb{R}^d$. If the limit

$$\lim_{N \to \infty} \int_{Q_N} f(x) \, dx$$

exists, we denote it by

$$\int_{\mathbb{R}^d} f(x) \, dx.$$

We deal with a special class of functions whose integrals over $\mathbb{R}^d$ exist. A continuous function f on $\mathbb{R}^d$ is said to be of **moderate decrease** if there exists $A > 0$ such that

$$|f(x)| \leq \frac{A}{1 + |x|^{d+1}}.$$

Note that if $d = 1$ we recover the definition given in Chapter 5. An important example of a function of moderate decrease in $\mathbb{R}$ is the Poisson kernel given by $\mathcal{P}_y(x) = \frac{1}{\pi} \frac{y}{x^2+y^2}$.

We claim that if f is of moderate decrease, then the above limit exists. Let $I_N = \int_{Q_N} f(x) \, dx$. Each I_N exists because f is continuous hence integrable. For $M > N$, we have

$$|I_M - I_N| \leq \int_{Q_M - Q_N} |f(x)| \, dx.$$

Now observe that the set $Q_M - Q_N$ is contained in the annulus $A(aN, bM) = \{aN \leq |x| \leq bM\}$, where a and b are constants that depend only on the dimension d. This is because the cube Q_N is contained in the annulus $N/2 \leq |x| \leq N\sqrt{d}/2$, so that we can take $a = 1/2$ and $b = \sqrt{d}/2$. Therefore, using the fact that f is of moderate decrease yields

$$|I_M - I_N| \leq A \int_{aN \leq |x| \leq bM} |x|^{-d-1}\, dx.$$

Now putting $\lambda = -d - 1$ in the calculation of the integral of the previous section, we find that

$$|I_M - I_N| \leq C \left(\frac{1}{aN} - \frac{1}{bM} \right).$$

So if f is of moderate decrease, we conclude that $\{I_N\}_{N=1}^{\infty}$ is a Cauchy sequence, and therefore $\int_{\mathbb{R}^d} f(x)\, dx$ exists.

Instead of the rectangles Q_N, we could have chosen the balls B_N centered at the origin and of radius N. Then, if f is of moderate decrease, the reader should have no difficulties proving that $\lim_{N\to\infty} \int_{B_N} f(x)\, dx$ exists, and that this limit equals $\lim_{N\to\infty} \int_{Q_N} f(x)\, dx$.

Some elementary properties of the integrals of functions of moderate decrease are summarized in Chapter 6.

3.2 Repeated integrals

In Chapters 5 and 6 we claimed that the multiplication formula held for functions of moderate decrease. This required an appropriate interchange of integration. Similarly for operators defined in terms of convolutions (with the Poisson kernel for example).

We now justify the necessary formula for iterated integrals. We only consider the case $d = 2$, although the reader will have no difficulty extending this result to arbitrary dimensions.

Theorem 3.1 *Suppose f is continuous on $\mathbb{R}^2$ and of moderate decrease. Then*

$$F(x_1) = \int_{\mathbb{R}} f(x_1, x_2)\, dx_2$$

is of moderate decrease on $\mathbb{R}$, and

$$\int_{\mathbb{R}^2} f(x)\, dx = \int_{\mathbb{R}} \left(\int_{\mathbb{R}} f(x_1, x_2)\, dx_2 \right) dx_1.$$

Proof. To see why F is of moderate decrease, note first that

$$|F(x_1)| \leq \int_{\mathbb{R}} \frac{A\,dx_2}{1+(x_1^2+x_2^2)^{3/2}} \leq \int_{|x_2|\leq|x_1|} + \int_{|x_2|\geq|x_1|}.$$

In the first integral, we observe that the integrand is $\leq A/(1+|x_1|^3)$, so

$$\int_{|x_2|\leq|x_1|} \frac{A\,dx_2}{1+(x_1^2+x_2^2)^{3/2}} \leq \frac{A}{1+|x_1|^3}\int_{|x_2|\leq|x_1|} dx_2 \leq \frac{A'}{1+|x_1|^2}.$$

For the second integral, we have

$$\int_{|x_2|\geq|x_1|} \frac{A\,dx_2}{1+(x_1^2+x_2^2)^{3/2}} \leq A''\int_{|x_2|\geq|x_1|} \frac{dx_2}{1+|x_2|^3} \leq \frac{A'''}{|x_1|^2},$$

thus F is of moderate decrease. In fact, this argument together with Theorem 2.1 shows that F is the uniform limit of continuous functions, thus is also continuous.

To establish the identity we simply use an approximation and Theorem 2.1 over finite rectangles. Write S^c to denote the complement of a set S. Given $\epsilon > 0$ choose N so large that

$$\left|\int_{\mathbb{R}^2} f(x_1,x_2)\,dx_1dx_2 - \int_{I_N\times I_N} f(x_1,x_2)\,dx_1dx_2\right| < \epsilon,$$

where $I_N = [-N,N]$. Now we know that

$$\int_{I_N\times I_N} f(x_1,x_2)\,dx_1dx_2 = \int_{I_N}\left(\int_{I_N} f(x_1,x_2)\,dx_2\right)dx_1.$$

But this last iterated integral can be written as

$$= \int_{\mathbb{R}}\left(\int_{\mathbb{R}} f(x_1,x_2)\,dx_2\right)dx_1 - \int_{I_N^c}\left(\int_{\mathbb{R}} f(x_1,x_2)\,dx_2\right)dx_1 - \int_{I_N}\left(\int_{I_N^c} f(x_1,x_2)\,dx_2\right)dx_1.$$

We can now estimate

$$\begin{aligned}\left|\int_{I_N}\left(\int_{I_N^c} f(x_1,x_2)\,dx_2\right)dx_1\right| &\leq O\left(\frac{1}{N^2}\right) \\ &\quad + C\int_{1\leq|x_1|\leq N}\left(\int_{|x_2|\geq N}\frac{dx_2}{(|x_1|+|y_1|)^3}\right)dx_1 \\ &\leq O\left(\frac{1}{N}\right).\end{aligned}$$

A similar argument shows that

$$\left| \int_{I_N^c} \left(\int_{\mathbb{R}} f(x_1, x_2)\, dx_2 \right) dx_1 \right| \leq \frac{C}{N}.$$

Therefore, we can find N so large that

$$\left| \int_{I_N \times I_N} f(x_1, x_2)\, dx_1 dx_2 - \int_{\mathbb{R}} \left(\int_{\mathbb{R}} f(x_1, x_2)\, dx_2 \right) dx_1 \right| < \epsilon,$$

and we are done.

3.3 Spherical coordinates

In $\mathbb{R}^d$, spherical coordinates are given by $x = r\gamma$, where $r \geq 0$ and γ belongs to the unit sphere S^{d-1}. If f is of moderate decrease, then for each fixed $\gamma \in S^{d-1}$, the function of f given by $f(r\gamma)r^{d-1}$ is also of moderate decrease on $\mathbb{R}$. Indeed, we have

$$\left| f(r\gamma)r^{d-1} \right| \leq A \frac{r^{d-1}}{1 + |r\gamma|^{d+1}} \leq \frac{B}{1 + r^2}.$$

As a result, by letting $R \to \infty$ in (3) we obtain the formula

$$\int_{\mathbb{R}^d} f(x)\, dx = \int_{S^{d-1}} \int_0^\infty f(r\gamma)\, r^{d-1}\, dr\, d\sigma(\gamma).$$

As a consequence, if we combine the fact that

$$\int_{\mathbb{R}^d} f(R(x))\, dx = \int_{\mathbb{R}^d} f(x)\, dx,$$

whenever R is a rotation, with the identity (3), then we obtain that

$$\int_{S^{d-1}} f(R(\gamma))\, d\sigma(\gamma) = \int_{S^{d-1}} f(\gamma)\, d\sigma(\gamma)\,. \tag{4}$$

Notes and References

Seeley [29] gives an elegant and brief introduction to Fourier series and the Fourier transform. The authoritative text on Fourier series is Zygmund [36]. For further applications of Fourier analysis to a variety of other topics, see Dym and McKean [8] and Körner [21]. The reader should also consult the book by Kahane and Lemarié-Rieusset [20], which contains many historical facts and other results related to Fourier series.

Chapter 1

The citation is taken from a letter of Fourier to an unknown correspondent (probably Lagrange), see Herivel [15].

More facts about the early history of Fourier series can be found in Sections I-III of Riemann's memoir [27].

Chapter 2

The quote is a translation of an excerpt in Riemann's paper [27].

For a proof of Littlewood's theorem (Problem 3), as well as other related "Tauberian theorems," see Chapter 7 in Titchmarsh [32].

Chapter 3

The citation is a translation of a passage in Dirichlet's memoir [6].

Chapter 4

The quote is translated from Hurwitz [17].

The problem of a ray of light reflecting inside a square is discussed in Chapter 23 of Hardy and Wright [13].

The relationship between the diameter of a curve and Fourier coefficients (Problem 1) is explored in Pfluger [26].

Many topics concerning equidistribution of sequences, including the results in Problems 2 and 3, are taken up in Kuipers and Niederreiter [22].

Chapter 5

The citation is a free translation of a passage in Schwartz [28].

For topics in finance, see Duffie [7], and in particular Chapter 5 for the Black-Scholes theory (Problems 1 and 2).

The results in Problems 4, 5, and 6 are worked out in John [19] and Widder [34].

For Problem 7, see Chapter 2 in Wiener [35].

The original proof of the nowhere differentiability of f_1 (Problem 8) is in Hardy [12].

Chapter 6

The quote is an excerpt from Cormack's Nobel Prize lecture [5].

More about the wave equation, as well as the results in Problems 3, 4, and 5 can be found in Chapter 5 of Folland [9].

A discussion of the relationship between rotational symmetry, the Fourier transform, and Bessel functions is in Chapter 4 of Stein and Weiss [31].

For more on the Radon transform, see Chapter 1 in John [18], Helgason [14], and Ludwig [25].

Chapter 7

The citation is taken from Bingham and Tukey [2].

Proofs of the structure theorem for finite abelian groups (Problem 2) can be found in Chapter 2 of Herstein [16], Chapter 2 in Lang [23], or Chapter 104 in Körner [21].

For Problem 4, see Andrews [1], which contains a short proof.

Chapter 8

The citation is from Bochner [3].

For more on the divisor function, see Chapter 18 in Hardy and Wright [13].

Another "elementary" proof that $L(1, \chi) \neq 0$ can be found in Chapter 3 of Gelfond and Linnik [11].

An alternate proof that $L(1, \chi) \neq 0$ based on algebraic number theory is in Weyl [33]. Also, two other analytic variants of the proof that $L(1, \chi) \neq 0$ can be found in Chapter 109 in Körner [21] and Chapter 6 in Serre [30]. See also the latter reference for Problems 3 and 4.

Appendix

Further details about the results on integration reviewed in the appendix can be found in Folland [10] (Chapter 4), Buck [4] (Chapter 4), or Lang [24] (Chapter 20).

Bibliography

[1] G. E. Andrews. *Number theory.* Dover Publications, New York, 1994. Corrected reprint of the 1971 originally published by W.B. Saunders Company.

[2] C. Bingham and J.W. Tukey. Fourier methods in the frequency analysis of data. *The Collected Works of John W Tukey*, Volume II Time Series: 1965-1984(Wadsworth Advanced Books & Software), 1984.

[3] S. Bochner. *The role of Mathematics in the Rise of Science.* Princeton University Press, Princeton, NJ, 1966.

[4] R. C. Buck. *Advanced Calculus.* McGraw-Hill, New York, third edition, 1978.

[5] A. M. Cormack. *Nobel Prize in Physiology and Medicine Lecture*, volume Volume 209. Science, 1980.

[6] G. L. Dirichlet. Sur la convergence des séries trigonometriques qui servent à representer une fonction arbitraire entre des limites données. *Crelle, Journal für die reine angewandte Mathematik*, 4:157–169, 1829.

[7] D. Duffie. *Dynamic Asset Pricing Theory.* Princeton University Press, Princeton, NJ, 2001.

[8] H. Dym and H. P. McKean. *Fourier Series and Integrals.* Academic Press, New York, 1972.

[9] G. B. Folland. *Introduction to Partial Differential Equations.* Princeton University Press, Princeton, NJ, 1995.

[10] G. B. Folland. *Advanced Calculus.* Prentice Hall, Englewood Cliffs, NJ, 2002.

[11] A. O. Gelfond and Yu. V. Linnik. *Elementary Methods in Analytic Number Theory.* Rand McNally & Compagny, Chicago, 1965.

[12] G. H. Hardy. Weierstrass's non-differentiable function. *Transactions, American Mathematical Society*, 17:301–325, 1916.

[13] G. H. Hardy and E. M. Wright. *An Introduction to the Theory of Numbers.* Oxford University Press, London, fifth edition, 1979.

[14] S. Helgason. The Radon transform on Euclidean spaces, compact two-point homogeneous spaces and Grassman manifolds. *Acta. Math.*, 113:153–180, 1965.

[15] J. Herivel. *Joseph Fourier The Man and the Physicist.* Clarendon Press, Oxford, 1975.

[16] I. N. Herstein. *Abstract Algebra.* Macmillan, New York, second edition, 1990.

[17] A. Hurwitz. Sur quelques applications géometriques des séries de Fourier. *Annales de l'Ecole Normale Supérieure*, 19(3):357–408, 1902.

[18] F. John. *Plane Waves and Spherical Mean Applied to Partial Differential Equations.* Interscience Publishers, New York, 1955.

[19] F. John. *Partial Differential Equations.* Springer-Verlag, New York, fourth edition, 1982.

[20] J.P. Kahane and P. G. Lemarié-Rieusset. *Séries de Fourier et ondelettes.* Cassini, Paris, 1998. English version: Gordon & Breach, 1995.

[21] T. W. Körner. *Fourier Analysis.* Cambridge University Press, Cambridge, UK, 1988.

[22] L. Kuipers and H. Niederreiter. *Uniform Distribution of Sequences.* Wiley, New York, 1974.

[23] S. Lang. *Undergraduate Algebra.* Springer-Verlag, New York, second edition, 1990.

[24] S. Lang. *Undergraduate Analysis.* Springer-Verlag, New York, second edition, 1997.

[25] D. Ludwig. The Radon transform on Euclidean space. *Comm. Pure Appl. Math.*, 19:49–81, 1966.

[26] A. Pfluger. On the diameter of planar curves and Fourier coefficients. *Colloquia Mathematica Societatis János Bolyai, Functions, series, operators*, 35:957–965, 1983.

[27] B. Riemann. Ueber die Darstellbarkeit einer Function durch eine trigonometrische Reihe. *Habilitation an der Universität zu Göttingen*, 1854. Collected Works, Springer Verlag, New York, 1990.

[28] L. Schwartz. *Théorie des distributions*, volume Volume I. Hermann, Paris, 1950.

[29] R. T. Seeley. *An Introduction to Fourier Series and Integrals.* W. A. Benjamin, New York, 1966.

[30] J.P. Serre. *A course in Arithmetic.* GTM 7. Springer Verlag, New York, 1973.

[31] E. M. Stein and G. Weiss. *Introduction to Fourier Analysis on Euclidean Spaces.* Princeton University Press, Princeton, NJ, 1971.

[32] E. C. Titchmarsh. *The Theory of Functions.* Oxford University Press, London, second edition, 1939.

[33] H. Weyl. *Algebraic Theory of Numbers*, volume Volume 1 of *Annals of Mathematics Studies.* Princeton University Press, Princeton, NJ, 1940.

[34] D. V. Widder. *The Heat Equation.* Academic Press, New York, 1975.

[35] N. Wiener. *The Fourier Integral and Certain of its Applications.* Cambridge University Press, Cambridge, UK, 1933.

[36] A. Zygmund. *Trigonometric Series*, volume Volumes I and II. Cambridge University Press, Cambridge, UK, second edition, 1959. Reprinted 1993.

Symbol Glossary

The page numbers on the right indicate the first time the symbol or notation is defined or used. As usual, $\mathbb{Z}$, $\mathbb{Q}$, $\mathbb{R}$ and $\mathbb{C}$ denote the integers, the rationals, the reals, and the complex numbers respectively.

$\triangle$	Laplacian	20, 185
$\|z\|$, $\overline{z}$	Absolute value and complex conjugate	23
e^z	Complex exponential	24
$\sinh x$, $\cosh x$	Hyperbolic sine and hyperbolic cosine	28
$\hat{f}(n)$, a_n	Fourier coefficient	34
$f(\theta) \sim \sum a_n e^{in\theta}$	Fourier series	34
$S_N(f)$	Partial sum of a Fourier series	35
D_N, $\mathcal{D}_R$, $\tilde{D}_N$, D_N^*	Dirichlet kernel, conjugate, and modified	37, 95, 165
P_r, $\mathcal{P}_y$, $\mathcal{P}_y^{(d)}$	Poisson kernels	37, 149, 210
O, o	Big O and little o notation	42, 62
C^k	Functions that are k times differentiable	44
$f * g$	Convolution	44, 139, 184, 239
σ_N, $\sigma_N(f)$	Cesàro mean	52, 53
F_N, $\mathcal{F}_R$	Fejér kernel	53, 163
$A(r)$, $A_r(f)$	Abel mean	54, 55
$\chi_{[a,b]}$	Characteristic function	61
$f(\theta^+)$, $f(\theta^-)$	One-sided limits at jump discontinuities	63
$\mathbb{R}^d$, $\mathbb{C}^d$	Euclidean spaces	71
$X \perp Y$	Orthogonal vectors	72
$\ell^2(\mathbb{Z})$	Square summable sequences	73
$\mathcal{R}$	Riemann integrable functions	75
$\zeta(s)$	Zeta function	98
$[x]$, $\langle x \rangle$	Integer and fractional parts	106
$\triangle_N$, $\sigma_{N,K}$, $\tilde{\triangle}_N$	Delayed means	114, 127, 174
H_t, $\mathcal{H}_t$, $\mathcal{H}_t^{(d)}$	Heat kernels	120, 146, 209
$\mathcal{M}(\mathbb{R})$	Space of functions of moderate decrease on $\mathbb{R}$	131

$\hat{f}(\xi)$	Fourier transform	134, 181
$\mathcal{S}, \mathcal{S}(\mathbb{R}), \mathcal{S}(\mathbb{R}^d)$	Schwartz space	134, 180
$\mathbb{R}^2_+$, $\overline{\mathbb{R}^2_+}$	Upper half-plane and its closure	149
$\vartheta(s)$, $\Theta(z\|\tau)$	Theta functions	155, 156
$\Gamma(s)$	Gamma function	165
$\\|x\\|$, $\|x\|$; (x, y), $x \cdot y$	Norm and inner product in $\mathbb{R}^d$	71, 176
x^α, $\|\alpha\|$, $\left(\frac{\partial}{\partial x}\right)^\alpha$	Monomial, its order, and differential operator	176
S^1, S^2, S^{d-1}	Unit circle in $\mathbb{R}^2$, and unit spheres in $\mathbb{R}^3$, $\mathbb{R}^d$	179, 180
M_t, $\widetilde{M}_t$	Spherical mean	194, 216
J_n	Bessel function	197, 213
$\mathcal{P}$, $\mathcal{P}_{t,\gamma}$	Plane	202
$\mathcal{R}$, $\mathcal{R}^*$	Radon and dual Radon transforms	203, 205
A_d, V_d	Area and volume of the unit sphere in $\mathbb{R}^d$	208
$\mathbb{Z}(N)$	Group of Nth roots of unity	219
$\mathbb{Z}/N\mathbb{Z}$	Group of integers modulo N	221
G, $\|G\|$	Abelian group and its order	226, 228
$G \approx H$	Isomorphic groups	227
$G_1 \times G_2$	Direct product of groups	228
$\mathbb{Z}^*(q)$	Group of units modulo q	227, 229, 244
$\hat{G}$	Dual group of G	231
$a\|b$	a divides b	242
$\gcd(a, b)$	Greatest common divisor of a and b	242
$\varphi(q)$	Number of integers relatively prime to q	254
χ, χ_0	Dirichlet character, and trivial Dirichlet character	254
$L(s, \chi)$	Dirichlet L-function	256
$\log_1\left(\frac{1}{1-z}\right)$, $\log_2 L(s, \chi)$	Logarithms	258, 264
$d(k)$	Number of positive divisors of k	269

Index

Relevant items that also arise in Book I are listed in this index, preceeded by the numeral I.